Foams

Foams

Structure and Dynamics

I. Cantat, S. Cohen-Addad, F. Elias, F. Graner, R. Höhler,
O. Pitois, F. Rouyer, and A. Saint-Jalmes

Translated by R. Flatman

Scientific editor (English edition) S.J. Cox

OXFORD
UNIVERSITY PRESS

Great Clarendon Street, Oxford, OX2 6DP,
United Kingdom

Oxford University Press is a department of the University of Oxford.
It furthers the University's objective of excellence in research, scholarship,
and education by publishing worldwide. Oxford is a registered trade mark of
Oxford University Press in the UK and in certain other countries

Original edition: *Les mousses – Structure et dynamique*
© Éditions Belin – Paris, 2010
English translation © Oxford University Press, 2013
Ouvrage publié avec le concours du Ministère français chargé de la culture – Centre national du livre

The moral rights of the authors have been asserted

First published 2013
First published in paperback 2018

All rights reserved. No part of this publication may be reproduced, stored in
a retrieval system, or transmitted, in any form or by any means, without the
prior permission in writing of Oxford University Press, or as expressly permitted
by law, by licence or under terms agreed with the appropriate reprographics
rights organization. Enquiries concerning reproduction outside the scope of the
above should be sent to the Rights Department, Oxford University Press, at the
address above

You must not circulate this work in any other form
and you must impose this same condition on any acquirer

Published in the United States of America by Oxford UniversityPress
198 Madison Avenue, NewYork, NY 10016, United States of America

British Library Cataloguing in Publication Data
Data available

Library of Congress Cataloging in Publication Data
Data available

ISBN 978-0-19-966289-0 (Hbk.)
ISBN 978-0-19-882433-6 (Pbk.)

Links to third party websites are provided by Oxford in good faith and
for information only. Oxford disclaims any responsibility for the materials
contained in any third party website referenced in this work.

Preface

The bubbly world of foams

A foam consists of a large number of bubbles packed together (FIG. 0.1), which gives this state of matter its remarkable properties. We describe here liquid foams, which consist of a gas and a liquid (often soapy water). They are in general opaque, remarkably stable, and even elastic, as is evident in whipped cream, shaving foam, and sea foam.

A great deal of excellent work has been conducted on these mixtures of soapy water and air: Boys' captivating book (published in 1890), for example, which arose from public lectures in which the author entertained the audience with his soap bubble demonstrations, or the book of Mysels, Frankel, and Shinoda that elucidates some of the many secrets of soap films, Bikerman's book on industrial applications of foams, and more recently the book of Weaire and Hutzler on the physics of foams, with its focus on structural problems.

In this book we set out the fundamentals of current knowledge on liquid foams and discuss some recent advances in our understanding of their structure, generation, ageing, and rheology. We have sought to provide information at several different levels: from simple, concrete explanations, perhaps for industrialists who find themselves interested in foams, to an in-depth discussion of scaling laws and detailed mathematical models, suitable for academics and students; all should, in addition, find the exercises and suggested experiments straightforward to implement.

This is certainly not an exhaustive volume: we consider solid foams only briefly and we avoid discussion of a number of unresolved questions. It is also the product of team-work, so that in places we find ourselves explaining and comparing different points of view, fundamental or applied, which is in itself a stimulating exercise. Such polyphonic writing requires extensive proofreading and polishing: we thank in particular Christian Counillon, Agnès Haasser, and Michel Laguës for having done this and for their excellent advice. We also thank with great pleasure all those colleagues and students who have helped us make progress thanks to their remarks and suggestions. Finally, we would like to acknowledge the CNRS-funded GDR "Mousses et Emulsions" and its directors M. Adler and C. Gay for their encouragement.

FIG. 0.1: A liquid foam some time after its creation. It is dry at the top, with large polyhedral bubbles, and wet at the bottom, with a smaller average bubble size. The bubbles become spherical when they come into contact with the liquid on which the foam floats. Photograph courtesy of S. Cohen-Addad, R.M. Guillermic, and A. Saint-Jalmes. See PLATE 1.

Experiments

The experiments presented at the end of each chapter aim to illustrate a number of points within that chapter. Some experiments can be carried out in minutes in a kitchen or bathroom, others require laboratory materials, but none are dangerous. Nonetheless, it is necessary to take the usual safety precautions and use common sense. At the beginning of each experiment, the theoretical prerequisites are indicated in the form of references to the pertinent paragraphs in the chapter. The level of difficulty and the time required for each experiment are classified according to four criteria:

Difficulty level: three levels of difficulty are identified for the experiment depending on the required materials:

- ☻ : Materials generally found in a kitchen or bathroom
- ☻☻ : Materials that can be bought or made at home
- ☻☻☻ : Laboratory materials

Cost: the cost of the experiment is divided into three levels of expense:

- $: cheap, less than about 5 dollars, euros, or pounds
- $$: moderate
- $$$: expensive, more than about 30 dollars, euros, or pounds

Preparation time: this refers to the time required to set up the experiment once the materials are assembled. Depending on the experiment this could take between five minutes and half a day.

Experimental time: once the experiment is set up, this is the actual experimental time, including the analysis, required to give a complete description of the phenomenon.

Contents

1 Uses of foams 1
   1 The foams around us 1
      1.1 Foams in mythology 1
      1.2 On your plate and in your glass 1
      1.3 Detergents and cosmetics 3
      1.4 Spontaneous or undesirable foams 4
   2 Foam identification 4
      2.1 Physico-chemical constituents 4
      2.2 Geometrical and physical properties 5
      2.3 Mechanical properties 6
   3 What are foams used for? 6
      3.1 Desirable functions 6
      3.2 Mineral flotation 8
   4 Solid foams and other cellular systems 9
      4.1 Solid foams 9
      4.2 Other cellular structures 11
   5 Experiments 13
      5.1 Three ways to make a foam 13
      5.2 Chocolate mousse 14
   References 15

2 Foams at equilibrium 17
   1 Description at all length-scales 17
      1.1 At the scale of a gas/liquid interface 17
      1.2 At the scale of a film 19
      1.3 At the scale of a bubble 21
      1.4 At the scale of a foam 22
   2 Local equilibrium laws 23
      2.1 Equilibrium of fluid interfaces 23
      2.2 Plateau's laws 26
   3 Dry foams 30
      3.1 Number of neighbours: topology 31
      3.2 Bubble geometry 35
      3.3 Topology and geometry 38
   4 Wet foams 45
      4.1 Modification of the structure 46
      4.2 Osmotic pressure 51
      4.3 Role of gravity 54

		5	2D and quasi-2D foams	55
			5.1 3D structure of a monolayer of bubbles between two plates	57
			5.2 A model for a dry 2D foam	58
			5.3 Two-dimensional liquid fraction	60
			5.4 2D foam flows	61
		6	Experiments	63
			6.1 Surface tension and surfactants	63
			6.2 Creation and observation of 2D and quasi-2D foams	65
			6.3 Giant soap films	66
			6.4 Kelvin cell	68
		7	Exercises	69
			7.1 Interfacial area of a foam	69
			7.2 Film tension and the Young–Laplace law	69
			7.3 Plateau's laws in 2D	70
			7.4 Euler's formula	71
			7.5 Perimeter of a regular 2D bubble	71
			7.6 Energy and pressure	72
		References		72
3	Birth, life, and death			75
	1	Foam evolution		75
		1.1 The competition between different processes		75
		1.2 Elementary topological processes		78
	2	Birth of a foam		82
		2.1 Foamability: introduction to the role of surfactants		82
		2.2 Interfacial properties and foamability		82
		2.3 Properties of liquid films and foamability		92
		2.4 Summary of the microscopic origins of foamability		98
	3	Coarsening		99
		3.1 Growth rate of a bubble in a dry foam		99
		3.2 Evolution of bubble distributions in a dry foam		104
		3.3 Effects of different parameters		109
	4	Drainage		113
		4.1 What is drainage?		114
		4.2 Free drainage		114
		4.3 Forced drainage		115
		4.4 Modelling flows in solid porous media		116
		4.5 Modelling the permeability of a liquid foam		119
		4.6 Drainage equations		127
		4.7 Comparison of theoretical predictions with experiments		128
		4.8 Summary and remarks		133
	5	Rupture and coalescence		134
		5.1 Rupture at the scale of a single film		134
		5.2 Rupture at the scale of a foam		140
		5.3 Defoamers and antifoams		140

6	Appendices		145
	6.1	Stabilizing agents	145
	6.2	Dissipation due to surfactant motion during the steady expansion of a film	151
7	Experiments		154
	7.1	Flow in a soap film	154
	7.2	Free drainage in a foam and the vertical motion of bubbles	156
	7.3	Forced drainage in a foam: observation of the wetting front	157
	7.4	Life and death of a foam measured by electrical conductivity	158
8	Exercises		161
	8.1	Exponent in the scale-invariant regime	161
	8.2	Frumkin equation of state	161
	8.3	Foam drainage and equilibrium height	161
	8.4	Drainage in the bulk and at the wall	162
	8.5	Free drainage: characteristic times and liquid fraction profiles	162
	8.6	The true 3D pressure and 2D surface pressure	162
References			162

4 Rheology 167

1	Introduction		167
2	Overview of the rheological behaviour of complex fluids		168
	2.1	Constitutive laws	168
	2.2	Shear tests	172
	2.3	Small and large strains	173
	2.4	Stress tensor in a complex fluid	174
3	Local origin of rheological properties		178
	3.1	Elastic shear modulus of a dry monodisperse foam	178
	3.2	The elastic limit of a dry foam	183
	3.3	Dissipative processes	187
4	The multiscale character of foam rheology		193
	4.1	Solid behaviour	194
	4.2	Transition from solid to liquid behaviour	208
	4.3	Foam flow	211
5	Appendix: From the discrete to the continuous		215
6	Experiments		217
	6.1	Observation of T1s	217
	6.2	Visualization of the yield stress	217
7	Exercises		218
	7.1	The Young–Laplace law and the stress in a spherical bubble	218
	7.2	Elasticity of a dry 2D foam	219
	7.3	Poynting's law	219
	7.4	Stress and strain in a square lattice	220
	7.5	Elasticity and plasticity	220
	7.6	Compressibility of a foam	221
References			221

5 Experimental and numerical methods — 225
- 1 Experimental methods — 225
  - 1.1 Methods used to study interfaces and isolated films — 225
  - 1.2 Methods for studying foams — 230
- 2 Numerical simulations — 242
  - 2.1 Predicting static structure — 242
  - 2.2 Predicting dynamics — 244
- 3 Methods of image analysis — 248
  - 3.1 Image treatment — 248
  - 3.2 Image analysis — 250
  - 3.3 Image analysis, liquid fraction, and stress in 2D — 254
- 4 Exercises — 255
  - 4.1 Measurement of the average liquid fraction of a foam — 255
  - 4.2 Pressure in the Potts model — 255
- References — 256

Notation — 259

Index — 263

1
Uses of foams

Are foams really useful? We smile when scientists announce that they study foams, because bubbles are considered child's play, but in this first chapter we will show that the converse is true: liquid foams exhibit a range of complex and unique properties, and are well-adapted to diverse applications in our daily lives and to numerous industrial processes.

1 The foams around us

1.1 Foams in mythology

In the *Mahabharata*, Indra, having been captured by the demon Namuchi, is only freed in exchange for the promise that he will not kill his kidnapper by day or night, with something neither wet nor dry. Indra keeps his word: when he returns to slay the demon (FIG. 1.1), it is twilight and he uses foam! Neither wet nor dry, foam meets Namuchi's contradictory demands.

In Greek mythology, Cronos castrates his father Uranus and throws his genitals into the sea. Hesiod recounts that "a white foam surrounded the immortal flesh, and in it grew a girl. At first it touched on holy Cythera, from where it came to Cyprus, circled by the waves. And there the goddess came forth..." [6]. This goddess is named Aphrodite, which translates literally as "born of the foam". *Aphron* means foam in Greek, so the subject of this book can be described as aphrology!

A Chinese myth describes how in order to thwart two dragons who claimed his throne, the emperor collects the foam from their drooling jaws and locks it in a casket. Although foams are not usually so stable, this foam remained in the casket for 2,000 years; when the casket is opened, the foam escapes and takes the form of a small dragon. The dragon touches a young girl who, several years later, gives birth to Pao Sze, the dragon princess renowned for her great beauty and sadness.

Although short of a full anthology, these examples show that foams, whether for their unique properties or their (supposedly) superficial nature, appear frequently in literature. A final example is Joyce's *Ulysses*, where foam appears in diverse forms: from the first sentence, in Buck Mulligan's shaving bowl, to its association with the barrels of beer flowing into the street.

1.2 On your plate and in your glass

Foam is, of course, associated with certain beverages. Beer foam, stabilized amongst other things by proteins, appeals both to the eyes and to the palate. Champagne foam, which is more fragile, is so distinctive that a discerning drinker is able to predict the

FIG. 1.1: The properties of foam inspired Ganesh, the mythical author of the *Mahabharata*, in the scene in which Indra attacks Namuchi. © Amar Chitra Katha Pvt. Ltd.

quality of the wine from the foam alone. In champagne and certain beers, carbon dioxide, which is a product of fermentation in the bottle, is concentrated by an overpressure (6 bars for champagne). The liquid is supersaturated at ordinary pressure but degasses when the bottle is opened; bubbles appear, rise, and form a foam. Some draught beers are dispensed with air or nitrogen in order to slow the ageing of the foam. Fizzy drinks and carbonated water are produced by adding carbon dioxide, but the resulting foams are not very stable due to the lack of proteins. The long-lived froth on a cappucino is stabilized by milk proteins. This leads us naturally to questions about foaming, drainage, and ageing which are dealt with later in this book (chap. 3).

Not only are foams highly drinkable, but their consistency and lightness make them a pleasure to eat as well. Chocolate or fruit mousses (FIG. 1.2) immediately spring to

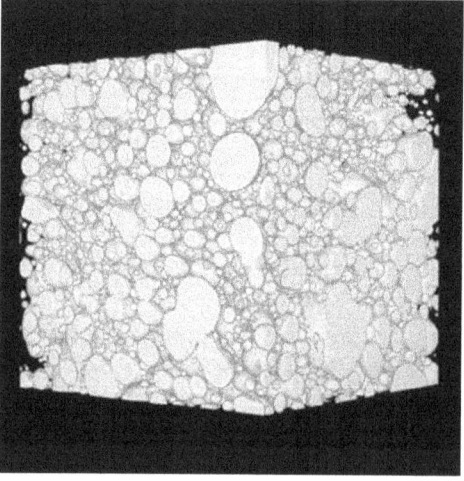

FIG. 1.2: Three-dimensional imaging experiment of the air bubbles in a chocolate mousse. The technique used is X-ray tomography, which is described in chap. 5. Image courtesy of P. Cloetens, ESRF.

mind, but air is also incorporated into ice cream, altering its texture and making it easier to serve, but without causing it to melt too quickly in the mouth. Air, of course, is a raw material which doesn't cost anything... whilst making a large contribution to the volume of the final product.

Edible foams are usually solid, obtained by solidifying a liquid foam either by refrigeration (chocolate mousse) or cooking (bread). The foaming or stabilizing agents for edible foams are natural, for example sugar (for meringue) or gelatine (for marshmallows, where it also serves as a gelling agent). Industrially produced edible foams tend to be stabilized by lecithin (from egg or soya) or xanthan gum, which makes them very viscous. Some foams, like bread and cake, have no need of a stabilizing agent because they are solidified before they collapse.

Baked alaska is an unusual combination of foams: ice cream is thermally stabilized by two other foams: the ice cream is placed on a solid foam, the sponge cake, and then covered by a liquid foam, an uncooked meringue, and placed in the oven. The ice cream is sufficiently well-insulated that it doesn't melt while the meringue is gently cooked!

There are also situations where foams appear without being sought, as a consequence of either boiling or simply agitation, as in the froth that is generated when cooking pasta or potatoes in water, where the proteins present enable the bubbles to persist. Stable milk foam can be seen when a baby is feeding from a bottle or when a glass of milk is shaken, while that which floats on coffee or tea is very short-lived.

1.3   Detergents and cosmetics

The presence of soap in a cleaning product has a consequence beyond the required property (removal of fat and solid particles): the appearance of a foam. It is widely believed that the presence of foam is a measure of cleaning efficacy but in fact the opposite is true: manufacturers try to suppress foams in mechanical cleaning (in washing machines and dishwashers). The addition of only a few drops of washing-up liquid to a dishwasher results in the machine overflowing due to a large quantity of stable foam. In order to avoid these problems, antifoam agents are added (cf. §5.3, chap. 3).

There are, however, advantages of using foams in some cases. For example, while tests show that the cleaning performances of non-foaming and foaming oven cleaners are hardly distinguishable, the latter will be desirable as the cleaner will cling to the vertical walls of the oven for longer. The ability of a foam to stick to a vertical surface, owing to its elasticity, also makes it useful for shaving; a small quantity of product covers the cheeks and stays there until the motion of the razor blade renders it sufficiently fluid to wash off. This leads to questions of rheology which will be discussed later in the book (chap. 4).

Generally speaking, foam products are very common in cosmetics: their ability to cover and cling to surfaces explains their use in hair gels or conditioners. When you wash your hair, the shampoo generates more foam in the second wash, indicating that the first application has succeeded in its role. But the shampoo foam is also there for your enjoyment. Similarly, it is in their role of providing comfort and pleasure that foaming bath products are successful.

1.4 Spontaneous or undesirable foams

As we have seen, foams arise from the presence of a gas in a liquid, for example the carbon dioxide in a fizzy drink or the alkanes in a shaving foam. A magma which degasses causes bubbles to accumulate, and therefore a foam forms in the magma chamber of a volcano. Depending on the conditions (gas flow rate, bubble size, cavity shape), the foam may collapse gradually or, as in Strombolian volcanoes, suddenly and spectacularly!

More common in nature is a foam that forms in a liquid that is agitated in such a way to allow air to be incorporated, as in a waterfall or a breaking wave. These foams are weak and transient if nothing stabilizes them, but in the presence of surfactants from pollutants or plankton, they become the long-lasting "white horses" seen along the coast where spectacular foam mountains can reach several metres high, as can be found for example in Australia. So the presence of a foam is a good indicator of contamination (industrial or natural) by surfactants.

In the same way, pouring washing-up liquid into a public fountain produces a splendid display of foam, which can be made more spectacular still if a few drops of food colouring are added. At the cinema, foams have a role in special effects, for example representing snow. In the final scene of *The Party* by Blake Edwards, a house, the scene of a lavish party, is overrun with soap bubbles and chaos ensues.

Undesirable foams are also encountered in the industrial world, where the focus is on developing products to get rid of them, a subject to which we will return later (§5.3, chap. 3). Such foams appear for example in settling tanks or during the manufacturing of steel, glass, pulp, sugar, watercolour paints, or products obtained by fermentation (like wine or penicillin). An over-acidic stomach secretes foam, and the recommended medicines contain an antifoam that allows them to reach the stomach wall before reacting.

2 Foam identification

As we have seen, liquid foams are commonly encountered all around us. We now describe the properties which make these materials unique and potentially useful in various industrial applications (see §3). Furthermore, a foam, under carefully controlled conditions, often serves as a model system for a cellular material or a biological tissue.

2.1 Physico-chemical constituents

A foam is a dispersion of gas in a liquid. The tightly packed gas bubbles occupy most of the volume (FIG. 0.1). The liquid phase, which consists of the soap films and their junctions, is continuous, unlike the gas phase. It contains special molecules, known as surfactants, which stabilize the bubbles by arranging themselves at the liquid/gas interfaces (see §2, chap. 3).

Different properties can be obtained by changing each constituent of a foam. Consider the gas: air, for example, which is easily available in large quantities and is non-flammable, is used to inflate the aqueous foams in fire extinguishers, enabling them to act as a barrier between the combustible material and oxygen from the atmosphere. The choice of gas also influences the ageing of the foam: carbon dioxide, which is soluble in water, makes drinks effervesce, while nitrogen or alkanes, which are less soluble, delay the ageing of beer or shaving foams (§3, chap. 3).

The liquid also exerts an influence on the ageing of a foam. It drains (because of gravity for example) through the network of films and their junctions, which has the effect of drying the foam (§4, chap. 3). The liquid presents countless possibilities. It can, for example, provide a vehicle for some active ingredient: varicose veins are treated by injecting a foam incorporating a sclerosing medicine, which has the same effect as for a traditional treatment but with much less liquid, and therefore less medication. Or the foaming liquid could be made to harden, through a chemical reaction or a change in temperature, to allow the generation of a solid cellular structure (a metallic or polyurethane foam for example) that is much more durable than a liquid foam.

The stabilizing molecules in a foam also influence its properties and, due to their particular chemical affinity for one material over another, foams are used in separation methods (like mineral flotation) or in detergency. As the impurities compete with the surfactants, they weaken and even destroy the liquid films (§5, chap. 3): the foam thus serves as a marker during cleaning (dishes, shampoo), since it only appears once the main contaminants have gone.

2.2 Geometrical and physical properties

The presence of a large quantity of gas makes a foam a markedly less dense material than the liquid it contains. Thus, a foam of water and air floats on water (as in white horses on the sea, or FIG. 0.1). It contains less material and thus fills the same volume or covers the same surface for a lower cost. When a foam is used to decontaminate a nuclear power station or asbestos, on the one hand less active product is used, but there is also much less waste to reprocess afterwards. In the same way, treating a material with a foam (to make it impermeable or grease-resistant, for example) rather than with a liquid not only enables you to economize on product but also to dry the material more quickly.

A foam also contains a large number of interfaces (§3, chap. 2), and therefore an enormous surface area per unit volume: with 50 or 100 grams of water, one can easily make a litre of foam with 10 m^2 of interface! The possibilities for molecular transfer are multiplied, which is valuable in foods where foams are strong flavour enhancers (of chocolate or spices for instance) and allows salt or sugar content to be reduced.

Its numerous interfaces make the foam refract and reflect light in all directions, preventing light from propagating in a straight line. A foam absorbs light (although liquid and gas are generally transparent), and it blurs vision as much as fog, frosted glass, or milk. Because of this, probing the bulk structure is made easier with 3D tomography. Sound travels more slowly (and less far) in an aqueous foam than in water or in air, due to the alternation between liquid and gas. Moreover, a liquid foam conducts electric current, particularly when the liquid fraction is high. These physical properties are exploited when probing foam structure (§1.2, chap. 5).

2.3 Mechanical properties

A foam's mechanical response is remarkably different from that of both the liquid and the gas from which it is made. In fact, a foam behaves either as a solid or a liquid (see chap. 4)! It can be a weak (visco-)elastic solid, capable of returning to its original shape if the deformation is not too great, or a (visco-)plastic solid, to be sculpted at leisure. It insulates against knocks and shock waves and therefore offers remarkable mechanical isolation.

A foam can also flow like a liquid, seeping into pores, filling cavities, and coating fibres. It can be poured into containers and run through tubes of various shapes. It behaves as a "yield stress fluid" and because its viscous resistance increases less quickly with flow rate than that of a normal fluid it is able to reduce frictional losses. It also keeps small particles in suspension, even when flow stops. All these characteristics are crucial for the use of foams in drilling for oil.

3 What are foams used for?

Foams have novel properties owing to their lightness, their very large specific surface area, and the fact that they exhibit both solid and liquid behaviour. These properties often meet the somewhat contradictory requirements of a number of industrial applications [1]. We characterize some of the sought-after functions (FIG. 1.3) and give examples of applications in which they are used.

3.1 Desirable functions

Why are foams preferable to other fluids in industrial processes? There are at least seven good reasons:

1. *Economize on product, reduce waste*
   Using a foam offers the possibility to employ a smaller quantity of active ingredient or raw material without diminishing its effect. It often reduces the amount of liquid that must be discarded or treated following an industrial operation, for example in decontamination and cleaning. So radioactive tanks could be decontaminated, following the decommissioning of a nuclear power station, by cleaning them with a specially formulated foam.
2. *Quickly fill large spaces*
   The high rate of expansion of a foam (the ratio of foam volume to the volume of liquid used to generate it) allows it to rapidly fill large spaces. For example, fire-fighting foams must cover large areas as quickly as possible. If an aeroplane's landing-gear fails, a layer of foam can be spread across more than a kilometre of runway within half an hour to allow the aeroplane to land. On a more playful note, "foam parties" also require large volumes of foam.

	Economize on product, reduce waste	Quickly fill large spaces	Isolate, confine, smother	Trap material	Absorb or exert pressure	Give a fluid the behaviour of a solid	Confer the structure of a foam on a solid
Cleaning	X					X	
Surface treatment	X		X			X	
Building materials							X
Reducing pollution	X	X	X	X		X	
Fire-fighting		X	X				
Extraction of natural resources				X	X	X	
Cosmetics	X					X	
Food	X					X	X
Army, Police		X	X	X	X		X

FIG. 1.3: The relation between the different functions that aqueous foams perform and their main uses and applications.

3. *Isolate, confine, smother*
   In fire-fighting it is necessary both to smother the fire and to isolate the source of combustion. Foams are able to extinguish burning hydrocarbons more effectively than water due to their low density and because they float and thus isolate the fuel from oxygen in the air. Other examples include the foam extinguishers kept in chemical factories in case of accidents, the sticky anti-riot foams used by police or soldiers, and even certain insects (like froghoppers) whose nymphs secrete a cocoon of foam around themselves to protect against drying out and temperature variations.

4. *Trap substances*
   The interfaces of a foam are able to capture very small particles of solid matter (for example in purification), ions (during water purification or in the separation of uranium), organic molecules (enzymes, proteins, polymers), or small molecules like the surfactants themselves. This forms the basis of numerous separation processes, like sewage treatment or mineral flotation, which we describe later in this chapter. Even a gas can be trapped, in the bubbles, as in methane recovery from coal mines.

5. *Absorb or exert pressure*
   Foams are used to muffle explosions (for example in mine clearance). A foam is injected into rock or concrete in a controlled way to fracture and excavate them. Conversely, foam also limits the problems of overpressure in the bore-holes used in drilling for oil.

6. ***Give a fluid the behaviour of a solid***
   Creating a foam is one way to increase the viscosity of a fluid to such an extent that it acquires the mechanical properties of a solid, as in the case of cosmetics or food. A foam which is sufficiently elastic doesn't flow under the effect of gravity, and can therefore be used to cover vertical surfaces, in order to clean or treat them for example. The elasticity of foams is also useful in displacing oil to extract it from a reservoir.
7. ***Confer the structure of a foam on a solid***
   Finally, liquid foams serve as precursors of solid foams for the construction of cellular materials, including bread and cakes, and polyurethane, glass, and metal foams [2]. The structure of the solid material is determined by that of the liquid foam, which is itself governed by the laws of capillary action (chap. 2), as long as it solidifies before ageing processes like drainage render it heterogeneous. Likewise, cellular concrete is obtained by mixing a foaming agent into the concrete, then degassing and solidifying it.

3.2 Mineral flotation

Among the many applications cited above, the process of mineral flotation is one of the most important. It has been used for more than a hundred years, and now uses close to 90% of the world's surfactant products. After it has been extracted from the ground, the ore-containing rock is ground up and then treated in order to separate the mineral, such as copper, zinc, nickel, talc, or carbon, from the gangue. This separation step must be inexpensive and rapid—typically of the order of one hundred tonnes of mineral per hour.

The principle of flotation is based on the wetting properties at the liquid/air interfaces. It is possible to optimize the physico-chemical conditions (by adjusting the pH, and the nature and concentration of the surfactants) so that the mineral grains are trapped by the interfaces while the gangue remains in the liquid. The attraction of using a liquid foam is of course that it has very high interfacial area.

The ground-up rock and mineral are placed in large tanks (height and diameter of several metres) (FIG. 1.4). Two types of surfactants (at least) are added: those that control the affinity of the mineral for the interface and those that cause the mixture to foam. Air is injected at the base while the mixture is vigorously mixed in a way that favours encounters between bubbles and particles. A foam is thus continually formed and rises to the surface of the tank before spilling out. It brings with it the mineral, attached to the bubbles (FIG. 1.4). During the ascent, a large proportion of the liquid falls back down, making the foam dry and fragile. During the collisions between bubbles, the gangue particles may be trapped in the liquid phase. They are discharged downwards with the draining liquid, a process that can be enhanced by spraying the tank with water from above. Other properties of the foam contribute to the efficacy of the process. Its elasticity limits convection in the column, its low density allows it to float on the surface, where it can be skimmed off, and its fragility and low water content are advantageous for recovering only the mineral.

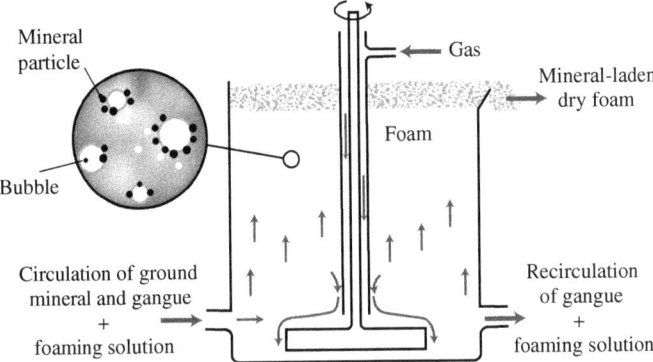

FIG. 1.4: Schematic diagram of the process of mineral flotation that is used to separate the gangue from the particles of the mineral, which are trapped at the air/water interfaces and then recovered at the top of the column.

4   Solid foams and other cellular systems
4.1   Solid foams

A closed-cell solid foam consists of pockets of gas surrounded by solid walls but, in contrast to liquid foams, solid foams can also exist if the walls have holes in them (so-called semi-open-cell foams), or are even entirely missing, as shown in FIG. 1.5.

4.1.1 Natural foams Many foam-like materials are found naturally in the world around us. Sea sponges are like open-cell foams, with a structure that allows water to penetrate, and an extremely large surface area which retains this water by capillary action. Radiolaria are single-celled organisms, typically a hundred microns in size, that multiply rapidly in marine plankton. Their skeletons (FIG. 1.5a) are ordered open-cell foams. Honeycombs also have very well-ordered structures and have become symbolic of a perfectly crystalline arrangement, although in reality they contain cells with different sizes. This, incidentally, poses problems for the bees when trying to join these cells together (FIG. 1.5b).

Closed-cell structures are also found naturally in bones, which are consequently light yet strong, and in cork, which because of its tiny air bubbles is light, supple, and provides good sound insulation. We make use of the lightness and roughness of pumice stones, that is rocks which contain bubbles, which usually have closed cells, but may be open, as in reticulite, an extreme form of pumice containing at least 95% air.

4.1.2 Artificial foams Since these properties are so desirable, many artificial foams have been created to mimic them. In the automobile industry, for example, decreasing weight leads to a reduction in manufacturing and running costs. Vehicles therefore contain polyurethane foams (cushions), rigid plastic foams (dashboard), and aluminium foams (bumper), to name just three.

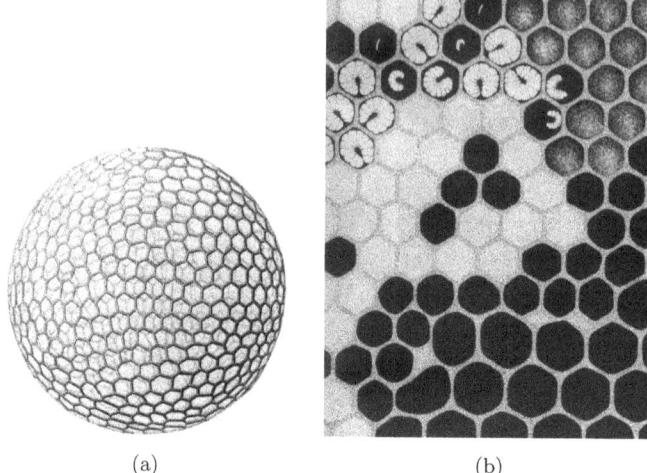

FIG. 1.5: Examples of natural solid foams. (a) The radiolarian skeleton can be seen as an open-cell solid foam (from [5]). (b) A bees' honeycomb is an example of a closed-cell solid foam. The cells which appear in white on the photo are filled with wax: they contain either honey or young bees undergoing metamorphosis. The others are either empty, occupied by larval worker bees, or full of pollen (in grey). The large cells are for the males (from [4]). © Pierre Deom, www.lahulotte.fr.

Some of these foams are obtained by solidifying a liquid foam. To better understand what controls properties such as the average bubble size, the bubble-size distribution, the volume fraction of solid, and the homogeneity of the material, it is necessary to model the liquid foam before solidification. This is a complicated problem which links many of the topics covered in this book: coarsening, drainage, rupture, rheology, and physical chemistry.

The solidification may follow either direct injection of a gas into a molten liquid (physical blowing agent), or the production of bubbles within a liquid through a chemical reaction which produces gas (chemical blowing agent), for example by adding a powder that decomposes at high temperature. Both aluminium foams and polymeric foams can be made like this. Another possibility is to mix air into the liquid by beating or whisking, as in chocolate mousse, which is solidified by cooling, or meringue, which is solidified by cooking (§1.2).

Polymeric foams typically contain 60% to 90% gas, and are found in artificial sponges, filters, membranes, and bottle closures (synthetic cork). Those which have an open-cell structure are generally obtained from thermosetting polymers, which polymerize and consequently harden as the foam expands. Polyurethane foam is probably the example most commonly used in industry, as it provides the filling for seats, cushions, and mattresses, is used in buildings for thermal insulation, and is injected into bicycle or car tyres to stop them deflating. Polymeric foams with a closed-cell structure are obtained by dispersing gas bubbles in a molten thermoplastic polymer (polystyrene, polyethylene, polyvinylchloride), which is already polymerized and hardens on cooling. As they are more rigid, they are used to make various light plastic objects. Expanded polystyrene is a good thermal insulator, so is used to make cups for hot drinks. There are also foams made from biodegradable polymers which are now used for packaging.

Metal foams have experienced a recent surge in popularity. Aluminium foam is lightweight yet strong, making it desirable as a structural material for walls, racing cars, and aircraft. It is very good at absorbing shocks (although it must be replaced after each impact), so is often used for bumpers on cars, trams, and even aeroplane nose-cones (to protect against bird strikes). Aluminium foam has a high surface area per unit volume, making it useful as a heat exchanger, for example in electronics, or as a catalyst and porous electrode in chemistry. Its diverse applications include uses in the military and aerospace industry, for example as weapon silencers and to stabilize (laminerize) flows in wind tunnels. Lead foam is obtained more easily than aluminium foam, because lead becomes molten at a lower temperature, but it is too soft for practical applications. On the other hand, steel foam is made at much higher temperatures, which makes it unprofitable, despite its strength and low density which would be valuable in making unsinkable ships. The large surface area, which facilitates more chemical reactions, in a small volume is put to good use in mobile phone batteries.

In many industrial processes, foam is an undesirable by-product; in the case of glass, it is particularly awkward because a homogeneous and transparent glass must not contain visible bubbles. However, manufacturers have recently turned this to their advantage and are now producing aerated glasses with around 20% gas. Foamed glass is light, cheap, watertight, does not pollute, is stable at temperatures up to 80 °C, and is also incredibly robust: in fact, a propagating fracture is stopped when it meets a bubble. It is already used for the foundations of houses and roads, and could become common in building for the same reasons as metal foams. It provides excellent thermal and acoustic insulation and doesn't contain fibres (unlike glass wool, rock wool, and asbestos). It exemplifies the importance to industry of producing aerated materials. Another foamed construction material is cellular concrete, which offers a compromise between mechanical strength and weight, with good thermal insulation.

4.2 Other cellular structures

Materials made of cells filled with a liquid or a solid, rather than gas, are visually similar to foams and in particular they tessellate space with non-overlapping cells without gaps between them.

Some of these systems are so similar to foams that they age in the same way (we tackle these aspects in chap. 3). This is true of concentrated emulsions (dispersions of one liquid in another, immiscible, one), of polycrystals (assemblies of monocrystals with different orientations in contact with each other) and magnetic garnets (assemblies of magnetic domains). We discuss this further on p. 77.

Vesicles are small water-filled sacs formed from lipids, typically around ten microns in size, which form a cellular pattern (FIG. 1.6a) when they assemble. As the lipid interfaces do not have the same tension, the contact angles are not necessarily 120°, unlike foams (see §2.2, chap. 2).

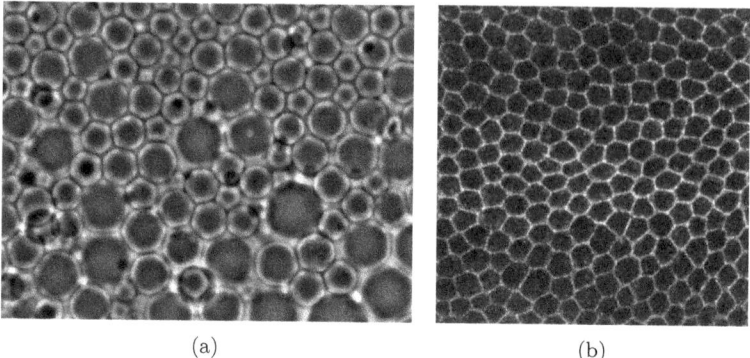

FIG. 1.6: **(a)** Vesicles are tiny droplets, typically of the order of 10 µm in size, enclosed by a lipid membrane. As they assemble, the competition between adhesion and curvature energies causes them to form a structure that resembles a foam. Photograph courtesy of D. Milioni. **(b)** Outlines of cells during the formation of a fruit fly (Drosophila) wing. Size of the image: 160 µm. Photograph courtesy of P.L. Bardet.

A biological tissue (epithelium) is a cellular arrangement in two dimensions. In a fly's eye, the facets are arranged in a hexagonal network, possibly even more perfectly than that of honeycomb. During the formation of the wings (FIG. 1.6b) or the thorax, the initially disordered pattern evolves towards a hexagonal network as the adult stage is approached, with the cells undergoing structural rearrangements similar to those seen in foams, which will be discussed in §1.2, chap. 3 and §3.2, chap. 4. These cell rearrangements are important in skin regeneration and wound healing. The cell assemblies are yield-stress materials, with energy barriers which can prevent them from relaxing to the global minimum of energy. In this sense they are closer to foams than liquids.

Finally, we give some more esoteric examples.

Carbon atoms in graphite make flat hexagonal arrangements which can roll up to form cylinders ("carbon nanotubes", FIG. 1.7a) or spheres ("buckyballs"). A liquid which solidifies changes density and cracks, forming patterns which are often cellular. Examples include salt lakes, basalt (like the Giant's Causeway in Northern Ireland), or simply a muddy puddle.

In cosmology, the universe consists of galaxies which enclose large regions of empty space between them in a structure which is often compared to a foam, although it is difficult to establish any link. Even more speculative is a recent measurement of cosmological radiation that suggests that the universe is finite but without boundary. One possible interpretation of this is that the universe is a dodecahedral cell with periodic boundary conditions, obtained by identifying the opposite faces pairwise (FIG. 1.7b). In other words, someone who sets off in a straight line comes back to his or her point of departure without ever having crossed the boundary (on earth, the analogy is of a journey that is always to the west, so that you finally return to your starting point from the east). This cell would have angles of 120° between faces and so to be able to

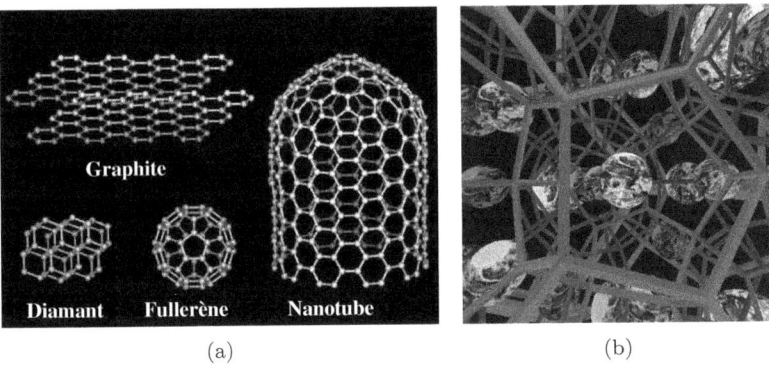

FIG. 1.7: Foams of all sizes. (a) Carbon atoms arranged in a regular hexagonal network, at a scale of the order of 10^{-10} m, which is the basis for fullerenes and nanotubes. (b) Some observations at the scale of the universe (10^{23} m) can be interpreted by assuming that the universe is an assembly of dodecahedra, with faces of constant curvature that meet at angles of 120° to form a periodic structure similar to a foam (cf. FIG. 2.21a). Image obtained using J. Weeks' *Curved Spaces* software [7]. See PLATE 2.

tessellate space with these cells, the space could not be Euclidean but instead would have positive curvature [3]. However, measurements by the Planck satellite suggest that this is not the case and indicate that the Universe is actually Euclidean.

5 Experiments

5.1 Three ways to make a foam

1) Whisking

difficulty level:	cost of materials: $
preparation time: 5 min	experimental time: 15 min

Phenomena demonstrated: *making a foam by whisking, foamability, liquid fraction, stability, loss of transparency, bubble size, appearance of elasticity.*

Materials:

- a bowl
- 3 eggs
- a whisk

1. Separate off the yolks and put the egg whites in the bowl.
2. Note the volume of egg white, its colour, and consistency.
3. Whisk the whites, stopping every 10 seconds to repeat the observations and, in addition, noting the bubble density and size.

2) Shaking

difficulty level:	cost of materials: $
preparation time: 5 min	experimental time: 1 hr

Phenomena demonstrated: *making a foam by shaking, foam disorder, loss of transparency, bubble size, appearance of elasticity.*

Materials:

- 9 plastic bottles
- water
- washing-up liquid

1. Mix air, water, and washing-up liquid, and shake.
2. For each of the three ingredients, put in a large amount, a little, or none at all.
3. Compare the foam produced by these nine combinations, noting the bubble density, the volume of foam, its transparency, colour, consistency, and stability.

3) Blowing

difficulty level:	cost of materials: $
preparation time: 5 min	experimental time: 1 hr

Phenomena demonstrated: *making a foam by beating and blowing, foam disorder, bubble size and geometry, appearance of elasticity.*

Materials:

- a large transparent bowl
- water
- washing-up liquid
- straws

1. Put the washing-up liquid in the water.
2. Make a foam either by agitating with a finger, blowing directly, or blowing through a straw.
3. Compare the foam produced by these three methods, noting the bubble density, the volume of foam, its transparency, colour, consistency, stability, and disorder (variety of bubble sizes).
4. When the bubbles are very large, observe their geometry, in particular that their faces meet in threes with an angle of 120° (see §2.2, chap. 2).

5.2 Chocolate mousse

difficulty level:	cost of materials: $$
preparation time: 5 min	experimental time: 20–40 min (then 2–3 hrs to cool)

Phenomena demonstrated: *making a foam by whisking, foamability, liquid fraction, stability, loss of transparency, bubble size, appearance of elasticity.*

Reference in the text: §1.2

Materials (for 6 people):

- 6 eggs
- 200 g chocolate
- three bowls
- a whisk

1. Separate the yolks from the egg whites.
2. Whisk the whites until they are very stiff: see §5.1.
3. Break the chocolate into small pieces and melt it in a microwave or in a saucepan containing a little hot water. Remove it from the heat, and allow to cool.
4. Mix the chocolate with the egg yolks, then mix all of this with the egg whites (try to be thorough, even if the foam almost completely collapses, as the mixture should be homogeneous).
5. Leave in the refrigerator for at least 2 hours (and not more than 24 hours).

Comments: The variations are countless. You could:

1. Put in more chocolate.
2. Add strong coffee or alcohol (to the melted chocolate), sugar (to the whites after whisking), crème fraîche (to the yolks), almonds (for decoration at the end), etc.
3. Replace two of the yolks with 150 g butter (mix thoroughly with the melted chocolate before adding to the yolks).
4. Or don't even separate the whites and the yolks! Whisk them together with icing sugar and rum, and add to the melted chocolate.

References

[1] J. AUBERT, A. KRAYNIK, and P. RAND, *Sci. Am.*, **254**, 74, 1986.
[2] J. BANHART and D. WEAIRE, *Phys. Today*, **July**, 37, 2002.
[3] J.-P. LUMINET, J.R. WEEKS, A. RIAZUELO, R. LEHOUCQ, and J.-P. UZAN, *Nature*, **425**, 593, 2003.
[4] *La Hulotte*, n° 28-29, 1er semestre 1997, p. 67 (inside cover).
[5] S. HILDEBRANDT and A. TROMBA, *Mathematics and Optimal Form*, Scientific American Library, W.H. Freeman, New York, 1985.
[6] D. Wender (trans.) *Hesiod* and *Theognis*, Penguin Classics, Harmondsworth, England, 1973.
[7] www.geometrygames.org/CurvedSpaces/index.html.fr

2
Foams at equilibrium

In this chapter we will discuss the structure and stability of foams at equilibrium. We start with a qualitative description of foam physics before turning to very dry foam structures. We will show how equilibrium properties are affected by the presence of a non-negligible amount of liquid in the foam, and we will describe the spatial distribution of this liquid.

1 Description at all length-scales

There are at least four length-scales at which we must consider the properties of a foam (FIG. 2.1):

- the observer's scale, of the order of a metre, at which a foam has the appearance of a soft, opaque solid (FIG. 2.1a).
- the millimetre scale (or even smaller, depending on the foam), at which bubbles can be distinguished. There are a small number of local geometric rules, called *Plateau's laws*, which describe how the bubbles pack together to form the foam skeleton (FIG. 2.1b).
- the micron scale, or slightly less, which shows how liquid is distributed between the bubbles (FIG. 2.1c).
- the nanometre scale, at which the molecular structure of the interfaces appears. The presence of particular molecules, like soap, which position themselves at the air/water interface, is essential to our understanding of the formation and stability of the interfaces (FIG. 2.1d).

In this introductory section we go through FIG. 2.1 in reverse, starting at the smallest scale and working our way up to the largest, to qualitatively describe the elementary physical properties that allow a foam to exist. In §2 we shall return to each of them to give a quantitative description.

1.1 At the scale of a gas/liquid interface

Surface tension and the Young–Laplace law Creating an interface between a liquid and a gas requires energy. This additional energy is γ, the energy per unit area of the interface, multiplied by the area of the interface created. It means that an interface tends to reduce its area, and it is therefore under tension. This is why γ is called *surface tension* or *superficial tension*. The word "superficial" refers here to the surface and not to something of little importance! On the contrary, this tension will be essential in what follows, because of the vast quantity of interface present in a foam (up to thousands of m² per litre, as discussed in exercise 7.1).

18 Foams at equilibrium

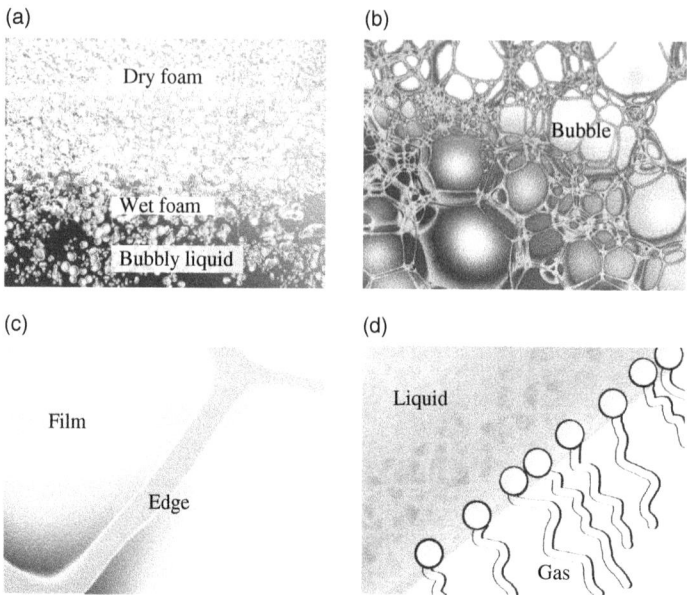

FIG. 2.1: Foam structure at different length-scales, with each image magnifying the previous one (scales are very approximate). (a) The whole foam, appearing wet at the bottom and dry at the top, typical scale $\sim 10^{-2}$–1 m. (b) Bubbles in a dry part of a foam, $\sim 10^{-4}$–10^{-2} m. (c) A Plateau border (liquid channel), $\sim 10^{-6}$–10^{-4} m, and a soap film (bubble wall), $\sim 10^{-8}$–10^{-6} m. (d) A gas/liquid interface containing molecules with long hydrophobic tails (in the gas) and hydrophilic heads (in the liquid) $\sim 10^{-10}$–10^{-8} m.

The shape of a liquid/liquid or gas/liquid interface is determined by the existence of this tension (see p. 35). For example, if the geometric constraints allow it, an interface is flat, while if an interface completely surrounds some fluid then (if the effect of gravity is negligible) the surface is spherical. In general, we determine the shape from the *Young–Laplace law: the pressure difference between the two sides of an interface is equal to the mean curvature of the interface multiplied by its surface tension* (see §2.1.3). The pressure is higher on the concave side. Surface tension acts to flatten the surface whilst the pressure difference tends to curve it.

Surfactants The additives used to make foams consist of molecules with a polar head and a long carbon chain for a tail. The head lowers this molecule's energy when it is surrounded by water molecules, so it is called *hydrophilic*, literally "water-loving". The tail increases the molecule's energy when it is surrounded by water molecules, so it is called *hydrophobic*. So the molecule both "likes" and "dislikes" water: it is *amphiphilic*. If such a molecule is dissolved in water, it therefore tends to adsorb at the air/water interface, with its head in the water and its tail in the air (FIG. 2.1d). It forms a layer which is one molecule thick, called a monolayer. Due to their effect on the surface tension of the interfaces to which they adsorb (see §2.1), amphiphilic molecules are also called *surface-active molecules*, or *surfactants* ("which act on the surface").

These molecules will adsorb at the surface until they cover it completely. If you then add more amphiphilic molecules, they remain in solution and don't play any significant role in the interfacial properties of a foam.

From interface to film: the role of surfactants Whether it is a foam in a river at the foot of a waterfall or a foam in a bottle of fizzy water that is shaken, there are many examples of bubbles which burst in less than a second. Each bubble which climbs to the surface is a small volume of gas enclosed in a film of water of several microns in thickness. In general, this type of thin film is unstable: under the effect of Van der Waals forces [22], the distance between the two air/water interfaces tends to diminish spontaneously. The film thins and then breaks.

If we instead add several drops of washing-up liquid to a bottle of water before shaking it, a similar foam forms, but this time it lasts. Why? Because certain amphiphilic molecules carry a small charge. For example, sodium dodecyl sulphate (commonly abbreviated to SDS, and known commercially as "lauryl sulphate", which is how it appears on the labels of washing-up liquids) is an *anionic* surfactant, meaning that it carries a negative electrostatic charge. The two air/water surfaces in a film are covered by charged monolayers that repel each other, stabilizing the film at the thickness at which electrostatic repulsion and van der Waals attraction cancel each other out (see §4.1.3). The repulsion between the interfaces is called the *disjoining pressure* (see FIG. 2.2a). We will see in §2.3, chap. 3 that this repulsion is not sufficient to guarantee the stability of a film, but that it contributes to it.

1.2 At the scale of a film

The familiar expression "soap film" is a two-fold abuse of language. On the one hand, rather than soap, which does not foam well, we often use washing-up liquid. Should we say "washing-up liquid film"? No! Because the film consists mainly of water, and there is only a small amount of surfactant.

Minimal surfaces It is easy to make a single film, for example by dipping a wire loop in a soapy liquid. When you take the wire out of the solution, the liquid is trapped between the wire and two liquid/gas interfaces. This system reduces its surface area as much as possible, under the constraints that the two interfaces mustn't touch each other (the surfactants prevent the formation of a hole) but must touch every part of the frame. The excess solution rapidly flows out, the two interfaces move closer together and adopt the same shape to leave a *thin film* of roughly constant thickness (much smaller than the dimensions of the frame or of a bubble). The film can therefore be described as a surface without thickness. Its shape obeys the Young–Laplace law for a single interface, but its surface tension is twice as large because of the two interfaces. Since it has the smallest possible surface area, it is called a *minimal surface* (see FIG. 2.2b). Since the two interfaces are open to the atmosphere, the pressure

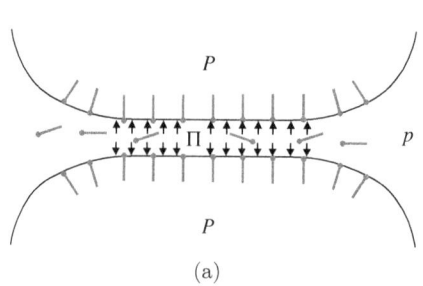 

FIG. 2.2: (a) The liquid in the thin films is at lower pressure than the gas, $p < P$. Mechanical equilibrium is determined by the surface tension where the interface is curved (Plateau borders) and by repulsion between two interfaces where the interfaces are very close (thin film). This repulsive force is fixed by the disjoining pressure Π, represented by arrows. (b) A minimal surface obtained with a soap film. The iridescence produced by the interference between the two interfaces of the film allow its thickness (of the order of microns) to be estimated. Photograph courtesy of F. Elias, see PLATE 3.

difference across the film is zero and the mean curvature of the film is also zero, a defining characteristic of minimal surfaces.

Film thickness A newly created film is not generally at equilibrium. It is a few microns thick and shows beautiful iridescence. If the film doesn't break, it thins due to gravity or *capillary suction* (see §1.3) until it reaches a thickness h of the order of a few tens of nanometres. The disjoining pressure then prevents further thinning of the film and ensures that it reaches mechanical equilibrium (see FIG. 2.2a).

From film to bubble A bubble is a soapy film of water which encloses a certain amount of gas. The presence of the gas is an additional constraint on the shape of the film, so that the film is no longer a minimal surface. A bubble assumes the smallest possible surface area to contain the gas, and if it is isolated it is spherical. The pressure in the bubble is slightly greater than atmospheric pressure, but not enough to compress the gas appreciably, and so its volume remains fixed. An isolated bubble ceases to be spherical when it is large enough to be sensitive to external forces, for example wind or gravity if it is several centimetres in size.

When two bubbles suspended in air come into contact, they spontaneously change their shape in order to share an interface and thus reduce the total interfacial area (see FIG. 2.8). The final shape results from the competition between the two bubbles each trying to minimize their surface area, so that they cannot both remain spherical. More generally, when a group of bubbles come together the conservation of volumes and the minimization of surface area leads to some simple laws which govern the local shape of the bubbles known as *Plateau's laws*, described in §2.2.

If the bubbles are suspended in water, putting two bubbles into contact leads to an increase in total surface area and so this does not occur spontaneously. The surface area of the film created between the two bubbles increases with the pressure with which they are squeezed together, known as the *osmotic pressure* (see §4.2).

1.3 At the scale of a bubble

Bubbles, films, and Plateau borders A bubble in the centre of a dry foam is polyhedral in shape because of its neighbours. Its faces are thin films that are gently curved either because of the pressure differences between the bubbles, or simply because its perimeter does not lie in one plane. The films intersect in threes along the edges (see FIG. 2.3a), which are liquid-carrying channels known as *Plateau borders* after Joseph Plateau (1801–1883). The curvature of the liquid/gas interfaces must remain finite (by the Young–Laplace law), which imposes a non-zero thickness on the Plateau borders (see FIG. 2.3b). The cross-section of each border is a small triangle with concave sides. Four Plateau borders intersect at the *vertices* (or nodes) of each polyhedral bubble.

From bubble to foam The amount of liquid contained in a foam is defined by the liquid volume fraction, $\phi_l = V_l/V_{\text{foam}}$, the ratio of the volume of liquid to the total volume of the foam. This quantity is linked to the density ρ of the foam by the relationship $\rho = \rho_l \phi_l + \rho_g(1 - \phi_l) \approx \rho_l \phi_l$, with ρ_l and ρ_g the densities of the solution and the gas respectively.

Different types of structure are obtained depending on the liquid fraction (FIG. 2.1a):
(1) for $\phi_l > \phi_l^*$ the bubbles are spherical and do not touch: this is a bubbly liquid;
(2) for $0.05 \lesssim \phi_l < \phi_l^*$ the bubbles touch and take the shape of a squashed sphere at each bubble/bubble contact: this is a wet foam;
(3) for $\phi_l \lesssim 0.05$ the bubbles are polyhedral and the Plateau borders have a negligible cross-section: this is a dry foam.

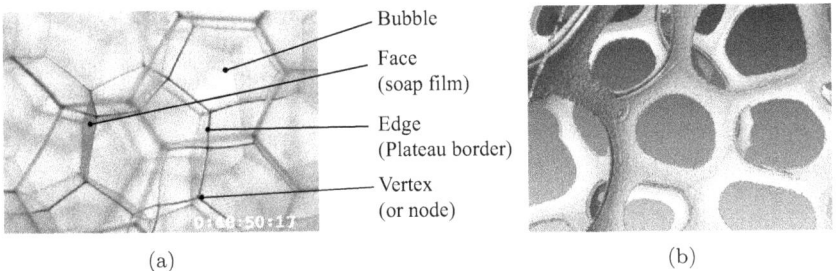

(a) (b)

FIG. 2.3: Bubble-scale structure of a foam. **(a)** Photo of a dry foam and definition of structural elements. Photograph courtesy of F. Elias. **(b)** The network of Plateau borders in a wet foam, obtained by X ray tomography at the ESRF (European Synchrotron Radiation Facility, see §1.2.4, chap. 5). The films are too thin to be detected. Image courtesy of R. Mokso.

The transition from bubbly liquid to wet foam occurs when the osmotic pressure vanishes (§4.2) [34], which corresponds to a precise value of liquid fraction, ϕ_l^*, close to 0.3 in 3D (see FIG. 4.17). Its value corresponds to the unfilled volume within a packing of hard spheres, as in a pile of oranges. It is a bit lower for an ordered foam structure than for a disordered structure, since if spheres are stacked in a regular fashion, rather than at random, more of them can be fitted in a given volume. The transition between wet foam and dry foam is less well defined. We speak of *dry foam* when the presence of the liquid doesn't play any role in the observed phenomenon, usually when the liquid fraction is less than 0.05.

The liquid phase is continuous: liquid circulates freely between the films and Plateau borders. At equilibrium, in the absence of gravity (see §1.4.1), the pressure in the liquid is thus constant. The lower the liquid fraction, the smaller the cross-section of the Plateau borders and the greater their curvature: they are therefore at lower pressure than the gas. In contrast, the curvatures of the films are small and of opposite sign for each interface. In the absence of disjoining pressure, i.e. as long as the film is at least a hundred nanometres thick, the pressure which exists between the interfaces is of the order of the gas pressure in the bubbles. That is why the liquid is drawn into the Plateau borders, a process known as capillary suction, which comes to a halt when the thickness of the film is sufficiently low that the disjoining pressure becomes non-negligible and balances it. At equilibrium the film is therefore thin and most of the liquid is found in the Plateau borders and in the vertices where they meet.

1.4 At the scale of a foam

1.4.1 Foam under gravity Below the millimetric scale, bubble pressures and surface tension are the only forces present. Above it, gravity causes the liquid to flow through the network of Plateau borders (see §4, chap. 3). At equilibrium, the pressure in the liquid is fixed by hydrostatic equilibrium, and so decreases with increasing height. The bubbles are not subject to gravity and the gas is, on average, at constant pressure. The pressure difference between the liquid and gas thus grows linearly with vertical position, allowing the radius of curvature of the Plateau borders and the liquid fraction to be deduced. A consequence of this is that at equilibrium there is more water at the bottom of the foam than at the top, as illustrated in FIG. 2.1a.

1.4.2 Quantity of interface At the scale of the bubble, we have seen that the quantity of liquid/gas interface is the smallest possible, subject to conserving the volume in the bubbles and, depending on the situation, the liquid volume or pressure. Yet at the scale of a foam the bubbles are never arranged in the optimal configuration. There are too many possible arrangements and too few bubble movements for it to be found. Instead, the foam is in a *local minimum* of energy: by moving a very small amount, it will increase its energy, but if it deforms further it may find a lower energy state.

2 Local equilibrium laws

Minimization of interfacial area, which governs the structure of a foam at equilibrium, is determined by simple geometric rules at the scale of a film and of a few bubbles. These laws, and the extent to which they are valid, are discussed in this section.

2.1 Equilibrium of fluid interfaces

2.1.1 Surface tension An increase in interfacial area by an amount dS induces a change in the energy (or the free energy, which we will make precise later) by an amount

$$dE_{\text{surf}} = \gamma dS, \qquad (2.1)$$

where γ is defined as the surface (or superficial) tension [22].

The tension γ can be considered an energy per unit area (J.m^{-2}) or a force per unit length (N.m^{-1}). One side of a line segment dl (see FIG. 2.4a) exerts a force $\gamma\,dl\,\hat{\mathbf{n}}$ on the other, with $\hat{\mathbf{n}}$ the unit normal to the segment in the plane of the film. For pure water, in air at room temperature, $\gamma_0 = 0.0727$ J.m^{-2} = 72.7 mN.m^{-1}.

A surface tension is defined for any interface that separates a liquid from a gas and, more generally, two immiscible fluids or a fluid and a solid. Unless indicated otherwise, we take the surface tension of a solution to be its surface tension in contact with air. The surface tension is always positive, since otherwise the interface would be unstable. It is only defined at a scale that is large compared to the size of the molecules, i.e. at a scale at which the interface can be treated as continuous. In this book we shall always assume that the size of the molecules at an interface is small compared to its characteristic size, given, for example, by its radii of curvature (see the definition in FIG. 2.6). So the energy depends only on the area of the interface and not on its shape: it is equal to the energy of a planar surface of the same area (although if the thickness h of the film is sufficiently low it may play a role; see §4.1.4).

At a microscopic level, the molecules of a liquid, which are attracted to each other, find an interface less favourable than when they are in the bulk [22]. This difference,

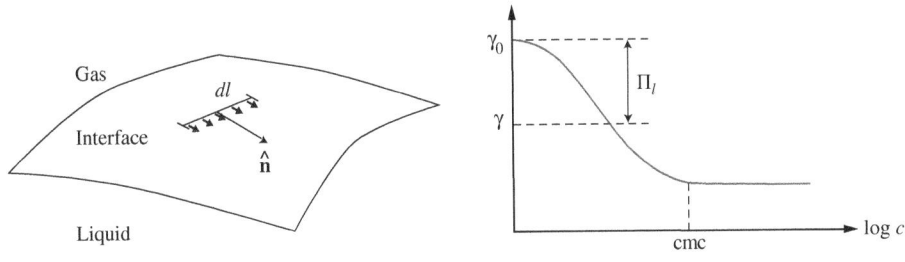

FIG. 2.4: Surface tension. (a) The surface tension is a force per unit length dl exerted along $\hat{\mathbf{n}}$ in the plane of the interface. (b) The change in surface tension of a solution with increasing surfactant concentration. The plateau marks the saturation of the interface by surfactant molecules and the corresponding concentration is called the critical micelle concentration (cmc).

which is a characteristic of the two fluids present, is the origin of surface tension. The surface tension depends on the temperature and what the liquid contains: in general, it is reduced when the temperature rises or when there are impurities. In particular, its value is significantly lowered by the presence of so-called *surface-active* or *amphiphilic* molecules (see §6.1, chap. 3). The more amphiphilic molecules are added, the more the surface tension is lowered, until a concentration, the *critical micelle concentration*, is reached at which the surface is saturated with amphiphilic molecules and the surface tension no longer varies. An example of the change in γ with surfactant concentration is shown in FIG. 2.4b. The difference between the surface tension of a pure liquid and the surface tension of the liquid in the presence of solutes is called the *surface pressure* or *Langmuir pressure*, denoted Π_l (see §2.2.1, chap. 3).

2.1.2 Film tension

FIG. 2.5 shows a thin film produced by drawing a wire frame from a bowl of soapy liquid. A horizontal slice through this film would have a cross-section of length L and thickness $h \sim 10^{-7}$ m $\ll L$, with perimeter $2L + 2h \sim 2L$ and area hL. The force exerted by the lower part of the film on the upper part has two contributions, one from the surface tension and the other from the pressure: $\mathbf{F}_\gamma = -\gamma\, 2L\, \mathbf{u}_z$ and $\mathbf{F}_p = p\, hL\, \mathbf{u}_z$, with $\mathbf{u}_z$ denoting the upwards vertical direction and p the pressure in the liquid.

The force exerted on this slice by the reference pressure p_{ref} (in general equal to P_{atm}) is $\mathbf{F}_{\text{ref}} = p_{\text{ref}} hL\, \mathbf{u}_z$ and so the excess force due to the presence of the film is $\mathbf{F} = [-2\gamma L + (p - p_{\text{ref}})hL]\mathbf{u}_z$ (see exercise 7.2). With $\gamma \sim 5 \times 10^{-2}$ N/m, $p - p_{\text{ref}} < 10^4$ Pa (see p. 48) and $h \sim 10^{-7}$m, we find that the second term is negligible. The excess force per unit length of film sliced, called the *film tension*, is therefore very well approximated by

$$\gamma_f \approx 2\gamma. \tag{2.2}$$

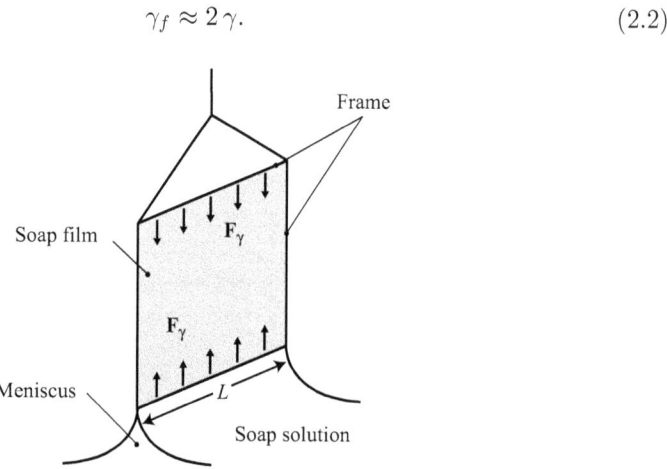

FIG. 2.5: Experimental measurement of surface tension. A soap film is formed by lifting a frame out of a soap solution. The film meets the solution in a liquid meniscus. Measuring the force $\mathbf{F}_\gamma$ necessary to keep the frame in equilibrium allows the surface tension to be determined (see §1.1.1, chap. 5).

Taking into account both the pressure term and a slight change in surface tension with thickness leads to variations of the order of 0.1%, which will be discussed in §4.1.4.

If the frame is moved vertically upwards by a distance dx, the associated energy change is, by eq. (2.1), $2\gamma L dx$. This quantity also represents the work done, $\mathbf{F}dx$.

2.1.3 The Young–Laplace law If an interface is not flat then the surface tension induces normal forces which are compensated, at equilibrium, by the pressures on each side (see FIG. 2.6b). There is a relationship between the pressures, the tension, and the shape of the interface, which we derive for an isolated bubble in the shaded section on p. 27, and state below for any curved surface.

We denote by S the area of an interface of tension γ between a domain A and a domain B; M denotes points on the interface, and $\mathbf{n}(M)$ is the normal at M oriented from A towards B. It is characterized locally by its principal radii of curvature R_1 and

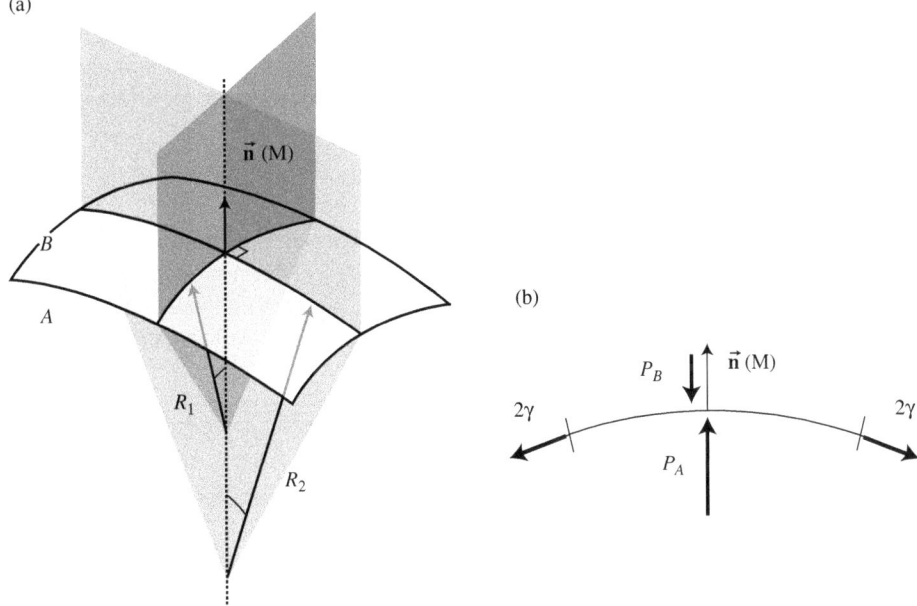

FIG. 2.6: The Young–Laplace law. (a) The interface separates space into two half-spaces A and B, with pressure P_A and P_B respectively. A plane parallel to the normal to the surface cuts it along a line with radius of curvature R, the radius of curvature of the surface in the direction given by the plane of the cut. By convention it is considered positive if the centre of curvature is in A. R takes different values as the plane of the cut rotates around the normal, and takes its extreme values in two particular directions, called the principal radii of curvature, R_1 and R_2, which turn out to be perpendicular. (b) Viewing a cross-section through the film illustrates the component of the force due to surface tension on an element of the surface in the direction of its normal $\mathbf{n}$. A pressure difference $P_A > P_B$ ensures that equilibrium is attained.

R_2, defined in FIG. 2.6a. By convention, the radius is counted positive if the centre of curvature is in A and negative if the centre of curvature is in B. We describe the changes in surface area and volume during a characteristic deformation in which each point M of the interface is moved a distance dx in the direction of the local normal $\mathbf{n}(M)$. The change in volume of A is thus $dV = Sdx$ and the change in surface area is $dS = HSdx$, with $H = 1/R_1 + 1/R_2$ the mean curvature of the surface.

At equilibrium, the energy change associated with this virtual transformation is zero (to first order in dx). Then $-\Delta P dV + \gamma dS = 0$ where $\Delta P = P_A - P_B$ is the difference in pressure between domain A and domain B. Finally, the curvature of the interface and the difference in pressure between the two regions separated by the interface balance:

$$\Delta P = P_A - P_B = \gamma H = \gamma \left(\frac{1}{R_1} + \frac{1}{R_2} \right). \tag{2.3}$$

This equation is called the *Young–Laplace law*, after Pierre-Simon de Laplace (1749–1827) and Thomas Young (1773–1829), or sometimes just the *Laplace law*. It indicates that surface tension tends to reduce the curvature of an interface, making it more planar, and that it is counterbalanced by a pressure difference, which tends to bend the interface.

2.2 Plateau's laws

Plateau's laws were formulated in 1873 by Joseph Plateau [52], but only fully proved in 1976 by Jean Taylor [2, 59]. They are based on an *ideal foam* model, defined below, which does not take into account all the details of the physics and chemistry of a foam. The laws stated in §2.2.2 remain valid for real foams as long as the hypotheses of the model are approximately satisfied (see §4).

2.2.1 Ideal foam
We define an ideal foam by making the following simplifications:

(1) The foam is *very dry*.
    The liquid occupies a negligible part of the total volume of the foam and we assume that the liquid volume fraction ϕ_l is zero (see its definition on p. 21). Experimentally, liquid fractions of $\phi_l \sim 10^{-4}$ to 10^{-2} are attainable.

(2) The foam is *at mechanical equilibrium* and is thus static.
    The forces within the foam are equilibrated and the foam is at rest, in a local energy minimum. At the scale of the bubbles, thermal fluctuations are negligible; other phenomena can alter the foam, however (see §1, chap. 3), and approximation (2) is only correct if the characteristic time of these changes is much greater than the time taken for the foam to return to mechanical equilibrium (typically a millisecond for a soap foam).

(3) The foam has an *energy proportional to the surface area of its bubbles* (or, for a monolayer of bubbles, proportional to the total perimeter; see §5 on 2D foams).
    The thin films separating the bubbles are "ideal": they have zero thickness and film tension 2γ (the factor 2 comes from the fact that each film has two interfaces, see

§2.1.2). This implies that each air/water interface has the same surface tension, as is the case for a foam at equilibrium if the monolayers of surfactant are themselves at equilibrium (see §2.2, chap. 3); equivalently, all other contributions to the energy (gravity, for example) are negligible, which is the case for very dry foams.

(4) The foam is *incompressible*.
The volume V of each bubble is fixed (or the area A for a monolayer of bubbles, see §5 on 2D foams). This approximation is valid over sufficiently short times that the transfer of gas between bubbles can be neglected (see §3, chap. 3), and remains true if we apply an overpressure to the foam, provided it remains small compared to atmospheric pressure (see exercise 7.6, chap. 4 and the shaded section on p. 27).

2.2.2 Local structure of an ideal foam

The Young–Laplace law for a single interface (eq. (2.3)) applies to an ideal film of film tension 2γ (see §2.1.2 and exercise 7.2). We consider two bubbles i and j, separated by a face S_{ij} with principal radii of curvature $R_{1,ij}(M)$ and $R_{2,ij}(M)$ counted as positive if the centre of curvature is in bubble i, M being any point of the surface. The mean curvature of the film $H_{ij} = 1/R_{1,ij} + 1/R_{2,ij}$ satisfies eq. (2.3) and is consequently *independent of M*:

$$P_i - P_j = 2\gamma H_{ij}. \tag{2.4}$$

The films are thus surfaces of constant mean curvature: this is **Plateau's first law**. For example, an isolated bubble has a spherical shape (see shaded section below).

We deduce from this that the sum of the curvatures of three adjacent faces is zero. To see this, note that each time a film is crossed, we add the pressure differences. Traversing a loop from bubble i to j to k and then back to i, the sum of the pressure differences is zero: $(P_i - P_j) + (P_j - P_k) + (P_k - P_i) = 0$. In other words,

$$H_{ij} + H_{jk} + H_{ki} = 0. \tag{2.5}$$

This is a strong constraint on the equilibrium structure of a foam.

Equilibrium of a bubble
The Young–Laplace law predicts that the overpressure in a bubble is a function of its radius (see eq. (2.4)). We demonstrate this directly in the following two examples.

a – Force balance on half a bubble
Imagine a spherical bubble with radius R that has been cut in half, as shown in FIG. 2.7. The gas is confined by a film of negligible thickness h ($h \ll R$) and of tension 2γ. Mechanical equilibrium is a result of the balance between the pressure forces acting on the internal and external surfaces $F_{\Delta P} = \pi R^2 \Delta P$ and the surface tension force $F_\gamma = 4\pi R\gamma$ acting on the cut perimeter of the bubble (see FIG. 2.7b).

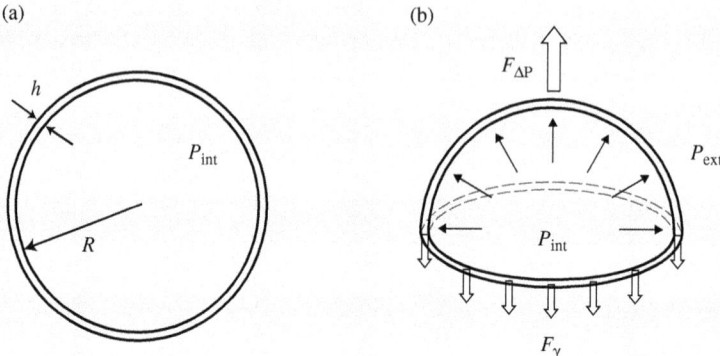

FIG. 2.7: (a) Cut through a soap bubble of radius R and film thickness h. (b) The forces acting on the upper half of the bubble.

These two forces are oriented perpendicular to the cut. This leads to the expected overpressure:

$$P_{\text{int}} - P_{\text{ext}} = 4\frac{\gamma}{R}. \tag{2.6}$$

b − Free energy balance over the whole bubble

In order to minimize the free energy E of the system consisting of the bubble and the external and internal gas, we can express how it changes with R, at constant temperature T, as:

$$dE = -P_{\text{int}}\, d\left[\frac{4}{3}\pi R^3\right] - P_{\text{ext}}\, d\left[-\frac{4}{3}\pi R^3\right] + 2\gamma\, d\left[4\pi R^2\right]$$

$$= 4\pi R\left[R(P_{\text{ext}} - P_{\text{int}}) + 4\gamma\right] dR. \tag{2.7}$$

The ideal gas law allows us to express P_{int} as a function of the radius R and of $nR_{GP}T$, with R_{GP} the ideal gas constant. Then integrating gives

$$E = -nR_{GP}T \ln R^3 + P_{\text{ext}}\frac{4}{3}\pi R^3 + 2\gamma 4\pi R^2 + \text{cst}.$$

The free energy increases when R becomes very large, because the external pressure and the surface tension both act to oppose growth. E also increases if R becomes very small, because the internal pressure prevents the bubble from collapsing. The curve $E(R)$ does indeed have a minimum when $dE/dR = 0$, given by eq. (2.6).

Up to bubble radii of the order of tens of microns, this overpressure is very small compared to atmospheric pressure. For a bubble of radius 2 mm and tension 50×10^{-3} N/m, $4\gamma/R \approx 100$ Pa $= 10^{-3}$ atm. The volume of gas may be considered fixed in most cases.

Plateau's second law concerns the way in which three films meet along an edge. Consider, for example, two bubbles of the same volume V touching each other (FIG. 2.8a). The system reduces its energy by sharing part of the interface; by symmetry the shared film is a circular disk, with radius denoted by a. We assume that each bubble is spherical, with a truncated region (which is the case in practice). The system is thus characterized by the radius R of the sphere and the radius a of the shared film. These two parameters are linked by conservation of volume: if a increases, R must likewise increase to be able to contain the volume driven out of the truncated region. Minimization of the surface area thus implies a compromise between the gain associated with the sharing of a part of the interface and the increase in radius of the sphere. Owing to the volume constraint, a single parameter must describe the structure and, rather than R or a, the appropriate parameter is the angle α defined in FIG. 2.8a, from which R and a can be deduced. An expression for the area, and thus the total energy of the system, as a function of α can be obtained analytically. The result is shown in FIG. 2.8b, and the minimum is obtained when $\alpha = 60°$.

The angle at which the three films meet is $\beta = 180 - \alpha = 120°$. This value can also be found from a different line of reasoning: the (vectorial) forces of surface tension exerted by each of the three films cancel out if and only if the angle between each pair of adjacent films is $120°$. When a third bubble is added, it minimizes the surface area of the system by sharing a film with each of the first two bubbles. These three films again intersect at $120°$.

Plateau's third law is obtained by adding a fourth bubble to the system: at the centre there are four edges connected at a *vertex* with tetrahedral symmetry (FIG. 2.9). This angle of $109.47°$ between the edges is the only possible choice that allows angles of $120°$ between the films. The angles found in these configurations are identical for more complex arrangements of bubbles and constitute a rule of assembly for films in a foam. In particular, a vertex at which more than four edges meet can always be separated into a number of four-fold vertices, reducing the energy (see exercise 7.3 in 2D).

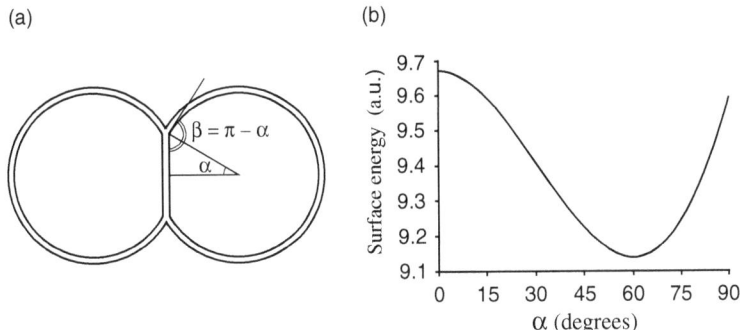

FIG. 2.8: The double bubble. (a) Two bubbles share a common film to minimize their surface area. (b) The surface energy of the double bubble as a function of the angle α. The only equilibrium solution is given by the energy minimum.

30 *Foams at equilibrium*

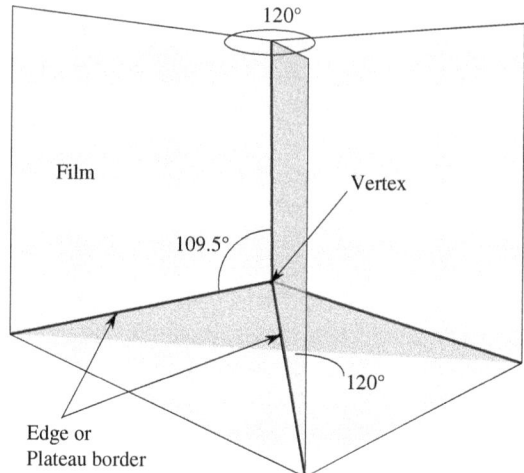

FIG. 2.9: Meeting of 4 bubbles: there are 6 films, 4 edges, and 1 vertex oriented at characteristic angles to each other, in a configuration that is found throughout a foam.

Summary of Plateau's laws

(1) Equilibrium of faces: the soap films are smooth and have a constant mean curvature which is determined by the Young–Laplace law (eq. (2.4)).
(2) Equilibrium of edges: the films always meet in threes along edges, forming angles of $120° = \arccos(-1/2)$.
(3) Equilibrium of vertices: the edges meet four-fold at vertices, forming angles of $\theta_a = \arccos(-1/3) \approx 109.5° \approx 1.91$ rad.

These three laws are a necessary and sufficient condition to ensure mechanical equilibrium of an ideal foam. If an infinitesimal perturbation (that preserves the bubble volumes) is imposed on the films in a foam structure that obeys these laws, the amount of interface, and thus the energy, is increased. Plateau's laws have two-dimensional equivalents, given on p. 59.

2.2.3 Condition at solid walls It is necessary to add to these laws a condition for contact with a solid wall: at equilibrium, the angle between a solid wall and a film in an ideal foam is 90°. In practice, this result is valid if the radius of curvature of the solid wall is large compared to the size of the Plateau border.

3 Dry foams

Despite their varied properties (chemical composition, surface tension, bubble size), all foams, and also highly concentrated emulsions, have a very similar appearance. In effect, the bubbles fill space in a disordered way. This leads to rules for the local arrangement of bubbles with their neighbours (topology), their shapes and sizes (geometry), and the relationship between topology and geometry. In order to understand the properties that all foams share, we consider again the simplified *ideal foam*, defined on p. 26, that is very dry, at mechanical equilibrium, and for which the energy is proportional to the interfacial area. Whenever it makes things simpler, we will consider a two-dimensional foam (see §5).

3.1 Number of neighbours: topology

3.1.1 Euler constant To a mathematician, a dry foam is a partition of space, that is, a division of space into many regions without overlaps or gaps. A 2D foam is a partition of the plane. Another example of a 2D partition is the division of a country into counties: each town belongs to exactly one county (no town belongs to two counties or to none). On a sheet of squared paper, the small squares partition the sheet.

We define a face to be an idealized surface separating two bubbles. There are as many faces as soap films in a foam. There is a relationship between the total numbers of bubbles, faces, edges, and vertices in space:

$$-N + N_{\text{faces}} - N_{\text{edges}} + N_{\text{vertices}} = \chi_{\text{Euler}} \; (= \text{constant}). \tag{2.8}$$

This formula is due to the mathematician Leonhard Euler (1707–1783); it is a topological relationship that is independent of the shape of the bubbles and is true even for a foam which is not at equilibrium.

The constant χ_{Euler} in eq. (2.8) is a small integer (check it for some simple examples!). It is typically equal to one, but may be zero or two if the foam is contained in a closed box or has a free surface (see shaded section on p. 43). The important thing is that it doesn't change if bubbles or faces are added or removed. For example, the transformation shown in FIG. 3.3 conserves the number of bubbles, increases the number of faces by one, the number of edges by two (one disappears and three appear), and the number of vertices by one (two disappear and three appear): the change in eq. (2.8) is thus $- \, 0 + 1 - 2 + 1 = 0$.

In two dimensions, Euler's formula is

$$N - N_{\text{edges}} + N_{\text{vertices}} = \chi_{\text{Euler}}. \tag{2.9}$$

For example, in FIG. 3.8 one bubble is lost, three edges are lost (five lost and two gained), and two vertices are lost, so $-1 - (-3) + (-2) = 0$. This can also be seen in FIGS. 3.4 and 3.7.

3.1.2 Average number of neighbours In the shaded section below we show a remarkable consequence of eq. (2.9) [30, 62]: in an ideal 2D foam with a large number of bubbles, the average[1] $\langle n \rangle$ of the number n of neighbours of the bubbles is always close to 6. More precisely, $\langle n \rangle$ is equal to 6, less a small correction that depends on the periphery of the foam (see shaded section on p. 43) and on N:

$$\langle n \rangle = 6 - \frac{\text{cst}}{N}. \tag{2.10}$$

---

[1] By convention in the whole of this chapter, angle brackets indicate an average over all bubbles in the foam, or, if given with a subscript, over a certain collection of bubbles.

In a real foam there are other, generally small, corrections, due for example to the non-zero liquid fraction. We emphasize that this rule does not fix the number of sides of each bubble, it is only the average which is fixed at 6. Indeed, the proportion of 5-sided bubbles is often greater than the proportion of 6-sided bubbles. If there are bubbles with 3, 4, or 5 sides, there must also be some with 7 or 8 sides, or even more.

To describe the difference from a hexagon, we define the quantity $q_t = 6 - n$, called the *topological charge* of a bubble. The average over the foam of the topological charge is zero for a large number of bubbles (its value is of the order of $1/N$) [28].

Consequences of Euler's formula and Plateau's laws in 2D [30, 62].
In two dimensions, n refers to the number of sides (or edges) of a bubble or, equivalently, its number of vertices. Thus $\langle n \rangle$ is the average number of sides per bubble, the average being taken over all N bubbles in the foam.

If N is sufficiently large that the effects of the periphery of the foam can be neglected, each edge is shared between two bubbles, $\langle n \rangle \approx 2 N_{\text{edges}}/N$. Further, each edge links two vertices and each vertex is shared between the three edges which meet there (see Plateau's laws in 2D on p. 59), so $N_{\text{vertices}}/2 = N_{\text{edges}}/3$. Putting this together, we can write

$$N = \frac{2}{\langle n \rangle} N_{\text{edges}} = \frac{3}{\langle n \rangle} N_{\text{vertices}}.$$

Then Euler's formula in 2D (eq. (2.9)) gives

$$1 - \frac{\langle n \rangle}{2} + \frac{\langle n \rangle}{3} = \frac{\chi_{\text{Euler}}}{N}. \tag{2.11}$$

The right-hand side tends towards 0 as the foam becomes large and we thus obtain $\langle n \rangle = 6$. This reasoning applies to any partition in which $\bar{z}$, the average number of edges which meet at a vertex, is known. Then $1/\langle n \rangle = 1/2 - 1/\bar{z}$. If all the vertices are four-fold, for example the squares on a sheet of squared paper, $\langle n \rangle$ and $\bar{z}$ are each equal to four and we verify that $1/4 = 1/2 - 1/4$.

Returning to three dimensions, we can show (see exercise 7.4) that for each *individual* bubble there is a link between its number of faces $f = N_{\text{faces}}$ and its average number of edges per face $\bar{n}$. A simple counting argument using Plateau's laws implies that

$$f = \frac{2}{\bar{n}} N_{\text{edges}} = \frac{3}{\bar{n}} N_{\text{vertices}}. \tag{2.12}$$

Inserting this into Euler's formula (eq. (2.8)) gives

$$6 - \bar{n} = \frac{12}{f}. \tag{2.13}$$

Whether the number of faces is very small or very large, $\bar{n}$ is thus always strictly less than 6. Put another way, no bubble has all hexagonal faces, and there must be some faces with 5, 4, or 3 edges.

Eq. (2.13) is easily tested on simple polyhedra such as those of FIG. 2.13, where $\bar{n} = 3, 4$, or 5, respectively. A familiar example is a football, which has 20 hexagonal and 12 pentagonal faces: here $\bar{n} = (6 \times 20 + 5 \times 12)/32$, and $6 - \bar{n} = 12/32$, which obeys eq. (2.13). Another example is the Kelvin cell (see the shaded section on p. 50), a truncated octahedron which tessellates 3D space (just as hexagons tessellate the plane). It has 8 hexagonal and 6 square faces, so $f = 14$, $\bar{n} = (8 \times 6 + 6 \times 4)/14$, and then $6f - \bar{n}f = 6 \times 14 - 8 \times 6 - 6 \times 4 = 12$, as required.

If eq. (2.13) is averaged over all the bubbles of a foam (cf. exercise 7.4) we find

$$\langle f \rangle = \left\langle \frac{12}{(6 - \bar{n})} \right\rangle. \tag{2.14}$$

In 3D, unlike 2D, the average number of faces in a foam is not fixed. If the foam is disordered, $\langle \bar{n} \rangle$ is often quite close to 5 (FIG. 2.10), and $\langle f \rangle$ is typically between 12 and 15 (and often between 13 and 14). On the other hand, the variance $\mu_2^f \equiv \langle f^2 \rangle - \langle f \rangle^2$ of the distribution of the number of faces in a foam is very variable: this is one measure of the disorder of a foam (see §3.2.1, chap. 5). An example of the distribution of f is given in FIG. 3.24.

3.1.3 Rivier's theorem For each bubble in a 3D foam, the number of faces that have an odd number of edges is even [57, 58]. To see this, assign to each face the number $(-1)^n$, where n is its number of edges. For any bubble, the product of $(-1)^n$ over all faces is $(-1)^{2N_{\text{edges}}}$, because each edge is counted twice, which is equal to 1. Hence the result.

Applying the same argument to any closed surface in a 3D foam, i.e. the outside of a group of connected bubbles, shows that the surface also has an even number of faces that have an odd number of edges. A corollary is that if a line is drawn

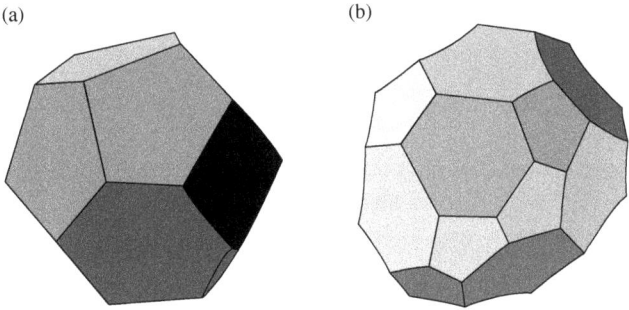

FIG. 2.10: Examples of simulated 3D bubbles. **(a)** A bubble with $f = 13$ faces that is commonly observed in foams [48]: it has one face with 4 edges, 10 with 5 edges and 2 with 6 edges. **(b)** A bubble with $f = 26$ faces, all either pentagonal or hexagonal; although it is at equilibrium and unstressed, it is irregular and elongated [18].

between the centres of any two bubbles that share a face with an odd number of edges, there is an even number of lines entering the closed surface. These "Rivier lines" can form loops, or end at the surface of the foam, but they never terminate in dangling ends.

3.1.4 Correlations between neighbouring bubbles

In disordered foams we observe that bubbles with more faces generally have neighbours with few faces.

More specifically, take a 3D bubble with f faces and write $\langle f \rangle_{\text{neighbours}}$ for the average number of faces of its f neighbours. Both experiments and simulations suggest that f and $\langle f \rangle_{\text{neighbours}}$ are inversely correlated, that is if one is large (compared to the average value) then the other is small.

It is not a general rule, but Lewis, Aboav, and Weaire have all observed in practice that $\langle f \rangle_{\text{neighbours}} \sim A + B/f$ where A is a constant slightly less than $\langle f \rangle$ and B is another positive constant [1, 56]; see FIG. 2.11 and eq. (5.13) for an alternative expression.

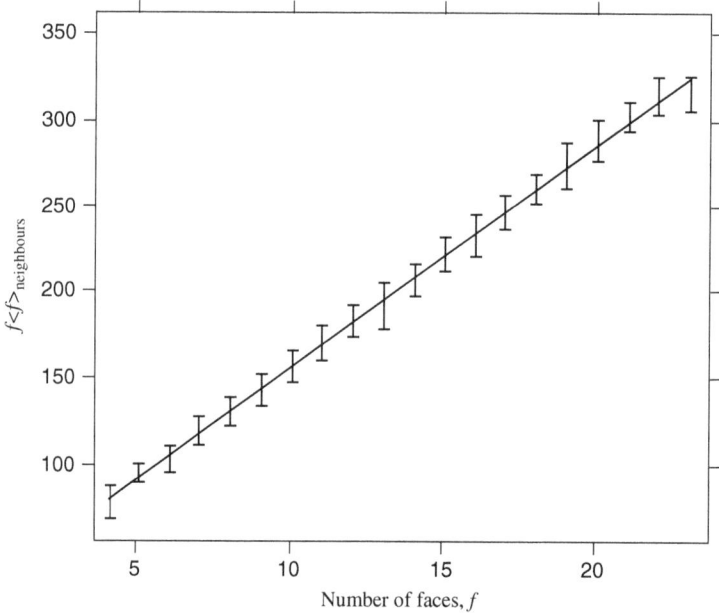

FIG. 2.11: Example of the relationship between the number of faces f of a bubble and the number of faces of its neighbours, in a foam obtained by numerical simulation with the *Surface Evolver* [38]. The midpoint of each error bar is the average of $f \langle f \rangle_{\text{neighbours}}$ over bubbles with the same f, and the length of each error bar represents the standard deviation. This representation underlines the affine variation of this term with f, as observed by Aboav and Weaire (solid line, eq. (5.13)).

3.2 Bubble geometry

3.2.1 Energy of a foam
Since each soap film has two gas/liquid interfaces, the surface energy (eq. (2.1)) of a foam can be written

$$E = 2\gamma \sum_{0 \leq i < j \leq N} S_{ij} = \gamma \sum_{0 \leq i,j \leq N} S_{ij} = \gamma \sum_{i=0}^{N} S_i = \gamma S_{\text{int}}, \quad (2.15)$$

where the index 0 means the region outside the foam, S_{ij} is the area of the face separating bubble i and j ($S_{ij} = 0$ if bubbles i and j are not neighbours, $S_{ij} = S_{ji}$, and $S_{ii} = 0$), $S_i = \sum_j S_{ij}$ is the surface area of bubble i, S_0 is the surface area of the outside of the whole foam, and $S_{\text{int}} = \sum_{i=0}^{N} S_i$ is the total amount of gas/liquid interface in the foam.

Eq. (2.15) can be non-dimensionalized by the average bubble volume, $\langle V \rangle = \frac{1}{N} \sum_{i=1}^{N} V_i$:

$$\frac{E}{\gamma \langle V \rangle^{2/3}} = \sum_{i=1}^{N} \frac{S_i}{\langle V \rangle^{2/3}}. \quad (2.16)$$

In 2D, the expression is similar:

$$\frac{E}{\lambda \langle A \rangle^{1/2}} = 2 \sum_{0 \leq i < j \leq N} \frac{\ell_{ij}}{\langle A \rangle^{1/2}} = \frac{L_{\text{int}}}{\langle A \rangle^{1/2}}, \quad (2.17)$$

where λ is the line tension, ℓ_{ij} and L_{int} are the 2D equivalents to S_{ij} and S_{int} in eq. (2.15), and $\langle A \rangle$ is the average bubble area (see §5.2). At equilibrium a foam is at a minimum of energy. Eq. (2.16) shows that the *existence* of surface tension is crucial in this minimization; its *value*, however, as well as the value of the average bubble size, which vary from one foam to another, don't play an important role. Instead, it is the shape of the bubbles that determines the minimum, and this is a purely geometric problem. That is, the equilibrium structures of foams are alike, irrespective of their average bubble size and chemical composition (although they do depend on the *polydispersity*, that is the width, or variance, of the bubble size distribution, see p. 252). Because of this universality, an expert can generally recognize from a photo if the foam is at equilibrium.

Energy minimization tends to reduce the surface area of the bubbles and would also reduce their volume if the bubble pressures did not oppose this. These pressures adapt in order to keep the volumes constant, and are slightly higher than the pressure of the outside atmosphere (see shaded section on p. 27). Foams which have a free surface, like the head on a glass of beer, obey (see exercise 7.6)

$$E = \frac{3}{2} \sum_{i=1}^{N} (P_i - P_{\text{atm}}) V_i. \quad (2.18)$$

Here P_{atm} is the gas pressure outside the foam, P_i is the pressure of bubble i, and E is the energy given by eq. (2.15). This expression, written here in three dimensions, generalizes to all dimensions (see exercise 7.6).

3.2.2 Global and local minimization Energy minimization causes *frustration*: an almost spherical bubble minimizes its surface area, but increases that of its neighbours, so that a foam must rapidly find a compromise. While it is out of equilibrium, after some external perturbation for example, everything that may be adjusted (curvature of the faces, position of the vertices, length of the edges) is adjusted while respecting the constraints (the number of bubbles and the bubble volumes). The total surface area thus decreases until Plateau's laws are satisfied throughout.

During this approach to equilibrium, it may happen that an edge shrinks without managing to satisfy Plateau's laws. Its length decreases and reaches zero, and its two ends meet. A vertex at which 6 edges meet is created (in 3D). This situation is unstable (see exercise 7.3 in 2D) and gives rise to the formation of a new face (a T1 process, see §1.2.2, chap. 3). The reverse process is equally likely. The number of faces and edges is consequently not fixed.

With or without T1 processes, when the surface area (or energy) of a foam reaches a minimum, it is at equilibrium, and stops moving. A foam cannot spontaneously overcome the energy barrier which separates one equilibrium configuration from another. In particular, thermal fluctuations with amplitude $k_B T \sim 4 \times 10^{-21}$ J, are negligible in comparison to the energy barriers between equilibrium states, which are much larger (see p. 80) and the foam is unable to change its configuration. The equilibrium configuration of a foam thus depends on its history, including, for example, the way in which the sample was prepared.

If the bubbles are generated randomly, there is consequently almost no chance of them spontaneously assuming, among all possible equilibrium configurations, the one with the smallest possible surface area or energy (FIG. 2.12). We say that a foam is

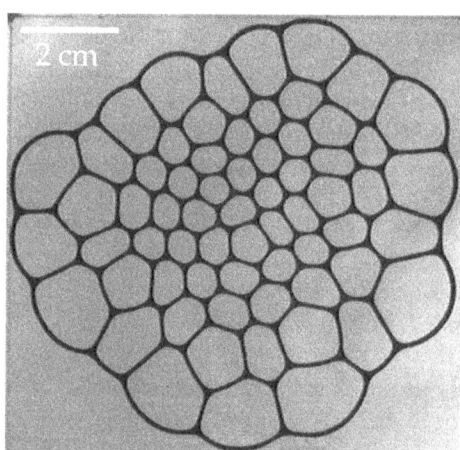

FIG. 2.12: Each bubble of this 2D foam has a similarly sized neighbour. Therefore each bubble is almost regular. It is a rare (and artificial) example where the foam might be expected to approach its minimum energy configuration [28]. Photograph courtesy of E. Janiaud [28]. Reproduced by kind permission of The American Physical Society.

in a *local, not a global, minimum of energy*. In fact, it is not proven (for a foam of more than three bubbles) which configuration corresponds to the global minimum of energy, even in the simple case where all the bubbles are exactly the same size (see the shaded section below).

Global minimization and the Weaire–Phelan structure

What is the *global* minimum of surface area, and therefore the global energy minimum, of a foam in which the bubbles are *all exactly the same size*? This is an elegant mathematical problem, made difficult because all candidates have comparable surface area (see eq. (2.19)). The theoretical lower limit is a sphere, which has $S/V^{2/3} \approx 4.836$.

In three dimensions, if all the bubbles have exactly the *same shape*, they have the same pressure and consequently every face has zero mean curvature. An idealized bubble with zero mean curvature (flat faces), defined on p. 44, which cannot be realized in practice, has $f = 13.4$ and $\bar{n} = 5.1$; its surface area is $S/V^{2/3} = 5.254$, a value which is probably impossible to beat. The best real bubble is conjectured to be the Kelvin cell (see shaded section on p. 50), with 14 faces and an average of 5.14 edges per face; its surface area is $S/V^{2/3} \approx 5.306$, believed to be the smallest possible value for a bubble which tiles space on its own.

Denis Weaire suggested that it might be possible to further lower the surface area by introducing as many pentagonal faces as possible (there are none in the Kelvin cell!), so that each face has close to $n = 5.1$ edges (and not just on average) and is consequently not very curved. As bubbles with only pentagonal faces are unable to tessellate space on their own, it is necessary to mix bubbles of at least two different shapes (but, in this case, with the same volume). Weaire proposed candidates inspired by known crystalline structures, and his student Phelan numerically simulated them using the *Surface Evolver* software. Their result, known as the Weaire–Phelan structure (FIG. 5.13), has $\langle S \rangle / V^{2/3} \approx 5.288$. This structure, which inspired the architect of the swimming pool for the 2008 Beijing Olympics, is the best result known for bubbles of the same volume. Does anything better exist? We don't know how much time will be required to answer this question...

In two dimensions the ideal bubble *can* be realized. It is possible to tile the plane with a bubble without curved edges: a tiling of regular hexagons, the hexagonal honeycomb, is the least perimeter division of the plane into bubbles with equal area (FIG. 4.10A). This was known to the Romans, more than two thousand years ago, as the solution found by bees to fill a beehive with cells using the smallest possible amount of wax (FIG. 1.5b), but was only rigorously demonstrated in 2001 [31].

3.2.3 Bubble shape Plateau's laws (see p. 30) are a necessary and sufficient condition for a collection of bubbles to form an equilibrium foam structure, but there are many different possible structures. In theory, bubbles of the same volume can have

(a) 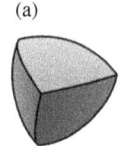 (b) 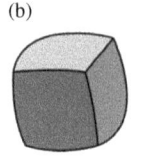 (c)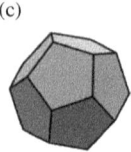

FIG. 2.13: In 3D, regular bubbles are possible only for a few values of f. For example: (a) $f = 4$, tetrahedron, (b) $f = 6$, cube, (c) $f = 12$, dodecahedron. The value of $S/V^{2/3}$ for these bubbles decreases with f and lies between about 5.3 (typical bubble, eq. (2.19)) and 4.83 (a sphere) [17].

very different surface areas, but simulations suggest a more limited range of shapes, which can be expressed in three equivalent ways [41]:

$$S \approx 5.3\, V^{2/3} \pm \text{a few \%}, \qquad \frac{S}{V} \approx \frac{3.3}{R_V}, \qquad R_S \approx 1.05 R_V, \qquad (2.19)$$

with $R_V = (3V/4\pi)^{1/3}$, the radius of a sphere with the same volume as the bubble, and $R_S = (S/4\pi)^{1/2}$, the radius of a sphere with the same surface area as the bubble. Regular bubbles, in which the faces are all identical (FIG. 2.13), have a slightly lower value of $S/V^{2/3}$.

The same can be said of 2D bubbles, where

$$\mathcal{P} \approx 3.72 A^{1/2} \pm \text{a few \%} \approx 6.6\, R_A, \qquad (2.20)$$

with $R_A = (A/\pi)^{1/2}$ and $\mathcal{P}$ the perimeter of the bubble. Highly deformed bubbles, which would have a very high ratio of perimeter to square-root of area, are almost never observed.

For regular 2D bubbles (those in which all the sides are identical, FIG. 2.14), the relationship (2.20) is true to within less than 1% (see FIG. 2.14b and exercise 7.5). This result is specific to foams, and is not true, for example, for polygons with straight edges and variable angles, for which the ratio $\mathcal{P}/A^{1/2}$ varies greatly with n (it is equal to 4 for a square, 3.72 for a hexagon, and 3.5 for a polygon with very large n, that closely resembles a circle).

3.3 Topology and geometry

3.3.1 Geometry–topology correlations
Experimentally, we observe that bubbles which have more neighbours are generally the largest. That is, edge lengths ℓ, face areas S, and bubble volumes V, each normalized by the average bubble size to the appropriate power, are correlated with the number of faces f. In 2D, this correlation is often characterized by the Desch–Feltham relationship, which gives the bubble perimeter in terms of its number of neighbours [56], or Lewis' relationship, which gives the surface area of the bubbles in terms of its number of neighbours [54]. In 3D, the curves $\langle V \rangle_f(f)$, $\langle A \rangle_f(f)$, and $\langle \ell \rangle_f(f)$, obtained from numerical simulations where measurement is easier than in experiments (FIG. 2.15), are not quite linear and

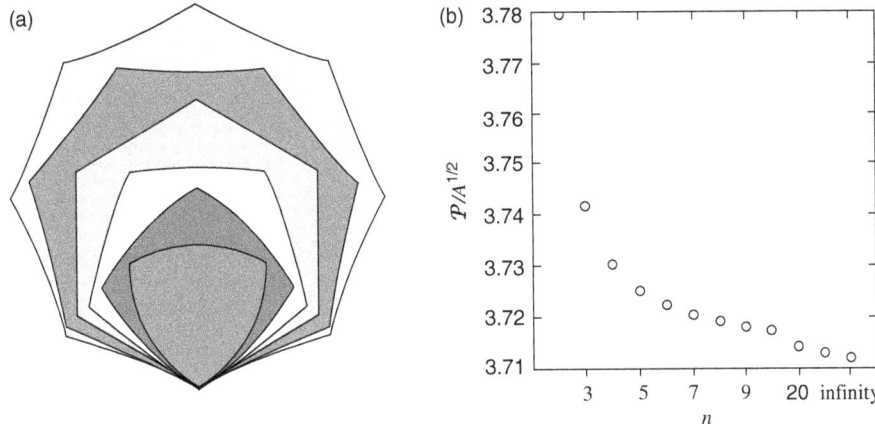

FIG. 2.14: Regular 2D bubbles with sides of equal lengths. (a) Bubbles with the same edge length superimposed one on top of another to compare their areas and curvatures ($n = 3, 4, 5, 6, 7$, and 8). Simulations by S. J. Cox. (b) Perimeter $\mathcal{P}/A^{1/2}$ for regular 2D bubbles with n sides [28], see exercise 7.5.

different fits are proposed in the literature. Nonetheless, it is significant that these curves are strictly monotonic.

This result is true for the whole foam *on average* (FIG. 2.15), but not for each individual bubble. It is currently *empirical*: we are unable to accurately predict the statistical correlation between bubble size and number of faces, although several models have been proposed [25, 49]. That the structure minimizes its energy is not sufficient to understand this correlation: recall that a foam is in a local, not a global, energy minimum, so to predict this correlation requires that we determine which "type" of local minimum is adopted most often, and this turns out not to be the global minimum. Rivier [56] suggested an explanation for this correlation by assuming that the value of the energy in a local minimum has no influence on the probability of being adopted by the foam, and that the foam can explore many different configurations. More recently, de Almeida [21] reconciled the minimization of surface energy with the maximizing of disorder during mixing by defining the analogue of a free energy for foams. The minimization of this energy functional enables the distribution of sizes and numbers of neighbours of bubbles, as well as their correlations, to be predicted statistically (FIG. 3.24).

There are different definitions of "disorder" in a foam. We can use a topological definition, for example μ_2^f (see p. 250), or a geometric one, for example μ_2^V (defined in a similar way by replacing the number of faces with the bubble volumes). These two measures of disorder usually vary in the same way from one foam to another [25, 54]: they both measure the difference between a given structure and a very organized structure. In a regular pattern in which all the bubbles are identical—the honeycomb in 2D and the Kelvin foam in 3D—both measures are identically zero. Nevertheless, it is possible to construct examples in which the two measures are very different. For

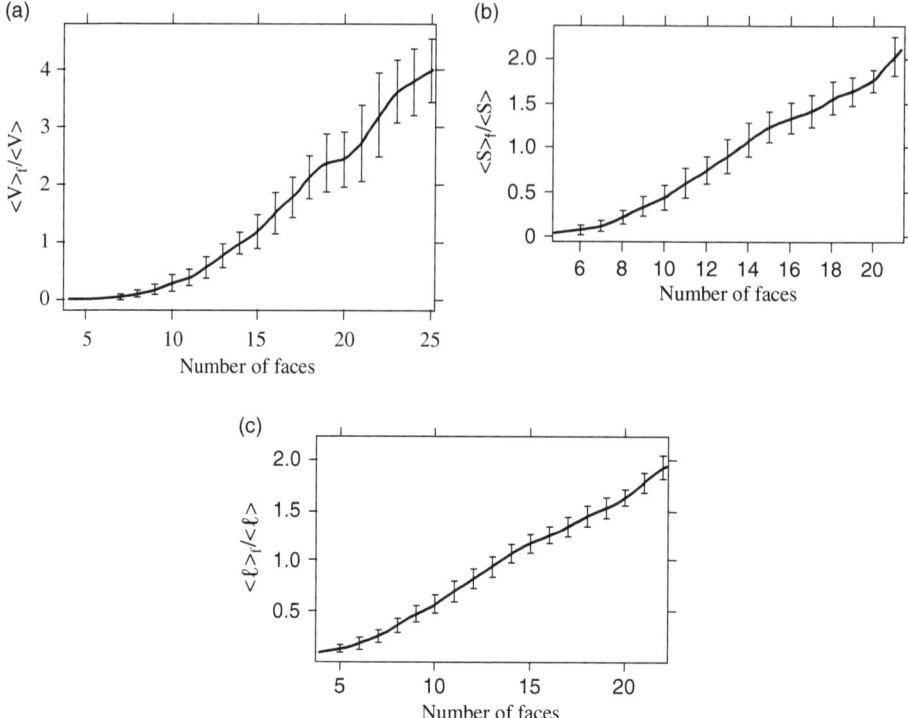

FIG. 2.15: Correlations between topology and bubble size in 3D [38] (Desch–Feltham and Lewis correlations). (a) Relationship between bubble volume and number of faces; (b) the same for surface area; (c) the same for edge lengths. To improve the statistics, the curves are averages taken from several images of a coarsening foam (during which these curves change very little, see §3.2.3, chap. 3). The continuous line is an average over all bubbles with given f (denoted $\langle \cdot \rangle_f$), normalized by the average value for all bubbles; the bars show the standard deviation for each value of f.

example, a 2D foam in which the bubbles are hexagonal but with very different sizes has a much greater geometric disorder than a topological one. Conversely, a monodisperse 2D foam can be made from heptagons and pentagons, with a much smaller geometric than topological disorder.

3.3.2 Number of neighbours, curvature, and pressure in 2D We describe a circuit around the perimeter of the 2D bubble shown in FIG. 2.16. The direction of motion keeps changing: first, from a to b we follow an arc of a circle: if R is its radius and ℓ its length, the angle through which we turn is ℓ/R. We then arrive at a vertex, which has internal angle $2\pi/3$; the angle through which we turn from b to c is called the external angle, equal to $\pi - 2\pi/3$, or $\pi/3$.

Continuing our circuit, we add the rotations c to g and $\ell_{ij}/R_{ij} = \kappa_{ij}\ell_{ij}$ along each edge ij (separating bubble i from bubble j). By convention, the radius R_{ij} and

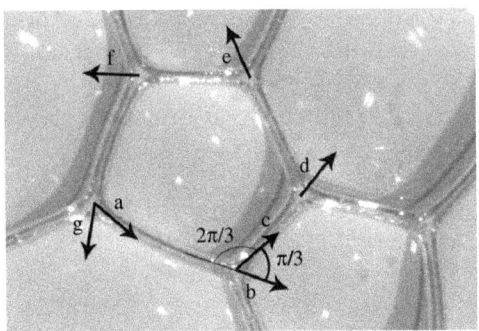

FIG. 2.16: A circuit around a bubble in 2D. We label this bubble i. It has n neighbours, here $n = 5$, labelled $j = 1, \ldots, n$. The edge separating bubbles i and j has length ℓ_{ij}, radius of curvature R_{ij}, and curvature $\kappa_{ij} = 1/R_{ij}$. In this example, all the R_{ij} are positive (bubble i is convex). Photograph courtesy of S. Courty [29]. Reproduced by kind permission of Springer Science + Business Media.

curvature κ_{ij} of the edge (ij) are counted as positive if the centre of curvature is towards i and negative otherwise (see FIG. 2.16). Therefore $R_{ij} = -R_{ji}$ and $\kappa_{ij} = -\kappa_{ji}$. Finally, we come back to our point of departure, a, after making a complete turn of 2π. Then

$$\sum_{j=1}^{n} \kappa_{ij}\ell_{ij} + n\frac{\pi}{3} = 2\pi, \qquad (2.21)$$

since we added $\pi/3$ at each of the n vertices. The quantity

$$q = \sum_{j=1}^{n} \kappa_{ij}\ell_{ij} \qquad (2.22)$$

is an angle. More precisely, it is the angle at which it is necessary to curve the sides of a polygon so that they meet at an angle of 120° and it can be closed. It is also called the *geometric charge*, and is proportional to the topological charge $q_t = 6 - n$ (see §3.1.2) which also describes the difference from an ordered foam. It is also used in other contexts, for example to describe the defects (*disclinations*) in a crystal or the curvature of space.

The definition in eq. (2.22) allows us to rewrite the relationship (2.21) in a form that links the geometry, on the left, and the topology, on the right:

$$q = \sum_{j=1}^{n} \kappa_{ij}\ell_{ij} = 2\pi - \frac{n\pi}{3} = (6-n)\frac{\pi}{3} = q_t\frac{\pi}{3}. \qquad (2.23)$$

This describes the link between the number of sides of a bubble and its shape. A bubble with 2, 3, 4, or 5 sides has, on average, positive curvatures and consequently convex sides (FIG. 2.14a). Bubbles with two sides are rare: they generally arise when a bubble with three sides loses one through film rupture, and we will disregard them in the following. Bubbles with one vertex are forbidden by Plateau's laws, since this would require its single side to curve back on itself with an angle of 120°, but that's impossible for an arc of a circle.

For bubbles with $n > 6$ sides, on the other hand, the geometric charge is negative, and on average the sides are concave (FIG. 2.14a). There is no upper limit for n, which may reach 40 or more.

A bubble may have all sides flat, i.e. without curvature ($\kappa_{ij} = 0$), and then it must be a hexagon ($n = 6$, $q = 0$), whether or not it is regular. The converse is not true: a hexagon may have both convex and concave sides (as long as the sum of $\kappa \ell$ remains zero).

If the geometric charges of all the bubbles are added, the sum $\sum_i q = \sum_{i,j=1}^n \kappa_{ij} \ell_{ij}$ counts each side twice (in the form (ij) and (ji)). Since $\ell_{ij} = \ell_{ji}$ and $\kappa_{ij} = -\kappa_{ji}$, the terms cancel each other out pairwise (except for the bubbles situated at the edge of a foam, see the shaded section on p. 43) and we recover the result (see p. 32) that in a (sufficiently large) dry 2D foam the average of the geometric (or topological) charge is zero.

In other words, there must be enough bubbles with 3, 4, or 5 sides, in which the charge is positive, to compensate for the negative charge of bubbles with more than 6 sides. We observe that in a disordered foam, the value most represented is $n = 5$, even if the average number of sides is $\langle n \rangle = 6$, because there can be faces with much more than $n = 6$ sides, but not much less.

Moreover, the pressure difference $P_i - P_j$ across an edge balances its curvature κ_{ij} (the Young–Laplace law, eq. (2.50)) and we obtain a fundamental relationship for 2D foams, discovered in 1952 by the mathematician John von Neumann (1903–1957) [56, 60]. It links the pressure in a bubble to its shape and to its number of neighbours:

$$\sum_{j=1}^n \frac{e}{2\lambda} (P_i - P_j) \ell_{ij} = \sum_{j=1}^n \kappa_{ij} \ell_{ij} = (6 - n) \frac{\pi}{3} = q, \qquad (2.24)$$

where e is the thickness of the 2D foam and λ the line tension (see p. 59). Bubbles with positive geometric charge (3, 4, or 5 sides) consequently have a higher pressure than the average (weighted by perimeter) of their neighbours. On the other hand, bubbles which have many sides have lower pressure compared to their neighbours. This plays a fundamental role in foam coarsening (see §3, chap. 3). In a way, the pressure is analogous to an electrostatic potential: it is maximal on bubbles which have a positive charge, and minimal on those with a negative charge. It is the curvature of the edges which allows us to visualize pressure differences between bubbles [28].

It is difficult to say which of the pressure, the geometric charge, and the curvature is the cause of the other two. Instead, we express the condition of equilibrium, and therefore the balance between all three, as the configuration of a foam which (locally) minimizes the total perimeter at fixed bubble areas.

All of the above is based on the fact that you turn an angle 2π when traversing a closed circuit. This is a very simple two-dimensional statement of a more general theorem in electrostatics, due to the mathematician Gauss, which has numerous implications for foams (2D and 3D). Rivier showed [3] that if you take any region of a 2D foam containing several bubbles, the sum of $\kappa\ell$ around the periphery of this region is always equal to the sum of the geometric charges contained within it [28]. The sum of the geometric charge over the whole foam is 6 (see shaded section below).

Geometric charge of a 2D bubble touching a wall
The argument associated with FIG. 2.16 is equally applicable to a bubble in contact with a wall. In this case, two of the edges meet the wall at a right angle (see p. 30). We recover eq. (2.23) if the wall is counted as two sides [28], i.e. a bubble is considered to have n sides if $n-2$ of them are free. It therefore has $n-3$ free vertices with external angle $\pi/3$ (see p. 40) and two vertices in contact with the wall with external angle $\pi/2$, and eq. (2.21) becomes

$$\sum_{j=1}^{n-2} \frac{\ell_{ij}}{R_{ij}} + (n-3)\frac{\pi}{3} + 2\frac{\pi}{2} = 2\pi. \tag{2.25}$$

Thus, a (pentagonal) bubble which has 4 free sides and one which touches a straight wall has zero geometric charge and all its sides can be straight. A (hexagonal) bubble which has 5 free sides and one which touches a straight wall has negative geometric charge and is concave.

With this convention, the sum of the geometric charges (eq. (2.22)) over the whole foam is 2π, both for a foam enclosed in a box (FIG. 5.12b) and for a free foam (FIG. 2.12) [28]. Equivalently, the total topological charge of a foam is 6, so the average topological charge is $6/N$ and the constant in eq. (2.8) is $\chi_{\text{Euler}} = 1$.

3.3.3 Number of neighbours, curvature, and pressure in 3D The topological charge can be generalized to three dimensions [58], as can the geometric charge:

$$q = \sum_{j=1}^{n} \frac{H_{ij} S_{ij}}{V_i^{1/3}}, \tag{2.26}$$

using the fact that each face is a surface of constant mean curvature ($H = 1/R_1 + 1/R_2$, see p. 27). Conventionally, curvatures $1/R_{ij,1}$ and $1/R_{ij,2}$ are counted as positive if the centre of curvature is closer to bubble i (see p. 27). The product $H_{ij}S_{ij}$ has the dimension of a length, so is normalized by bubble volume V_i to the power $1/3$.

When defined in this way, the geometric charge doesn't play exactly the same role as in 2D. In particular, the von Neumann relationship between the curvature of the faces of a bubble and its number of neighbours (eq. (2.24)) has no equivalent in 3D, since:

- $H_{ij} = -H_{ji}$ and $S_{ij} = S_{ji}$ but $V_i \neq V_j$, so even in the case of a very dry foam without walls, the terms of eq. (2.26) do not cancel each other out pairwise on neighbouring bubbles (as it does in 2D, see p. 42), except in the case of monodisperse foams (where all the V_i are equal).
- There are no bubbles with all faces planar, which would provide a reference shape, as the hexagon does in 2D. Since in 3D the edges meet at an angle $\theta_a \approx 109.5° \approx 1.91$ radians (see p. 30), this would impose on a flat face with n straight edges the condition $n(\pi - \theta_a) = 2\pi$ (cf. §3.3.2), so that $n = 2\pi/(\pi - \theta_a) \approx 5.104\ldots$ [4, 35]. That is, this face would have a non-integer number of edges! Nonetheless, we can determine some of the properties of this *idealized bubble with exactly zero curvature*. Since the average number of edges per face $\bar{n}$ and the number of faces per bubble f are related (eq. (2.13)), it would have $f = 12/(6 - \bar{n}) \approx 13.4$ faces. This is also not a whole number, but it is very close to the average number of faces observed in real foams (see p. 33) [4].
- The geometric charge of a bubble does not depend only on its number of faces, i.e. two bubbles with the same f do not necessarily have the same q. This observation is particularly significant for coarsening (see §3, chap. 3). We observe in practice that the differences are small: on average, bubbles which have less than 12 faces are rather convex ($q > 0$), those which have more than 15 faces are rather concave ($q < 0$), and the geometric charge of bubbles with f faces varies little (FIG. 2.17). Nonetheless, if a bubble has faces which all have approximately the same number of edges and the same shape, we observe that it has a lower q (it is more concave) than a more asymmetric bubble with the same f (cf. the dashed line in FIG. 2.17).
- The function $q(f)$ is not linear (FIG. 2.17), unlike $q(n)$ in 2D. By assuming that bubbles are almost regular and that the faces are not very curved, Mullins was able to derive analytically the approximate formula [51]:

$$q_{\text{Mull}}(f) \approx -\left(\frac{3}{4\pi}\right)^{1/3} G_1(f)\, G_2(f),$$

with $\quad G_1(f) = \dfrac{\pi}{3} - 2\arctan\left[1.86\, \dfrac{(f-1)^{1/2}}{f-2}\right],$

and $\quad G_2(f) = 5.35\, f^{2/3} \left(\dfrac{f-2}{2\,(f-1)^{1/2}} - \dfrac{3}{8} G_1(f)\right)^{-1/3}. \quad (2.27)$

This expression changes sign at $f = 13.3$; for large f it increases (in absolute value) as the square-root of f. The central role of q in foam coarsening (see FIG. 3.22 and eq. (3.30)) has motivated an intensive search for a simple expression for $q(f)$ (see [38]). Theoretical calculations focus on regular bubbles [33], while numerical simulations give detailed information about the equilibrium configuration of a foam [42] (or even a single bubble, in order to reduce the calculation time [17]).

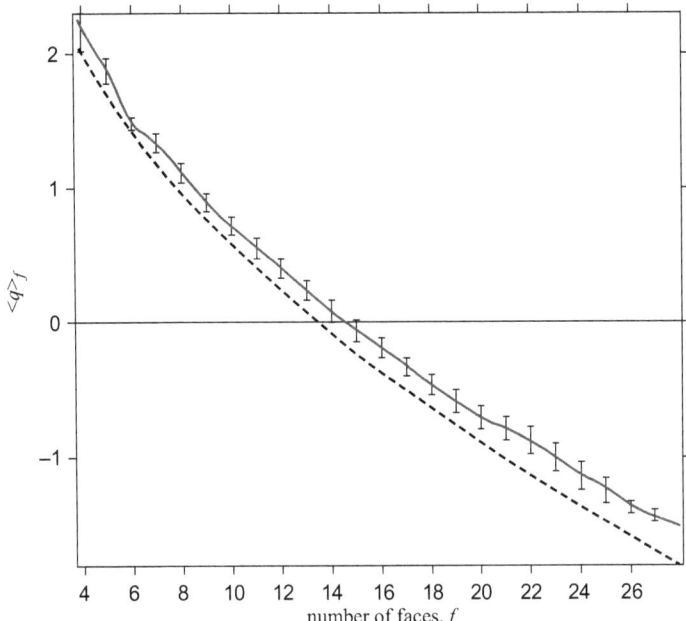

FIG. 2.17: Average geometric charge of 500 bubbles in a dry 3D foam as a function of the number of faces, obtained with *Surface Evolver* [38]. The geometric charge q of a bubble is the mean curvature integrated over the surface, normalized by $V^{1/3}$ (see eq. (2.26)), and $\langle q \rangle_f$ is its average for bubbles with f faces. The error bars represent the standard deviation for each f and the dashed line shows the theory for regular bubbles, eq. (2.27).

These different approaches confirm Mullins' law for 3D dry bubbles (eq. (2.27) and FIG. 2.17). More recently, MacPherson and Srolovitz [45] showed that

$$\sum_{j=1}^{n} H_{ij} S_{ij} = 2\pi \mathcal{L} - \pi \frac{\mathcal{A}}{3}, \qquad (2.28)$$

where $\mathcal{L}$ is a (complicated but exact) function of the shape of a bubble, with a value close to twice its diameter, and $\mathcal{A}$ the sum of the lengths of its edges. The geometric charge can be deduced from this expression (see eq. (2.26)).

4 Wet foams

We now describe the equilibrium configuration of a foam which contains a non-negligible quantity of liquid, so that the parts of an ideal foam (films, edges, and vertices) must all inflate to accommodate foaming solution. We determine the distribution of liquid in such a foam from the equilibrium laws established in §2 and we show that this equilibrium can be established even in the presence of gravity.

4.1 Modification of the structure

The quantity of liquid in a foam is characterized by the liquid fraction $\phi_l = V_l/V_{\text{foam}}$ (see p. 21). A foam which has just been made usually has $\phi_l \simeq 1\text{--}10\%$, and its liquid fraction decreases over time because of gravity-driven drainage (see §4, chap. 3). To understand the structure of a wet foam, we do the opposite: we take a dry foam structure and add solution.

FIG. 2.18 shows how the liquid modifies the structure of a foam. As long as ϕ_l is small, typically $\phi_l < 5\%$, we can describe a wet foam in terms of:

- *bubbles*, with gas volumes equal to those in the reference dry structure;
- *soap films*, the surface area of which decreases as ϕ_l increases, so that their thickness h, and therefore their volume, remains negligible;
- *Plateau borders*, the liquid channels which form between the films, replacing the edges in a dry foam (see FIGS. 2.3 and 2.19), with characteristic length ℓ and cross-sectional area proportional to r_{PB}^2, which will be determined below;
- *vertices*, the junctions of four Plateau borders, of characteristic volume r_{PB}^3, which replace the vertices of a dry foam.

At higher liquid fractions, when $r_{PB} \sim \ell/2$, neighbouring vertices meet and the Plateau borders are no longer clearly defined. The foam then becomes a collection of almost-spherical bubbles which are slightly deformed where they press against one another.

Above a critical liquid fraction, denoted ϕ_l^*, the bubbles no longer touch each other. When $\phi_l > \phi_l^*$, we no longer speak of a foam, but rather of a suspension of bubbles, or a bubbly liquid. The value of ϕ_l^* depends on the structure (see FIG. 4.17).

4.1.1 Validity of Plateau's laws
In the rest of this section we consider a foam with low but non-vanishing liquid fraction, i.e. with $h \ll r_{PB} \ll \ell/2$ (the film thickness, Plateau border size, and Plateau border length respectively). Plateau's laws should

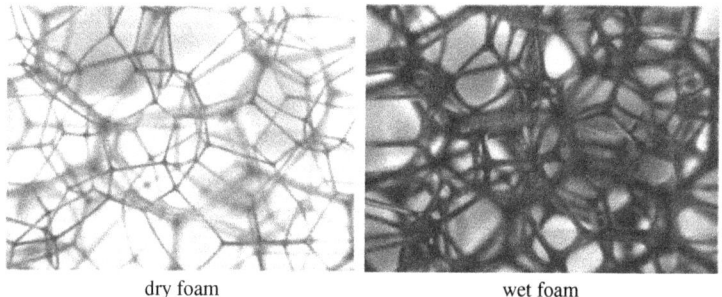

dry foam   wet foam

FIG. 2.18: When liquid is added to a dry foam (left, initial state) it accumulates mainly in the vertices and Plateau borders (right, final state). Photographs courtesy of F. Elias.

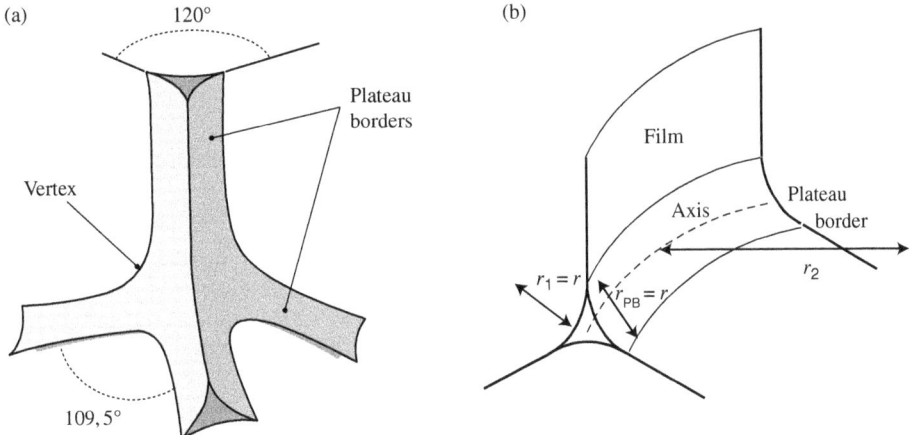

FIG. 2.19: **(a)** A vertex in a moderately wet foam. **(b)** A Plateau border, indicating the notation used. The lengths r_1 and r_2 are the radii of curvature of the liquid/gas interfaces and r_{PB} is the length of one side of the cross-section of the Plateau border. We show in the text that $r_{PB} = |r_1|$, denoted r in the following.

remain valid, at least approximately (see FIG. 2.19a). If the three thin films attached to each Plateau border are extrapolated inwards they almost meet, at an angle close to 120°, along *the Plateau border axis* (see FIG. 2.19b). Similarly, the axes of four Plateau borders almost meet at a point that we call *the centre of the vertex*, and intersect at close to 109.5°. The structure of this foam of non-negligible liquid fraction is thus very close to that of an ideal dry foam, with a simple thickening of the edges. In two dimensions these results are exact (see the decoration theorem on p. 61). In three dimensions we could be more precise about the angles by taking into account a line tension which acts along the axis of the Plateau borders, and whose value depends on the liquid fraction [26, 40].

4.1.2 Plateau border geometry The liquid in a wet foam is confined between gas/liquid interfaces, so it is the Young–Laplace law (eq. (2.3)) which imposes the local curvature and determines the shape of the Plateau borders and vertices. If we ignore gravity (§4.3), at equilibrium the pressure p of the liquid is uniform (the liquid phase is continuous) and the Young–Laplace law is

$$P - p = \gamma \left(\frac{1}{r_1} + \frac{1}{r_2} \right), \tag{2.29}$$

where P is the bubble (gas) pressure and r_1 and r_2 are the radii of curvature defined in planes perpendicular and parallel to the axis of the Plateau border. r_1 is of the order of the Plateau border size r_{PB}, and r_2 is of the order of the radii of curvature of the soap films. The difference in pressure between the two phases is the capillary pressure, denoted P_c. For the values of ϕ_l considered here, the Plateau borders are sufficiently long that $r_1 \ll |r_2|$. Writing $r = r_1$, eq. (2.29) thus simplifies to

$$P_c = P - p \simeq \frac{\gamma}{r}. \tag{2.30}$$

The film curvatures are at most of the order of $1/\ell$ and the pressure variations from one bubble to the next are consequently at most of the order of $\gamma/\ell \ll \gamma/r$. We can deduce from this that in the absence of gravity, for foams of moderate liquid fraction, the radius of curvature r of the Plateau borders is almost uniform within the foam. Moreover, the pressure difference between the liquid and the gas is much greater than the pressure differences between bubbles, and the lower the liquid fraction, the lower the pressure of the liquid phase. For $r = 0.1$ mm and $\gamma = 50 \times 10^{-3}$ N.m^{-1}, we find $P - p = 500$ Pa.

To a good approximation (see §4.1.4), the corners of the Plateau borders meet the films tangentially, the angles between the films are 120°, and the three radii of curvature r_1 are equal. So the triangle in which the cross-section of a Plateau border is inscribed is equilateral with side length $r_{PB} = r$ (see FIG. 2.19b). The area of the Plateau border cross-section is

$$s = \left(\sqrt{3} - \frac{\pi}{2}\right) r^2. \tag{2.31}$$

4.1.3 Film geometry The curvature of the films is smaller than that of the Plateau borders, which are therefore at a lower pressure than the films. Then capillary suction acts to suck much of the water from the films into the Plateau borders. Nevertheless, equilibrium must be reached, and in order to understand this we need to look at a smaller scale, the thickness of a film, of several tens of nanometers, at which the interfaces of the film begin to interact (see §2.3.1, chap. 3). This interaction is associated with a new constraint, oriented perpendicular to the interfaces, known as the *disjoining pressure* [8, 36]. Its value, and especially its sign, provide information about the interaction of the two interfaces: a positive disjoining pressure is a result of interfaces which repel each other. As discussed in §1.2, this is a necessary condition for films to be stable (and therefore for foams to exist).

If we neglect film curvature, the pressure in the gas, P, is the same on each side of the film. In the liquid, both the disjoining pressure Π_d (see FIG. 2.2a) and the bulk liquid pressure p act. The balance of normal forces on an interface is then

$$\Pi_d(h) = P - p, \tag{2.32}$$

where h is the film thickness. Equilibrium is reached when h is such that the jump in pressure at the interface, $P - p$, is the equal to the jump P_c at the Plateau border surface (eq. (2.30)), thus

$$\Pi_d(h) \simeq \frac{\gamma}{r}. \tag{2.33}$$

The variation of Π_d with h will be presented in §2.3, chap. 3. The equilibrium thicknesses are very small, of the order of tens of nanometres. Most of the liquid phase is therefore situated in the network of Plateau borders. The liquid fraction then fixes r (see shaded section on p. 50) and we deduce h from eq. (2.33).

4.1.4 Transition from a film to a Plateau border

The surface tension γ_f of a film is the energy necessary to increase its surface by one unit of area. We have accepted up until now that this tension has a value of 2γ (see §2.1.2). A careful thermodynamic calculation (see [8, 36]) shows however that γ_f depends slightly on the disjoining pressure through the relationship

$$\gamma_f(\gamma, h) = 2\gamma + \int_h^\infty \Pi_d(h')\, dh' + h\Pi_d(h). \tag{2.34}$$

Since Π_d decreases quickly when h becomes large, both correction terms vanish for thick films. But for a thin film the surface tension γ_f is slightly lower than 2γ. The magnitude of the correction can be estimated by assuming that $\Pi_d \sim \gamma/r$ for lengths of the order of $h \sim 10^{-7}$ m. For $r \simeq 10^{-4}$ m, we find $(2\gamma - \gamma_f)/\gamma \simeq h/r \simeq 10^{-3}$. The relative variation in γ is consequently very small and, in this book, we do not distinguish γ_f from 2γ.

Nevertheless, the difference between these two values is large enough to create an observable angle θ_{macro} between the film and the Plateau border where they meet. The mechanical equilibrium of this "transition region" is illustrated in FIG. 2.20. We determine the balance of forces along a line tangent to the film and perpendicular to the axis of the Plateau border (horizontal direction on the figure). To the left of the transition region the film pulls with a tension $\gamma_f(h)$; this is quite far from the Plateau border so the film thickness is constant and equal to h. On the right, the thickness of the film is quite large so that Π_d is negligible and the Plateau border pulls with a tension $2\gamma \cos \theta_{\text{macro}}$. We deduce that $\cos \theta_{\text{macro}} = \gamma_f(h)/(2\gamma)$. With the previous numerical values, this suggests $\cos(\theta_{\text{macro}}) = 0.999$ and $\theta_{\text{macro}} \sim 3°$.

In the transition region neither the curvatures nor the interactions between the interfaces can be neglected. Looking more closely, we observe that a film and a Plateau border join tangentially. We can estimate the length of the transition region, ξ, by dimensional analysis. In this region, the order of magnitude of the curvature is h/ξ^2,

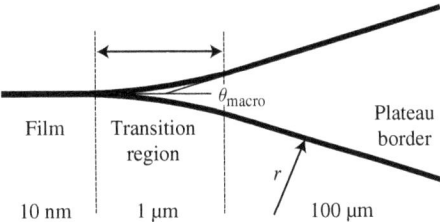

FIG. 2.20: Transition from a film to a Plateau border. From a distance, the Plateau border appears to intersect the film at an angle θ_{macro} along the contact line. A more precise examination of this line reveals a "transition region" in which the Plateau border curvature varies rapidly, owing to the disjoining pressure. When viewed at the film scale, the Plateau border thus meets the film tangentially.

which must equal the curvature of the Plateau border $1/r$. Then $\xi \simeq \sqrt{hr}$, which corresponds to a length of the order of a micron, i.e. much greater than h but smaller than r.

The Kelvin cell: an example of an ordered foam

An ordered monodisperse foam with a liquid fraction less than 6.3% forms the Kelvin structure (body-centred cubic, bcc).

In the dry case, the Kelvin cell has 8 hexagonal faces, 6 square faces, and 36 edges, each of length ℓ (see FIG. 2.21 and experiment (6.4)). Its surface area and volume are

$$S_K = 8\frac{3\sqrt{3}}{2}\ell^2 + 6\ell^2, \qquad V_K = 8\sqrt{2}\ell^3. \qquad (2.35)$$

The radius of a sphere of the same volume is

$$R_V = \ell(6\sqrt{2}/\pi)^{1/3}. \qquad (2.36)$$

The characteristics of a Kelvin cell with low but finite liquid fraction are obtained from the dry case and from the size r of the Plateau borders. The volume of a

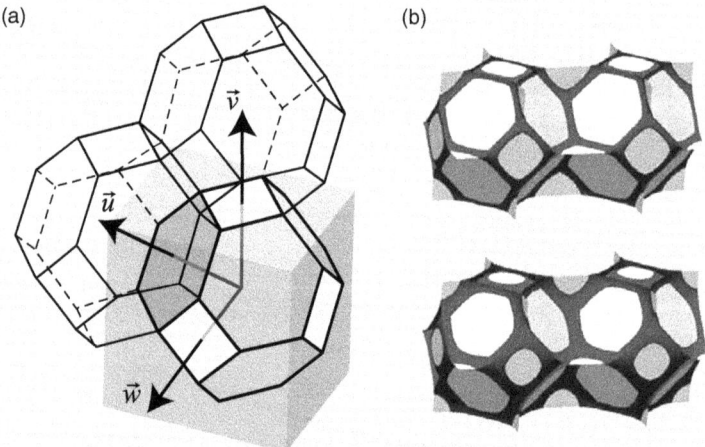

FIG. 2.21: The Kelvin cell is the unit cell of an ordered monodisperse foam with low liquid fraction. (a) Dry Kelvin cells (shown schematically with straight edges and flat faces). The arrows represent the three basis vectors of the body-centred cubic (bcc) lattice. Each bubble is inscribed in a cube. It shares one six-sided face with its 8 nearest neighbours, placed at the vertices of the cube (vector $\vec{u}$), and one four-sided face (within each face of the cube) with its 6 nearest neighbours, placed at the centre of the neighbouring cubes (vectors $\vec{v}$ and $\vec{w}$). (b) Kelvin cells for liquid fractions of 1% (top) and 3% (bottom), obtained with the *Surface Evolver*.

Plateau border is roughly $s\ell$, where $s = (\sqrt{3} - \pi/2)r^2$ is the cross-section of the Plateau borders (see FIG. 2.19b). There are 12 Plateau borders per bubble (36 edges shared by 3 bubbles), with a total volume $V_{PB} = 12s\ell \approx 1.94 r^2 \ell$.

The volume of a vertex can be approximated by considering four spheres of radius r packed tetrahedrally. The tetrahedron connecting the sphere centres has side length $2r$ and volume $2\sqrt{2}r^3/3$; it overlaps each sphere over $1/24$ of its volume, that is $\pi r^3/18$. So the volume of the cavity between the four spheres is $2\sqrt{2}r^3/3 - 4\pi r^3/18$. For all 6 vertices of a Kelvin cell, the total volume is $V_n = 1.47 r^3$. A more precise numerical calculation in *Surface Evolver* (see §2.1.2, chap. 5) suggests that the volume of the Plateau borders and vertices is equal to $V_{PB} + V_n = 1.935\, r^2\ell + 2.2627\, r^3$. The result is a little higher than our approximation because the surface of a vertex is not really part of a sphere of radius r.

A rough estimate of the volume of liquid in the films, not taking into account the reduction in surface associated with the non-zero size of the Plateau borders, gives $V_f = hS_K/2$. If $r \gg \sqrt{h\ell}$ then this volume is negligible compared to the volume of the Plateau borders. The liquid fraction is then approximately

$$\phi_l \simeq (V_{PB} + V_n)/V_K \simeq 0.171\frac{r^2}{\ell^2} + 0.2\frac{r^3}{\ell^3}. \qquad (2.37)$$

For $\phi_l < 1\%$, we have $r \ll \ell$ and it is reasonable to neglect the second term of this expression:

$$\phi_l \simeq 0.171\frac{r^2}{\ell^2} \simeq 0.33\frac{r^2}{R_V^2}; \quad \text{so } r \simeq R_V\sqrt{3\phi_l}. \qquad (2.38)$$

These results are quite close to those predicted by numerical simulations of disordered foams, and they serve as a reference throughout the book.

4.2 Osmotic pressure

If a solution (of sugar, for example) is separated from its solvent by a semi-permeable membrane, a flow of solvent is established towards the solution. The pressure that must be applied to arrest this flow is called the osmotic pressure of the solution. As the foam is a "solution" of bubbles in water, we can ask if a similar phenomenon occurs in foams. We show here that it does, but for a very different reason.

4.2.1 Physical origin
Princen first demonstrated the existence of an osmotic pressure in concentrated emulsions and foams [53]. In contrast to the case of solutions, for which this pressure is due to the entropy of the solute, in foams it arises because a wet foam has lower surface energy than a dry foam.

We consider a foam in contact with a reservoir of soapy water at a semi-permeable membrane through which liquid but not bubbles can pass (FIG. 2.22a). In order to

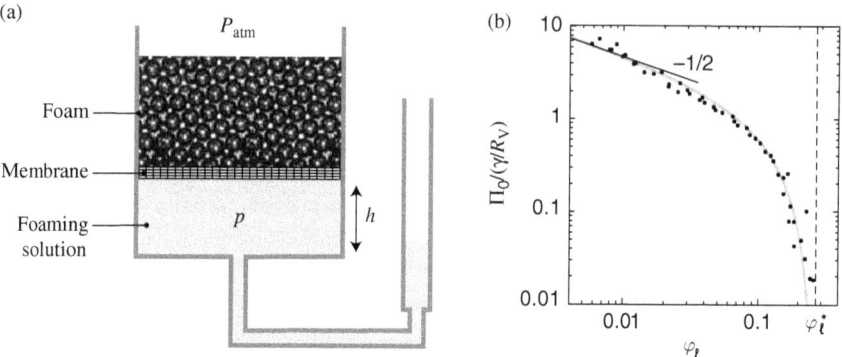

FIG. 2.22: (a) Diagram of an experiment measuring the osmotic pressure Π_o of a foam using a U-tube connected to the solution reservoir ($\Pi_o = \rho_l g h$). (b) The change in Π_o with liquid fraction for an ordered monodisperse foam with bubble radius $R_V = 150$ μm, $\gamma = 37$ mN.m^{-1}, and $\phi_i^* = 0.26$ (see FIG. 4.17). The continuous line is eq. (2.46) and the line of slope $-\frac{1}{2}$ shows the limit for dry foams (eq. (2.45)). Data from [34].

prevent the liquid being sucked up by the foam, we must impose a pressure difference Π_o between the reservoir (at pressure p) and the atmosphere above the foam (at P_{atm}):[2]

$$\Pi_o = P_{\text{atm}} - p. \qquad (2.39)$$

The osmotic pressure Π_o is deduced from the work required to extract from the foam a volume dV_l of liquid. Neglecting the gravitational potential energy and the compressibility of the gas, this work is determined by the change in interfacial, or surface, energy:

$$-\Pi_o \, dV_l = \gamma \, dS_{\text{int}}, \qquad (2.40)$$

where S_{int} is the total area of the interfaces. However, a variation in dV_l changes the liquid fraction $d\phi_l$ according to

$$d\phi_l = d\left(\frac{V_l}{V_l + V_g}\right) = \frac{(1-\phi_l)^2}{V_g} dV_l, \qquad (2.41)$$

with V_g the volume of gas in the foam. Then Π_o varies with ϕ_l according to

$$\Pi_o = -\gamma \, (1-\phi_l)^2 \frac{d}{d\phi_l}\left(\frac{S_{\text{int}}}{V_g}\right). \qquad (2.42)$$

---

[2] Note that the difference between the osmotic pressure Π_o and the capillary pressure P_c (defined on p. 48) is small compared to atmospheric pressure P_{atm}.

At the volume fraction corresponding to close packing (denoted ϕ_l^*, see FIG. 4.17) the bubbles are spherical, the interfacial energy density $\gamma S_{\text{int}}/V_g$ is a minimum, and consequently $\Pi_o(\phi_l^*) = 0$. As the liquid fraction is lowered, the bubbles deform (they become polyhedral as $\phi_l \to 0$) and their surface area is therefore greater than for spheres of the same volume. In this case, the foam sucks up liquid from the reservoir, in order to enlarge the Plateau borders and make the bubbles rounder, with a pressure Π_o. The osmotic pressure becomes larger as the foam gets drier (see FIG. 2.22b).

4.2.2 Variation of osmotic pressure with liquid fraction

We now consider an ordered monodisperse foam at equilibrium. If its liquid fraction doesn't exceed a few percent, it forms the body-centred cubic Kelvin structure (see shaded section on p. 50). At higher liquid fractions, it takes a face-centred cubic (fcc) or hexagonal close-packed (hcp) structure. Given the structure of a foam we can predict the osmotic pressure for a given bubble size and liquid fraction.

Dry foam In the dry limit, where the Plateau borders are long and slender, the jump in pressure between the gas in a bubble and the liquid around it depends simply on the radius of curvature r of the Plateau borders (see §4.1.2):

$$P - p = \frac{\gamma}{r}. \tag{2.43}$$

The overpressure in a bubble is of the order of $4\gamma/R_V \ll P_{\text{atm}}$. Consequently, $P \approx P_{\text{atm}}$ and the osmotic pressure is

$$\Pi_o = \frac{\gamma}{r}. \tag{2.44}$$

By expressing r as a function of ϕ_l using eq. (2.38), we find

$$(\text{bcc}) \quad \Pi_o \simeq \frac{\gamma}{R_V} \frac{1}{\sqrt{3\phi_l}}. \tag{2.45}$$

A similar result holds for disordered foams. Thus, for a surface tension of 30 mN.m^{-1}, bubbles of radius $R_V = 150$ µm, and a liquid fraction $\phi_l = 0.01$, we predict $\Pi_o = 1150$ Pa, which corresponds to a height of water $h = \Pi_o/\rho_l g$ of the order of 11 cm, easily measurable using the device shown in FIG. 2.22a. Note that eq. (2.45) agrees with experimental data in the dry limit, $\phi_l \lesssim 1\%$ (FIG. 2.22b) and indicates that the smaller the bubbles, the more significant the role of the interfaces, and therefore the larger the value of Π_o.

Wet foam In order to predict the osmotic pressure of a wet foam, we first need to know how the interfacial energy density varies with ϕ_l (eq. (2.42)), e.g. from numerical simulations (*Surface Evolver*, §2.1.2, chap. 5). As ϕ_l^* is approached, the osmotic

54 Foams at equilibrium

pressure falls rapidly: FIG. 2.22b shows that Π_o decreases by an order of magnitude when ϕ_l doubles (around $\phi_l \sim 0.1$). This happens because the contact area between adjacent bubbles tends to zero at $\phi_l = \phi_l^*$ (cf. the shaded section on p. 197). At equilibrium, an ordered wet monodisperse foam has a face-centred cubic configuration and so $\phi_l^* = 0.26$. For $\phi_l \leq \phi_l^*$, the measurements and simulations of osmotic pressure are well-described by the empirical expression

$$\text{(fcc or hcp)} \quad \Pi_o(\phi_l) = 7.3 \frac{\gamma}{R_V} \frac{(\phi_l - \phi_l^*)^2}{\sqrt{\phi_l}}, \tag{2.46}$$

shown in FIG. 2.22b. This expression coincides with eq. (2.45) in the dry limit.

Thus given the bubble size and the surface tension, measuring Π_o indicates the liquid fraction of the foam. This expression also describes the osmotic pressure of concentrated emulsions, with ϕ_l corresponding to the volume fraction of the continuous phase [47, 53].

4.3 Role of gravity

We have seen that the geometry of the liquid-carrying parts of a foam is dictated by the liquid pressure, and have until now assumed that the liquid pressure is uniform throughout the foam. In fact, at equilibrium the pressure in the liquid is determined by hydrostatics: as a function of height z, we have $p(z) = p(0) - \rho_l g z$. On the other hand, the film curvatures are not modified by the presence of the liquid, and so the bubble pressures are not sensitive to liquid fraction or gravity, and can be assumed to be equal, on average, to a constant P, with weak variations compared to $P - p$.

Eq. (2.30) links liquid pressure to Plateau border curvature and therefore radius of curvature r to height:

$$\frac{\gamma}{r(z)} = P - p(0) + \rho_l g z = \frac{\gamma}{r(0)} + \rho_l g z. \tag{2.47}$$

Since r is connected to liquid fraction (through eq. (2.38)), we have

$$\phi_l^{-1/2}(z) - \phi_l^{-1/2}(0) = \frac{\sqrt{3} z R_V}{\lambda_c^2} \simeq \frac{z d}{\lambda_c^2}, \tag{2.48}$$

where $d \sim 2 R_V$ is a characteristic bubble diameter, $\lambda_c = \sqrt{\gamma/\rho_l g}$ is the capillary length, and $\phi_l(0)$ is the liquid fraction at the bottom of the foam. The latter is equal to ϕ_l^*, the liquid fraction of a packing of spherical bubbles (see FIG. 4.17). Eq. (2.30) is valid for dry foams and should not be applied at the bottom of a foam, although any error in liquid fraction is restricted to this part of the profile. Eq. (2.48) is known as the equilibrium liquid fraction profile. This profile is shown in FIG. 2.23. The maximum equilibrium foam height is fixed by the disjoining pressure of the films and the pressure in the Plateau borders.

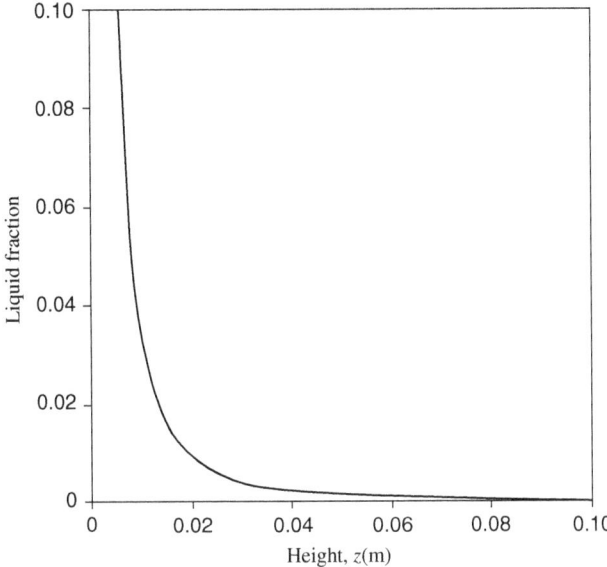

FIG. 2.23: The equilibrium profile of liquid fraction (see FIG. 0.1). The liquid fraction, obtained from eq. (2.48) (with $\gamma = 40$ mN.m^{-1} and $d = 2$ mm), changes from ϕ_l^* at the bottom of the foam ($z = 0$) to a value close to zero at the top of the foam.

5   2D and quasi-2D foams

In this section we present detailed information on bubble monolayers (quasi-2D foams). We describe their structure, including the 3D aspects (§5.1), a strictly 2D model of their equilibrium structure (§5.2 and §5.3), and aspects of rheology which are specifically 2D (§5.4).

A foam usually extends in three directions in space. In a foam that is not too polydisperse, it is possible for the bubbles to be arranged in a single layer (FIG. 2.24). These bubble monolayers are obtained by forcing bubbles to spread out over a (generally flat and horizontal) surface, or between two surfaces; the three geometries used are represented in FIG. 2.25a–c. We talk about a *bubble monolayer* or a *quasi-2D foam* when each bubble touches both the upper and lower surfaces.

A horizontal quasi-2D foam is not significantly affected by drainage (see §4, chap. 3), and when protected from evaporation (FIG. 2.25a and 2.25b) shows little coalescence (see §5, chap. 3). As quasi-2D foams are much more stable than 3D foams, and also much easier to visualize, they are widely used as model systems for research. Numerous results obtained in 2D are equally valid in 3D, at least qualitatively.

Viewed from above (see FIG. 2.24), only the boundaries between two bubbles are visible: they form a continuous network of curved lines, easily found by image analysis (see §3.1, chap. 5). Structural information is purely two-dimensional, and this is

56   *Foams at equilibrium*

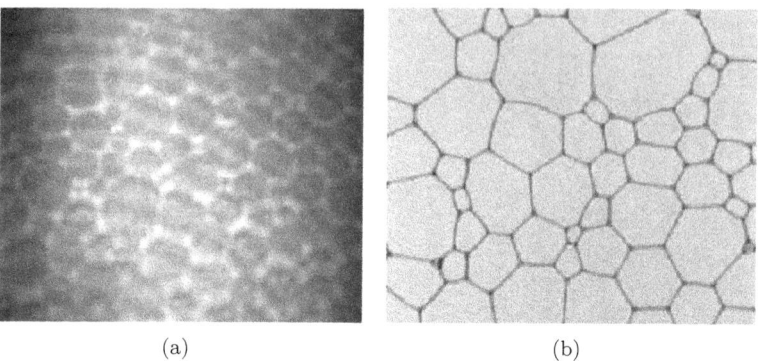

FIG. 2.24: 2D foams. **(a)** Langmuir foam (pentanoic acid on the surface of water) observed in polarized light at the Brewster angle (bubbles of around 50 µm). The reflected intensity cancels out on a bare interface, and 2D bubbles, which are not very dense, then appear darker than the dense continuous phase (see shaded section on p. 58) [16]. Photograph courtesy of S. Courty. **(b)** Dry foam between two plates, viewed from above (bubbles of several mm). Photograph courtesy of O. Lordereau [44], reproduced by kind permission of Springer Science + Business Media.

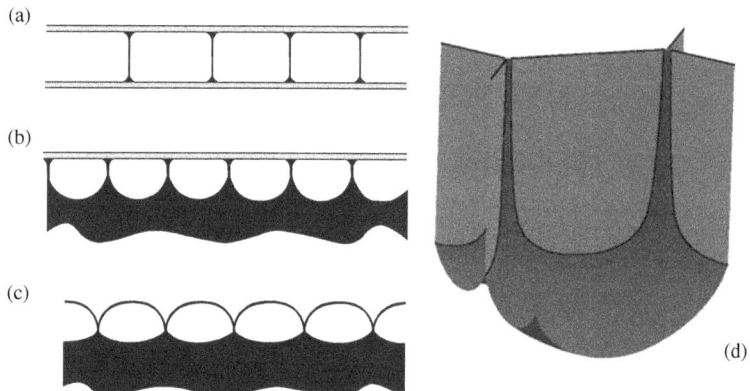

FIG. 2.25: Three possible configurations for a 2D foam (after [14]): **(a)** between two solid flat plates (Hele–Shaw cell or *glass–glass*), **(b)** between the foaming solution and a solid flat plate (*liquid–glass*), **(c)** above the foaming solution (Bragg bubble raft). **(d)** The shape of the liquid/gas interface in a foam of type (b), obtained by numerical simulation [55].

sufficient to interpret most of the observed phenomena. Nevertheless, analysing these 2D images does require knowledge of the complete, three-dimensional, structure of a 2D foam. For example, the shape of a bubble between liquid and plate is shown in FIG. 2.25d.

5.1 3D structure of a monolayer of bubbles between two plates

We next describe the 3D structure of the simplest quasi-2D foam, i.e. a foam between two plates. It consists of two types of films: wetting films covering the surface of the plates, and films separating bubbles (simply referred to as films in the following). The latter appear clearly in FIG. 2.26 as elongated rectangles. A more precise examination shows that the films, when viewed from above, are gently curved. For an isolated dry bubble between two plates, the cylindrical shape represents a minimum of surface energy and, by analogy, we conjecture that the films between the bubbles are parts of a cylinder with axis perpendicular to the plates. That is, the curvature viewed from the side is zero, but the curvature viewed from above adapts itself to the pressure difference between two neighbouring bubbles. This property is essential, for it enables us to relate the bubble pressures to the curvature of the films, as measured on the images (see eq. (2.50)).

There are also two types of Plateau borders (FIG. 2.26). Firstly, the intersections between three inter-bubble films, which have the same cross-sectional shape as in a 3D foam and are oriented perpendicular to the plates. In addition, *surface Plateau borders* are the intersections between the wetting films and the inter-bubble films that appear on an image when viewed from above. In a 2D foam between two plates there are two identical sets of surface Plateau borders: the network that contacts the top plate is superimposed on the network on the bottom plate (see FIG. 2.24b). The cross-section of these surface Plateau borders is visible in FIG. 2.25a. The thickness of the lines formed by the surface Plateau borders on the images gives a qualitative idea of their radius of curvature, although the precise relationship is difficult to determine because the curved walls of a surface Plateau border deflect light (see §1.2.3, chap. 5).

The monolayer is stable if the plates are not too far apart. If the liquid fraction is significant, or if the plates are very close together, the surface Plateau borders at the top and at the bottom touch each other, forming a liquid bridge, and the films disappear. If, on the other hand, the distance, e, between the plates increases, the

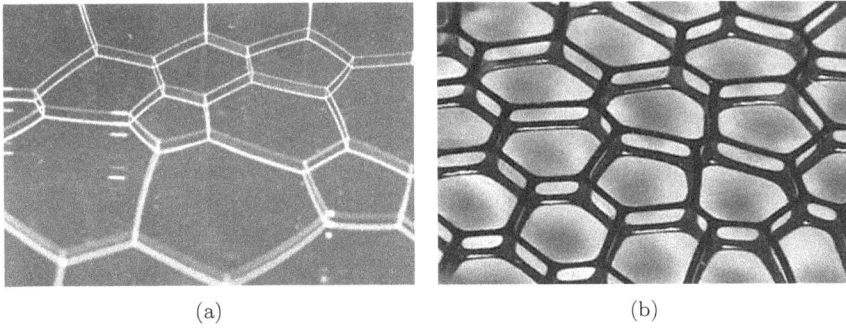

(a) (b)

FIG. 2.26: (a) Oblique view of a foam between two plates, illustrating both types of Plateau border described in the text. Photograph courtesy of I. Cantat. (b) A ferrofluid foam with slightly higher liquid content, illuminated from below. Photograph courtesy of E. Janiaud.

bubbles are slightly stretched at first, then finally, beyond a certain distance $e_{\max}$, the smallest bubbles become detached from one of the plates and the 2D structure is lost. For a single bubble of volume V between two plates, we have, in the dry limit, $e_{\max} = (\pi V)^{1/3}$. Beyond that, the cylindrical form becomes unstable, leading to a hemisphere in contact with one of the two plates [20]. This stability limit decreases with the size of the surface Plateau borders.

The structure of a fast-flowing foam between two plates is more complicated. In fact, when surface Plateau borders move along the plates, they experience friction (FIGS. 4.13 and 5.16), and so the film curvature viewed perpendicular to the plates is no longer zero. Then the pressure difference between two neighbouring bubbles is no longer directly related to the curvature measured on an image.

Truly 2D foams: Langmuir foams
Langmuir foams deserve more than any others to be called two-dimensional since they are only one molecule thick [43].

Water-insoluble surfactants are dissolved in chloroform and deposited on a liquid surface to form a Langmuir monolayer. They can exist in different phases: 2D gas (molecules well-separated and disordered), 2D liquid (molecules in close proximity and disordered), 2D liquid crystal (molecules with a certain order; also called a mesophase), 2D solid (resists shearing in the plane), or a 2D crystal (molecules completely ordered). Under very high pressures, there is an irreversible return to three dimensions. The surface pressure (or Langmuir pressure) $\Pi_l \equiv \gamma_{\text{water}} - \gamma$ depends on the area per molecule a_m. The isotherm $\Pi_l(a_m)$, measured by compressing the surface with barriers, gives rise to coexistence plateaux between phases, although less markedly than in 3D.

In the liquid/gas coexistence domain, the monolayer spontaneously organizes in the form of a 2D foam composed of bubbles of gaseous phase, separated by a continuous liquid phase (or the reverse, depending on the confinement). A Langmuir foam coarsens [7], and its structure is directly visible with a microscope at the Brewster angle (see FIG. 2.24a), or by adding fluorescent markers to increase the contrast between the liquid phase and the gas phase.

One particular property of this foam is that it is not necessary to add a supplementary chemical species to stabilize the liquid interfaces: the electrostatic dipolar interactions between the surfactants are sufficient. We also find this property in ferrofluid foams, which are stabilized by magnetic interactions (see §6.1.4, chap. 3).

5.2 A model for a dry 2D foam

The energy of a 2D foam at equilibrium depends only on the shape of the network of lines viewed from above, a property that enables a model of bubble monolayers (or Langmuir foams) to be developed.

Energy and line tension In the dry limit, the structure of a 2D foam between two plates is unaffected by translation in the vertical direction over the distance e which separates the two plates.

This translational invariance is lost in geometries (b) and (c) of FIG. 2.25, and also in geometry (a) as soon as the surface Plateau borders are no longer negligible. However, as long as the vertices can be ignored, the curvature r of the surface Plateau borders (in the plane of their cross-section) can be considered as constant everywhere. The quantity of interface per unit length of surface Plateau border is thus a well-defined constant dependent upon the precise shape of the cross-section of the surface Plateau borders. In this case the energy is

$$E = 2\lambda \sum \ell_{ij} = \lambda L_{\text{int}}, \qquad (2.49)$$

where λ is the line tension, defined for a single interface analogously to surface tension, and L_{int} is the total length of interface when the foam is viewed from above. The value of λ is approximately $\gamma(e + r(\pi - 2))$ and is slightly greater than γe. In the very dry limit for a foam between two plates, $\lambda \approx \gamma e$.

Plateau's laws and the structure of dry 2D foams The behaviour of a 2D foam is best described in terms of a cross-section through a horizontal cut, as shown in FIG. 2.27. The films are lines, the Plateau borders are small concave triangles and the surface Plateau borders are no longer taken into account. We can now use a 2D vocabulary: the line tension 2λ (in Newtons) replaces the surface tension 2γ of a film, the energy is proportional to the edge lengths, and we refer to bubble area rather than volume.

We consider first the case of an ideal dry foam (see definition on p. 26). The principle of energy minimization gives the 2D version of Plateau's laws (cf. p. 30):

- edges are arcs of circles;
- edges meet in threes at vertices;
- angles at vertices are 120°;
- edges meet solid walls at 90°.

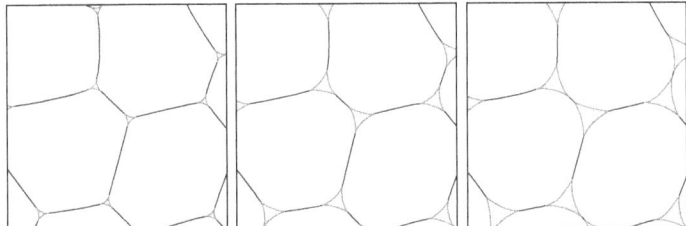

FIG. 2.27: 2D representation of a bubble monolayer at different liquid fractions [19]. This mathematical idealization of a 2D foam has given rise to many theoretical predictions and numerical simulations. The three images illustrate the decoration theorem (§5.3): when the liquid fraction increases, the structure of the foam remains unchanged. In particular, the film curvatures are not modified as long as the Plateau borders do not touch each other.

60 *Foams at equilibrium*

The first property results directly from the equilibrium of pressures along an edge separating two bubbles. In fact, with P_i and P_j the pressures in two neighbouring bubbles i and j, e the thickness of the foam, and R_{ij} the radius of curvature of the edge, counted as positive if the centre of curvature is towards bubble i, the pressure difference is

$$P_i - P_j = \frac{2\lambda}{eR_{ij}}. \tag{2.50}$$

Since $P_i - P_j$ is the same everywhere along the edge, its radius of curvature is constant: it is an arc of a circle.

The third property expresses the equilibrium of a vertex under the effect of three forces (line tensions) of the same magnitude, tangent to the edges. The instability of vertices at which four edges meet is dealt with in exercise 7.3. As in 3D (eq. (2.5)), the algebraic sum of the curvatures of three edges which meet at a vertex is zero [46]. In addition, the three corresponding centres of curvatures are aligned [50].

The fourth property arises from symmetry: it is independent of the actual contact angle of the foaming solution, which affects only the matching between the Plateau border and the wall.

5.3 Two-dimensional liquid fraction

The liquid fraction is the ratio of the volume of liquid to the total volume of the foam: it is well-defined for a foam in a Hele Shaw cell (FIG. 2.25a), but no longer makes sense when the foam consists of a monolayer of bubbles floating on the foaming solution (FIGS. 2.25b and c). The lower boundary of the latter systems is not well-defined and therefore the liquid fraction is no longer a useful quantity (although the osmotic pressure, §4.2, is still well-defined).

Instead, we define the *two-dimensional liquid fraction* $\phi_{l,2D}$ as the area fraction of liquid in a horizontal cut through the monolayer (FIG. 2.27). Unfortunately, this value depends on the position of the cut, as the cross-section of the vertical Plateau borders varies with height, more so as the foam becomes wetter. The section that is chosen is based on the following rule: the minimum edge length at which a T1 is triggered (see p. 78) must be the same in the real foam as in its idealized 2D representation [55]. This simple rule gives an unambiguous definition of the liquid fraction in a 2D foam and it enables comparisons between experimental, theoretical, and numerical results.

The actual (3D) liquid fraction of a foam between two plates is always greater than the two-dimensional liquid fraction, as it takes into account the presence of liquid in the surface Plateau borders. In the following, the liquid fraction of a monolayer of bubbles will refer to its two-dimensional liquid fraction and $\phi_{l,2D}$ will be written ϕ_l.

The smallest accessible liquid fractions vary greatly from one geometry to another. All the accessible liquid fractions in 3D are also accessible in a Hele–Shaw cell (FIG. 2.25a): it is sufficient to take a 3D foam with the required liquid fraction and confine it between two plates. It is possible to reach liquid fractions as low as $\phi_l \sim 10^{-4}$ in this geometry. In the liquid glass set-up (FIG. 2.25b), the liquid fraction can be reduced by increasing the distance between the surface of the solution and the plate, until bubbles start to detach from the plate around $\phi_l \sim 10^{-2}$. For the bubble raft (FIG. 2.25c), the liquid fraction decreases when the lateral pressure exerted on the bubbles increases. However, if the pressure increases too much, the bubbles slide on top of one another and the structure becomes three-dimensional. Low liquid fractions are inaccessible and $\phi_l \geq 10^{-1}$.

Decoration theorem The decoration theorem states that the 2D structure of a wet foam is the structure of a dry foam in which each vertex is decorated by a Plateau border [63]. This implies that if the three edges meeting at a Plateau border are extended inwards (while respecting their curvature), they intersect at a single point (the centre of the Plateau border) at 120°. In 3D, this is a good approximation (see §4.1.1) but no longer strictly true.

A 2D Plateau border is the cross-section of a 3D Plateau border, the shape of which is discussed in §4.1.2. They are small triangles with concave sides, whose radius of curvature r is fixed by the pressure p of the liquid phase and the pressure P of the neighbouring bubble according to $P - p = \gamma/r$. If all three neighbouring bubbles have the same pressure, the area s of the Plateau border is given by eq. (2.31). In a hexagonal network there are two Plateau borders per hexagon, of area A_h, and so the liquid fraction is

$$\phi_{l,\text{hex}} = \frac{2s}{A_h} = \left(2\sqrt{3} - \pi\right) \frac{r^2}{A_h}. \tag{2.51}$$

In this case the Plateau borders come into contact at a critical liquid fraction of 9% (cf. FIG. 4.17).

For a large number of bubbles, the average number of Plateau borders is always equal to two in a 2D foam (see eq. (2.10)). In a disordered foam, the area s of a Plateau border remains close to the value given by eq. (2.31), and eq. (2.51) is therefore a good approximation:

$$\phi_l \approx \left(2\sqrt{3} - \pi\right) \frac{r^2}{\langle A \rangle}, \tag{2.52}$$

with $\langle A \rangle$ the total area of the foam divided by the number of bubbles.

5.4 2D foam flows

Both 2D and 3D *quasi-static* flows are a priori very similar, and a 2D study is likely to highlight in a simple way the 3D response. On the other hand, *rapid* 2D flows are not a model for 3D because friction with the bounding plates (in a glass–glass or liquid–glass geometry) is a form of dissipation with no 3D equivalent (see p. 190). In a bubble raft, where friction is absent, rapid flows are unstable and the bubbles do not stay in a monolayer. Nonetheless, confined foam flows are interesting, for example for their applications in microfluidics or in the oil industry (flow in porous and fractured rocks).

Thanks to their geometry, direct visualization of 2D foams gives access to many flow properties. With the help of image analysis (see §3.2, chap. 5), we can identify the outline of individual bubbles and follow their movement. We can then find the stress and strain fields (tensors), strain rates, and time averages of velocity, pressure (see §3.3.3, chap. 5), and T1 density (see definition on p. 78), as described below. This information is more difficult to obtain in 3D.

62 *Foams at equilibrium*

Quasi-static flow around an obstacle FIG. 2.28 illustrates quantities that can be found through image analysis [24]. Here, a foam confined between liquid and glass flows continuously along a channel containing a circular obstacle (FIG. 2.28a; for the 3D case see FIG. 4.34). The bubbles are deformed as they pass around the obstacle (elongated parallel to the flow downstream and stretched perpendicular to the flow upstream) and store elastic energy in a way that can be quantified with a texture tensor (see §3.2.3, chap. 5) derived from the images. The velocity and pressure fields are shown in FIGS. 2.28b and 2.28c. The map of T1 density in FIG. 2.28d illustrates the number of T1 events per unit time and unit area and their average orientation. This measure of the plasticity field shows upstream–downstream asymmetry. Taken together, these observations enable rigorous testing of models of foams [11].

Shear between two walls The quasi-static flow of a 2D foam between coaxial cylinders, one fixed and the other rotating (cylindrical Couette flow [23]), is not always homogeneous (cf. §4.3.2, chap. 4) since the stress is heterogeneous [13], and localized shear bands are often observed near the moving wall. If a foam is sheared between two straight parallel walls [61], the stress is a priori homogeneous, and the way in which

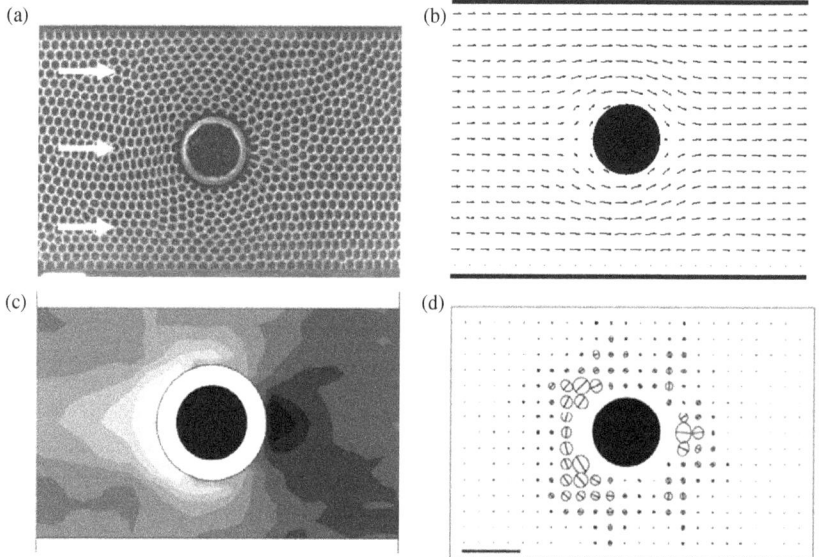

FIG. 2.28: Quasi-static flow of a monodisperse 2D liquid–glass foam around a cylindrical obstacle [24]. (a) Photograph of the experiment (courtesy of B. Dollet). Maps of the (b) velocity field, (c) pressure field (light grey represents high pressure, dark grey represents low pressure), and (d) T1 density. The latter is represented by an ellipse in which the axis shown represents the link between neighbouring bubbles before the T1 (the other axis, not shown, is for the link between bubbles that become neighbours after the T1), and the axis length is proportional to the frequency of T1s oriented in that direction.

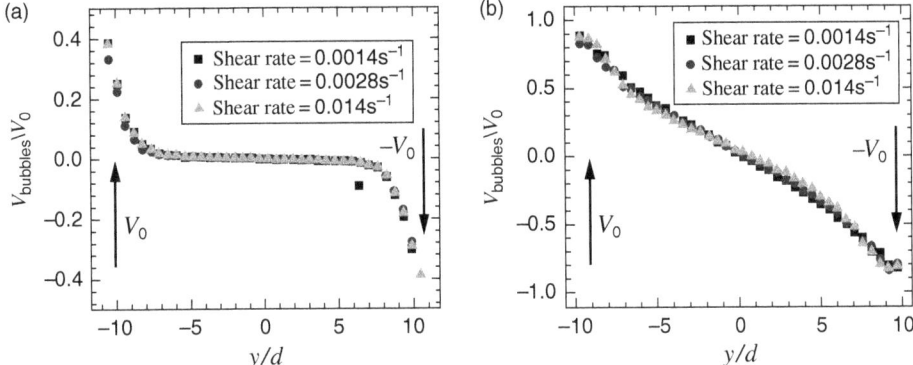

FIG. 2.29: Velocity profile of a monodisperse 2D foam with bubble diameter d in simple shear between parallel walls travelling at velocity $-V_o$ and V_o [61]. **(a)** Liquid–glass experiment; **(b)** bubble raft. In the first case, only the bubbles near the walls move, whereas in the second case the velocity profile is almost linear.

the foam is confined becomes significant [13]: if the foam is confined between two glass plates or between liquid and glass, the flow is localized close to the walls; on the other hand, the flow of a bubble raft is homogeneous, as illustrated in FIG. 2.29 [61], although structural disorder can also play a role [39], as well as the way in which the foam is prepared [6, 12].

Migration of a large bubble Consider the flow of a foam through a Hele-Shaw cell that consists of one large bubble surrounded by small ones. At low flow rate all the bubbles move at the same speed but at high flow rate the large bubble migrates more quickly than the others, without rupture of the films, through a succession of T1s in front and behind the large bubble [10]. This migration is due to the lower wall friction felt by the large bubble.

6 Experiments

6.1 Surface tension and surfactants

1) Pepper displaced by soap

difficulty level: ☺	cost of materials: $
preparation time: 5 min	experimental time: 5 min

<u>Phenomenon demonstrated</u>: *spreading of surfactant molecules at an interface.*

<u>References in the text</u>: §2.1.1 and §2, chap. 3.

<u>Materials</u>:
- a bowl
- water

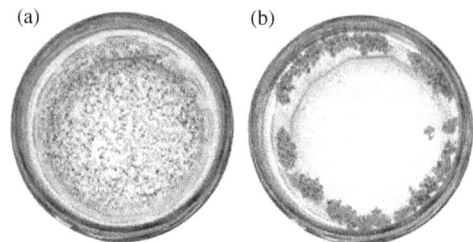

FIG. 2.30: Pepper displaced by soap: **(a)** the pepper is sprinkled uniformly on the surface of the water; **(b)** a drop of washing-up liquid is placed at the centre and pushes the pepper to the edges. Photographs courtesy of F. Elias and F. Rouyer.

- a small amount of washing-up liquid diluted in water (about 5%)
- black pepper

1. Fill the bowl with water.
2. Sprinkle the pepper evenly to form a thin layer at the surface of the water (FIG. 2.30a).
3. Carefully place a drop of the detergent solution at the centre.

The pepper is pushed to the edges of the bowl (FIG. 2.30b) by the surfactants spreading out on the surface of the water. Repeat the experiment with a drop of pure water to emphasize the role of the surfactants.

2) My cup runneth over

difficulty level:	cost of materials: $
preparation time: 5 min	experimental time: 5 min

<u>Phenomenon demonstrated</u>: *reduction of surface tension by the addition of surfactant.*

<u>References in the text</u>: §2.1.1 and §2, chap. 3.

<u>Materials</u>:

- a transparent, plastic cup
- water
- a small amount of washing-up liquid diluted in water (about 5%)

1. Fill the cup to the brim, until the liquid surface is higher than the top of the glass (FIG. 2.31a).
2. Carefully place a drop of detergent solution right at the centre of the surface.

The water runs over the sides of the glass. The surfactant lowers the surface tension and changes the contact angle between the glass and the liquid. Then the interface becomes flatter and the excess is lost (FIGS. 2.31b and 2.31c).

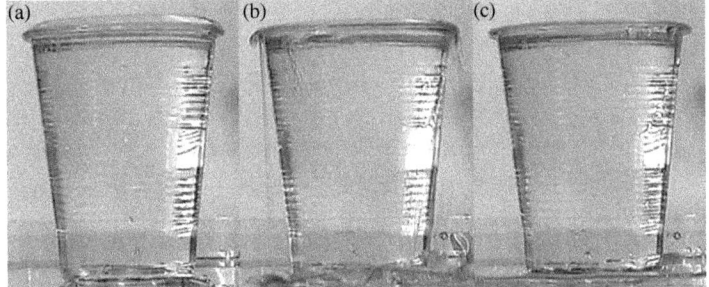

FIG. 2.31: One more drop of liquid makes the cup overflow: **(a)** the interface is raised above the rim of the cup because of a high contact angle; **(b)** adding a little surfactant lowers the contact angle and causes the liquid to overflow; **(c)** the excess water is lost. Photographs courtesy of F. Elias and F. Rouyer.

6.2 Creation and observation of 2D and quasi-2D foams

difficulty level: ⊛⊛	cost of materials: $$
preparation time: 10 mins	experimental time: 30 mins

<u>Phenomena demonstrated</u>: *Structure of a 2D foam: hexagonal arrangement, Plateau's laws.*

<u>References in the text</u>: §2.2, §3 and §5.

<u>Materials</u>:

- a transparent container with a large open surface (for example a large plastic food container)
- water
- washing-up liquid
- a straw
- a transparent flat plate, for example a flat-bottomed glass dish
- an empty transparent CD case (i.e. without the plastic CD support)

1. Dilute a small amount of detergent in the container filled almost to the top with water. The three steps below describe three ways of forming a 2D foam (see FIG. 2.25).
2. **Bubble raft**: with one end below the surface, blow air through the straw. Bubbles rise, spread out, and arrange themselves on the surface.
3. **Liquid–glass**: cover this foam with the pre-wetted (i.e. dipped in the detergent solution) plate. The bubbles now appear polygonal from above.
4. **Hele–Shaw (glass–glass)**: lower the closed CD case vertically into the soapy water, until the hole for the CD support is underwater. Put one end of the straw

into this hole and blow air in. You should create a 2D foam inside the case, with polygonal bubbles and a well-defined liquid fraction.

In each case, the harder you blow, the larger and more polydisperse the bubbles. Observe the curvature of the films separating neighbouring bubbles. If you blow more gently and smoothly, the foam will be less polydisperse, with ordered regions. It is sometimes possible to obtain a honeycomb pattern, with perhaps a few defects (dislocations, grain boundaries, etc.).

Comments:

1. You can also create a 2D foam by squeezing a small amount of shaving foam between two transparent plates. In this case, the bubbles are smaller and polydisperse.
2. Once the 2D foam has been made, it is possible, with good lighting and a good camera, to illustrate the properties presented here: Plateau's laws (§2.2), topology (§3.1, §3.3), coarsening and von Neumann's law (§3.1.1, chap. 3), and 2D flow (§5.4).

6.3 Giant soap films

difficulty level: ☺☺	cost of materials: $$
preparation time: 30 min to 2 hrs	experimental time: 30 s

Phenomena demonstrated: *liquid flow, film tension, iridescence, robustness and rupture of soap films.*

References in the text: §1.2 and §1.1.3, chap. 5.

Materials:

- patience
- a fairly high point of attachment (e.g. a ceiling hook) above a cleanable floor (inside a building)
- nylon fishing line (or cotton thread)
- a plastic bottle (e.g. a 2 litre soft-drink bottle)
- a bucket
- several litres of water
- good quality washing-up liquid (see comments below)
- a weight (100–200 g)
- a paper clip
- a bradawl (or a cork-screw or large needle)

1. Pierce a small hole in the cap of the bottle with the bradawl, small enough to allow a liquid flow rate of only a few drops per second.
2. Thread two lengths of fishing line (better is a single length folded in half) through the hole and attach them to the paper clip (which then rests in the neck of the bottle).

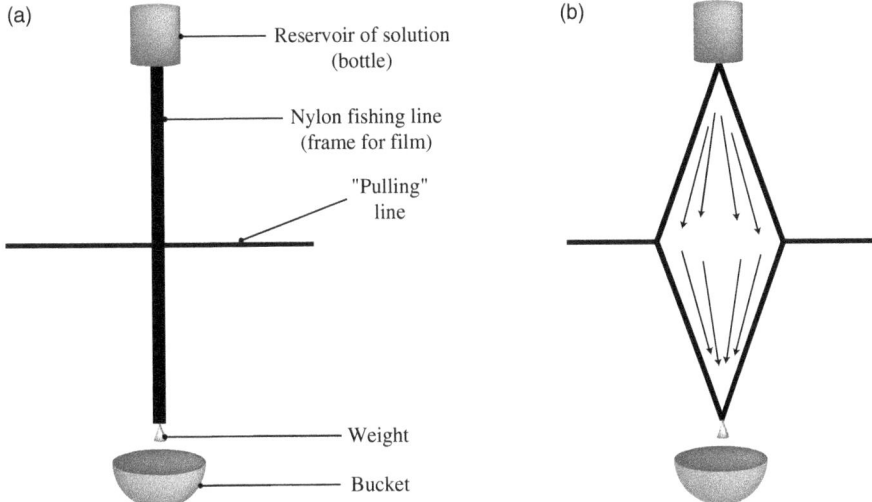

FIG. 2.32: Giant soap film: **(a)** At the start the two lengths of fishing line are touching. **(b)** When they are pulled apart they enclose a giant film.

3. Make two large holes (a few cm in diameter) at the base of the bottle, thread a second piece of fishing line through it and hang the bottle from the ceiling, neck downwards.
4. Attach the other ends of the two pieces of fishing line to the weight and check that they are firmly attached (FIG. 2.32a); place the bucket beneath the apparatus.
5. Attach to each of these vertical pieces of line a horizontal "pulling" line (the other end of which can be fixed so that it does not get tangled).
6. Pour into the bottle (through a hole in its base) a small amount of soapy water (5% washing-up liquid).
7. When the liquid has descended to completely cover the lines, pull them apart (FIG. 2.32b).

Most films last only a few seconds or perhaps a minute, but if they rupture they can be regenerated many times while there is still soap-solution in the bottle. The visible iridescence of the soap film indicates its micron thicknesses (see §1.1.3, chap. 5), so a film of 10 m in height and a metre in width contains only a few centilitres of liquid! Blow on the film, and observe how deformable it is. Objects that have been dipped in the soap solution can be passed through the film. Moving an object within the film allows you to observe its wake, showing the 2D flow of liquid in the film. It is even possible to use objects with different shapes (for example an aerofoil) in order to visualize the influence of the shape of the object on the flow of the fluid [15].

68   *Foams at equilibrium*

FIG. 2.33: With a good washing-up liquid, this giant film reached 18 m in height and showed spectacular colours. Photograph courtesy of F. Mondot [5]. See PLATE 4.

Comments:
1. For more details see [5].
2. For this activity the best washing-up liquid seems to be sold by Procter and Gamble under the label Dreft, Dawn, or Fairy. It enables the generation of films greater than 10 m in height which may last for more than ten minutes.
3. By feeding liquid into the film at several points, you can improve its longevity. In the laboratory, some last for hours.

6.4   Kelvin cell

difficulty level: 🌀	cost of materials: $
preparation time: 15 min to 2 hrs	experimental time: infinite

Phenomenon demonstrated: *3D visualization of a Kelvin cell.*

Reference in the text: shaded section on p. 50 on the Kelvin cell.

Materials:
- scissors
- glue
- photocopier (or tracing paper)

1. Copy FIG. 2.34 using a photocopier or tracing paper.
2. Cut around the edge.
3. Fold.
4. Glue the tabs around the faces and stick the structure together.

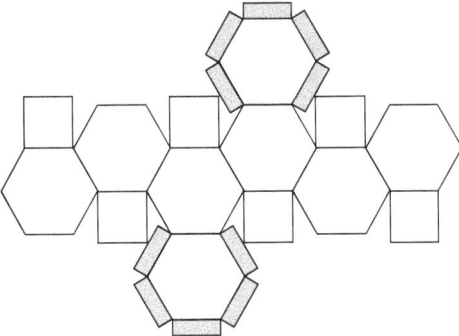

FIG. 2.34: Kelvin cell, to cut, fold, assemble, and stick, then to keep on your desk (diagram taken from the Ph.D. thesis of E. Janiaud) [37].

Comments:

1. Increasing the size of FIG. 2.34 makes the task easier.
2. Why stop at one? Make several cells to see how they fit together and amuse yourself indefinitely!

7 Exercises

Solutions to the exercises are available on request through the following webpage: http://www.oup.co.uk/academic/physics/admin/solutions.

7.1 Interfacial area of a foam

(a) Find, from the data on p. 50, the total interfacial area per unit volume of a dry Kelvin foam, as a function of the radius R_V of the bubbles. Compare this with a packing of spherical bubbles of the same radius. We will ignore this difference in the following, which deals only with orders of magnitude.

(b) A monolayer of SDS has a thickness of about 1.5 nm, and the surface area per molecule at the surface is about 0.20 nm². Consider a litre of foam made with a solution of SDS (see p. 19), with bubbles of 1 mm in diameter. Calculate the volume of SDS required, if it is assumed to be pure and of the same density as in the monolayer.

(c) If the concentration of the SDS solution, measured before foaming, is 8×10^{-3} mol.L^{-1} and the liquid fraction of the foam is 5%, what is the volume of solution required to produce the litre of foam? How are the surfactants distributed between the liquid phase of the foam and the surface?

7.2 Film tension and the Young–Laplace law

This exercise illustrates, for a cylinder, relationships (2.3) and (2.4).

70 Foams at equilibrium

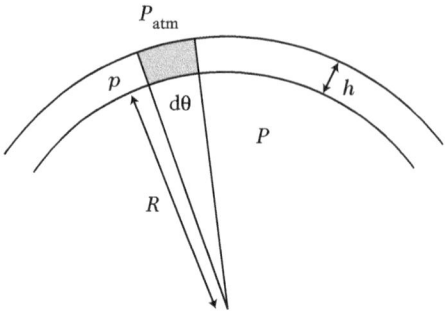

FIG. 2.35: Determination of the film tension.

(a) Write down the balance of forces exerted on the shaded film element in FIG. 2.35, including surface tension, the thickness h of the film, a constant pressure p in the liquid phase, and pressures P_{atm} and P on either side. The film is part of a cylinder of internal radius R and external radius $R + h$.

(b) Show that you recover the Young–Laplace law (eq. (2.3)) by defining the film tension as $\gamma_{\text{film}} = 2\gamma - (p - P_{\text{atm}})h$, which justifies the expression for the film tension in §2.1.2. By estimating the order of magnitude of each term, show that $\gamma_{\text{film}} \sim 2\gamma$ to recover relationship (2.4).

7.3 Plateau's laws in 2D

This exercise illustrates Plateau's laws, listed on p. 30. It is continued in exercise 7.5, chap. 4.

Consider a system consisting of two parallel plates held apart by four pins positioned at the corners of a rectangle $a \times b$. A soap film is attached to each pin, and is surrounded by the pin, the two plates (along which it slides freely), and a central film, as illustrated in FIG. 2.36. The pressure is P_{atm} on both sides of each film, which is thus straight (when viewed from above).

Fix the lengths a and b and allow the system a single degree of freedom: the length d of the central film, which can be oriented in the x or y direction.

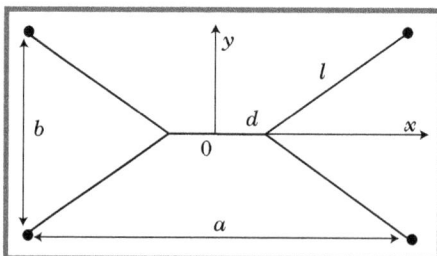

FIG. 2.36: Plateau's laws. The four black dots are the pins viewed from above.

(a) Calculate the total edge length, as a function of d, for the two directions of the central film.
(b) Calculate the energy of the configuration in which all four pinned films meet at the centre ($d = 0$). Deduce from this that, depending on the value of a/b, the four-fold vertex is unstable in one or both directions, and thus always unstable.
(c) From the energy, show that as a function of a/b, there are only one or two equilibrium positions (in the latter case, one is stable and the other metastable). Confirm that the angles between the films are 120° for these equilibrium structures.

7.4 Euler's formula

(a) A 3D bubble is analogous to a 2D foam which is wrapped onto a sphere; in this case eq. (2.9) has $\chi_{\text{Euler}} = 2$. By replacing the number of bubbles by the number of faces in eq. (2.9), and using eq. (2.12), demonstrate eq. (2.13).
(b) Starting from eq. (2.9), repeat the argument presented in 2D in the shaded section on p. 32 for a 3D foam. Then deduce the relationship between the average number of faces per bubble and the average number of edges per face in a 3D foam.

7.5 Perimeter of a regular 2D bubble

Consider a 2D bubble which has n identical sides, that is the same length b, the same curvature $\kappa = 1/\rho$, and which meet each other at an angle of 120°, as in FIG. 2.37.

Calculate the area of a sector of the bubble as a function of b and n, i.e. calculate the area of the triangle OBB' joining the centre of the bubble to two adjacent vertices, and then calculate the correction due to the curvature of the side (concave or convex) by determining the area of the circular segment between the straight line BB' and the arc BB'. Deduce from this the perimeter $\mathcal{P}$ and the area $\mathcal{A}$ of the bubble, as a

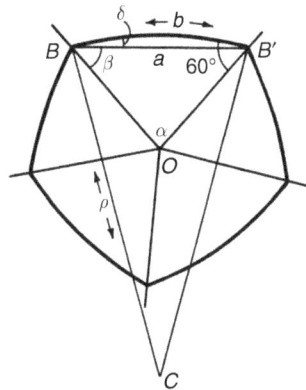

FIG. 2.37: A 2D bubble which has $n = 5$ identical sides, adapted from [27]. It is like a regular polygon except that its sides are curved.

function of b and n. Tabulate the dimensionless quantity $\mathcal{P}/\sqrt{\mathcal{A}}$ as a function of n, and show that its numerical value depends only weakly on n.

7.6 Energy and pressure

Consider an equilibrium foam with free boundary (i.e. which doesn't touch the walls of a box). Suppose that we expand all the lengths by multiplying them by a certain factor k (whilst conserving the pressure and the shape of each bubble). How do the surface areas and volumes of each bubble change?

Since the foam is at equilibrium (i.e. it minimizes some judiciously chosen thermodynamic potential) before expansion, demonstrate eq. (2.18), linking the surface energy of the entire foam with the pressures in each of the bubbles. Deduce from it the equation of state of the foam. Show that these equations also apply to a two-dimensional foam.

References

[1] D.A. ABOAV, *Metallography*, **3**, 383, 1970.
[2] F.J. ALMGREN Jr and J.E. TAYLOR, *Sci. Am.*, **235**, 82, 1976.
[3] T. ASTE, D. BOOSÉ, and N. RIVIER, *Phys. Rev. E*, **53**, 6181, 1996.
[4] J.E. AVRON and D. LEVINE, *Phys. Rev. Lett.*, **69**, 208, 1992.
[5] P. BALLET and F. GRANER, *Eur. J. Phys.*, **27**, 951, 2006.
[6] S. BÉNITO, F. MOLINO, C.-H. BRUNEAU, T. COLIN and C. GAY, *Eur. Phys. J. E*, **35**, 51, 2012.
[7] B. BERGE, A.J. SIMON, and A. LIBCHABER, *Phys. Rev. A*, **41**, 6893, 1990.
[8] V. BERGERON, *J. Phys.: Cond. Matter*, **11**, R215, 1999.
[9] B.P. BINKS, *Curr. Opin. Colloid Interface Sci.*, **7**, 21, 2002.
[10] I. CANTAT, C. POLONI, and R. DELANNAY, *Phys. Rev. E*, **73**, 011505, 2006.
[11] I. CHEDDADI, P. SARAMITO, B. DOLLET, C. RAUFASTE, and F. GRANER, *Eur. Phys. J. E*, **34**, 1, 2011.
[12] I. CHEDDADI, P. SARAMITO, and F. GRANER, *J. Rheol.*, **56**, 213, 2012.
[13] I. CHEDDADI, P. SARAMITO, C. RAUFASTE, P. MARMOTTANT, and F. GRANER, *Eur. Phys. J. E*, **27**, 123, 2008.
[14] R.J. CLANCY, E. JANIAUD, D. WEAIRE, and S. HUTZLER, *Eur. Phys. J. E*, **21**, 123, 2006.
[15] Y. COUDER, *J. Phys. Lett.*, **45**, 353, 1984.
[16] S. COURTY, B. DOLLET, F. ELIAS, P. HEINIG, and F. GRANER, *Europhys. Lett.*, **64**, 709, 2003. See erratum: B. DOLLET, F. ELIAS, and F. GRANER, *Europhys. Lett.*, **88**, 69901, 2009.
[17] S.J. COX and M.A. FORTES, *Phil. Mag. Lett.*, **83**, 281, 2003.
[18] S.J. COX and F. GRANER, *Phys. Rev. E*, **69**, 031409, 2004.
[19] S.J. COX, *J. Non-Newton. Fluid Mech.*, **137**, 39, 2006.
[20] S.J. COX, D. WEAIRE, and M.F. VAZ, *Eur. Phys. J. E*, **7**, 311, 2002.
[21] R.M.C. DE ALMEIDA and J.C.M. MOMBACH, *Physica A*, **236**, 268, 1997.

[22] P.-G. DE GENNES, F. BROCHARD-WYART, and D. QUÉRÉ, *Capillarity and Wetting Phenomena: Drops, Bubbles, Pearls, Waves*, Springer, New York, 2004.
[23] G. DEBRÉGEAS, H. TABUTEAU, and J.-M. DI MEGLIO, *Phys. Rev. Lett.*, **87**, 178305, 2001.
[24] B. DOLLET and F. GRANER, *J. Fluid Mech.*, **585**, 181, 2007.
[25] M. DURAND, J. KÄFER, C. QUILLIET, S. COX, S.A. TALEBI, and F. GRANER, *Phys. Rev. Lett.*, **107**, 168304, 2011.
[26] J.-C. GÉMINARD, F. CAILLIER, and P. OSWALD, *Phil. Mag. Lett.*, **84**, 199, 2004.
[27] J.A. GLAZIER, *Dynamics of cellular patterns*, Ph.D. thesis, University of Chicago, 1989. http://biocomplexity.indiana.edu/jglazier/docs/dissertation/Glazier-Dissertation.pdf
[28] F. GRANER, Y. JIANG, E. JANIAUD, and C. FLAMENT, *Phys. Rev. E*, **63**, 011402, 2001.
[29] F. GRANER, "Two-dimensional fluid foams at equilibrium", in *Morphology of Condensed Matter—Physics and Geometry of Spatially Complex Systems*, K. Mecke and D. Stoyan (eds.), pp. 187–214, Lecture Notes in Physics 600, Springer, Heidelberg, 2002.
[30] W.C. GRAUSTEIN, *Ann. Math.*, **32**, 149, 1931.
[31] T.C. HALES, *Disc. Comp. Geom.*, **25**, 1, 2001.
[32] S. HILGENFELDT, A.M. KRAYNIK, S.A. KOEHLER, and H.A. STONE, *Phys. Rev. Lett.*, **86**, 2685, 2001.
[33] S. HILGENFELDT, A.M. KRAYNIK, D.A. REINELT, and J.M. SULLIVAN, *Europhys. Lett.*, **67**, 484, 2004.
[34] R. HÖHLER, Y. YIP CHEUNG SANG, E. LORENCEAU, and S. COHEN-ADDAD, *Langmuir*, **24**, 418, 2008.
[35] C. ISENBERG, *The Science of Soap Films and Soap Bubbles*, Dover, New York, 1992.
[36] I.B. IVANOV, *Thin Liquid Films: Fundamentals and Applications*, Surfactant Sciences Series Vol. 29, CRC Press, Boca Raton, 1988.
[37] E. JANIAUD, *Élasticité, Morphologie et Drainage Magnétique dans les Mousses Liquide*, Thèse de l'Université Paris 7, 2004. http://tel.archives-ouvertes.fr/tel-00007740.
[38] S. JURINE, S. COX and F. GRANER, *Colloid Surf. A.*, **263**, 18, 2005.
[39] G. KATGERT, M.E. MÖBIUS, and M. VAN HECKE, *Phys. Rev. Lett.*, **101**, 058301, 2008.
[40] N. KERN and D. WEAIRE, *Phil. Mag.*, **83**, 2973, 2003.
[41] A.M. KRAYNIK, D.A. REINELT, and F. VAN SWOL, *Phys. Rev. E*, **67**, 031403, 2003.
[42] A.M. KRAYNIK, D.A. REINELT, and F. VAN SWOL, *Phys. Rev. Lett.*, **93**, 208301, 2004.
[43] M. LÖSCHE, E. SACKMANN, and H. MÖHWALD, *Ber. Bunsen.-Ges. Phys. Chem.*, **87**, 848, 1983.
[44] O. LORDEREAU, *Les mousses bidimensionnelles: de la caractérisation à la rhéologie des matériaux hétérogènes*, Ph.D. thesis, Université de Rennes, 2002.
[45] R. MACPHERSON and D. SROLOVITZ, *Nature*, **446**, 1053, 2007.

[46] M. MANCINI and C. OGUEY, *Eur. Phys. J. E*, **22**, 181, 2007.
[47] T.G. MASON, M.-D. LACASSE, G.S. GREST, D. LEVINE, J. BIBETTE, and D. A. WEITZ, *Phys. Rev. E*, **56**, 3150, 1997.
[48] E.B. MATZKE, *Am. J. Bot.*, **33**, 288, 1939.
[49] M.P. MIKLIUS and S. HILGENFELDT, *Phys. Rev. Lett.*, **108**, 015502, 2012.
[50] C. MOUKARZEL, *Physica A*, **199**, 19, 1993.
[51] W.W. MULLINS, *Acta Metall.*, **37**, 2979, 1989.
[52] J.A.F. PLATEAU, *Statique expérimentale et théorique des liquides soumis aux seules forces moléculaires*, Gauthier-Villard, Paris, 1873.
[53] H.M. PRINCEN and A.D. KISS, *Langmuir*, **3**, 36, 1987.
[54] C. QUILLIET, S. ATAEI TALEBI, D. RABAUD, J. KÄFER, S.J. COX, and F. GRANER, *Phil. Mag. Lett.*, **88**, 651, 2008.
[55] C. RAUFASTE, B. DOLLET, S. COX, Y. JIANG, and F. GRANER, *Eur. Phys. J. E*, **23**, 217, 2007.
[56] N. RIVIER, in *Disorder and Granular Media*, edited by D. Bideau and A. Hansen, Elsevier, Amsterdam, p. 55, 1993.
[57] N. RIVIER, *Phil. Mag. A*, **40**, 859, 1979.
[58] J.-F. SADOC and R. MOSSERI, *Geometrical Frustration*, Cambridge University Press, Collection Alea-Saclay, 1999, chap. 4.
[59] J.E. TAYLOR, *Ann. Math*, **103**, 489, 1976.
[60] J. VON NEUMANN, in *Metal Interfaces*, American Society for Metals, Cleveland, p. 108, 1952.
[61] Y. WANG, K. KRISHAN, and M. DENNIN, *Phys. Rev. E*, **73**, 031401, 2006.
[62] D. WEAIRE and N. RIVIER, *Contemp. Phys.*, **25**, 59, 1984.
[63] D. WEAIRE and S. HUTZLER, *The Physics of Foams*, Clarendon Press, Oxford, 1999.

3
Birth, life, and death

1 Foam evolution

We first introduce qualitatively the different mechanisms by which a foam forms, ages, rearranges, and then collapses, which collectively determine foam stability and lifetime. Later on in the chapter each of these stabilizing and destabilizing mechanisms is described in detail.

1.1 The competition between different processes

Aqueous solutions don't all foam in the same way. When you try to produce a foam—for example by shaking a closed flask containing liquid and gas—you may observe different behaviour: a few short-lived bubbles, a fragile foam which rapidly disappears, or a more stable one lasting several hours. The fact that a foam forms at all is the result of several different mechanisms which tend either to produce and stabilize it (§1.1.1) or to destroy it (§§1.1.2, 1.1.3 and 1.1.4). Each of these effects has its own characteristic time- and length-scales and a particular dependence on the liquid fraction and the bubble size. The competition between them determines the foam's lifetime, its macroscopic behaviour, and how it evolves in time, or ages. The complexity of the phenomena that are observed comes in part from the combined effects of these different mechanisms.

1.1.1 Stabilizing mechanisms A foam doesn't form spontaneously. Energy is required to disperse the gas in the liquid, but energy alone is evidently not sufficient. The *foamability* or *foaming capacity* of a solution is a qualitative measure of its capacity to produce a foam when it is shaken, or when bubbles are injected into it. Surfactants have a strong influence on the foamability of a solution, as described in §2. They adsorb at liquid interfaces and films and modify their properties. Both the surface concentration at equilibrium and the speed with which they cover an interface are significant. In addition to the static properties of a surfactant-laden interface, we will see that its viscoelasticity is equally important.

1.1.2 Drainage In response to gravity the liquid flows through the foam (FIG. 3.1) and the foam is said to *drain*. In §4 we describe a theoretical approach to drainage that is similar to the description of flow in porous media, in which the foam is treated as a continuous medium with an effective permeability. Nevertheless, a foam has characteristics which distinguish it from a porous solid: the size of the pores (the cross-sectional

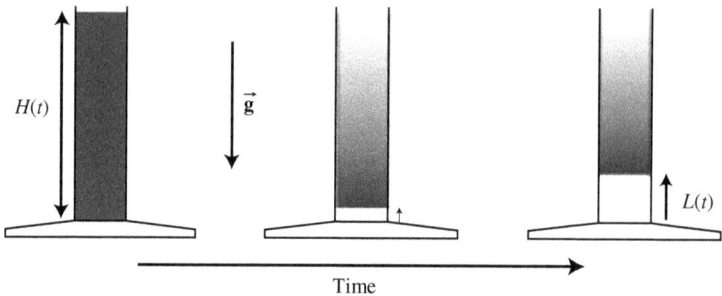

FIG. 3.1: Liquid accumulates beneath a draining foam over time, and it becomes drier and drier (see experiment 7.2).

area of the Plateau borders) depends on the liquid fraction, and thus on the flow; moreover, the walls of these pores (the gas/liquid interfaces) are viscoelastic (see §2.2).

The drainage rate depends on the difference in density between the gas and the liquid. In emulsions—that is, a dispersion of one liquid (oil) in another (water), or vice versa—drainage is less spectacular and markedly slower, because the densities are similar, and the process is called *creaming*.

Note that drainage is important in the early stages of solid foam formation, when the liquid is often still molten. It must be avoided if a homogeneous product is required.

1.1.3 Coarsening Any gas is to some extent soluble in any liquid. As a consequence, the liquid films do not act as absolute barriers to the gas. As a result of the pressure difference between bubbles, the gas may diffuse through the liquid phase from one bubble to another. As we explain on p. 77, this results in a reduction in the number of bubbles and an increase in their average size, and hence this process, called *coarsening*, influences the life of a foam. In §3 we describe how the volume of a bubble varies as a function of its topology and its size, and discuss the effects of physico-chemical parameters and the liquid fraction.

The laws of equilibrium, established in §2.2.2, chap. 2, assume that the volume of each bubble is fixed. Coarsening modifies the bubble volumes, but sufficiently slowly that the conclusions of that chapter are correct: coarsening induces just a slow variation in the control parameter (the volume of the bubbles) and the foam structure changes little by little. In other words, as a foam slowly coarsens, it is (at almost every instant) very close to mechanical equilibrium.

It is important to distinguish bubbly liquids from foams. In the first case, the bubbles are well separated from each other; the diffusion of the gas is between the bubbles and the solution, which acts as a reservoir of gas. The pressure of a bubble is proportional to its curvature (eq. (2.4)). Then the pressure of a spherical bubble is inversely proportional to its radius (eq. (2.6)), so that the smallest bubbles have a greater overpressure relative to the liquid and they lose more gas than they receive from it. On the other hand, the largest bubbles receive more gas from the liquid than they lose to it, so the small bubbles shrink and the largest continue to grow. Small bubbles eventually disappear, and since no new bubbles are created, the total number of bubbles decreases, and so the average size increases. This process, which shows an increase in average bubble radius with $t^{1/3}$ (both in 2D and 3D), is known as *Ostwald ripening* (or Lifshitz–Slyozov–Wagner ripening), and we will consider it no further.

In the case of dry foams, gas exchange occurs directly, from bubble to bubble. Here too, a bubble's pressure is proportional to its curvature (eq. (2.4)), but now the curvature depends on the bubble shape (eqs. (2.26) and (2.27)). For a dry foam, the average bubble radius increases with $t^{1/2}$ (both in 2D and 3D). This process, called *von Neumann–Mullins coarsening*, is described in detail in §3. In a dry 2D foam, the curvature of an edge depends directly on a bubble's number of neighbours (eqs. (2.21)–(2.24)). This simplifies the theoretical study of coarsening: the change in area is thus determined by the number of neighbours (and no longer by size as in Ostwald ripening). It is generally the case that for a given liquid fraction the behaviour of the foam lies somewhere between the two extremes.

There are other structured materials which, like foams, are organized into cellular domains of a discrete phase dispersed in a second continuous phase. If the domains can exchange matter (generally by diffusion), then the material coarsens in a similar fashion: some domains disappear and the average size of the remaining domains increases. This coarsening reduces the amount of energetically-costly interfaces.

This principle leads to various physical processes that drive coarsening. Foams coarsen due to pressure-driven diffusion. Emulsions, in which the structure resembles a foam, coarsen in the same way (it is this which causes mayonnaise to separate, for example). A crystal consists of small monocrystalline domains, called grains, in which the crystalline orientations are different (FIG. 3.2). Their walls, or *grain boundaries*, have an energy-cost proportional to their surface area. The crystals coarsen in a similar fashion to dry foams. In fact it was the metallurgist Cyril Stanley Smith who launched the modern study of foam physics in 1952: after one of Smith's lectures on crystal structure, von Neumann established the 2D growth law described in §3. Under certain conditions the thin films of a magnetic garnet arrange themselves into domains of the same magnetization, typically with dimensions of about ten micrometers, separated by walls (Bloch walls) and this cellular structure coarsens. Finally, during a first-order

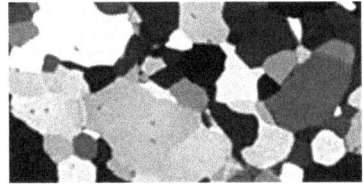

FIG. 3.2: The grains of a polycrystal of ice formed in the Antarctic, observed here between crossed polarizers. Each colour corresponds to an orientation of the crystal lattice. As in a dry foam, the grains coarsen and deform. Image courtesy of G. Durand [19]; the size is 5×2.5 cm. See PLATE 5.

1.1.4 Film rupture

The study of the rupture of a single film (which depends on the dimensions of the film and the liquid fraction) and the interactions between films (highly significant for rupture) will be presented in §5. We have already noted that, in comparison with drainage and coarsening, it is the mechanism of foam evolution that is least understood.

Any process which induces the rupture of a film between two bubbles (see §5.1) tends to destroy a foam. Since the thickness of a film (<1 µm) is small compared to its surface area (of the order of mm^2), the film between two bubbles is a fragile thing. It is easily ruptured, and then the two bubbles are said to *coalesce* (FIG. 3.8). The result of the coalescence is a reduction in the total number of bubbles (which is the same effect as with coarsening but brought about by a completely different mechanism).

A film breaks if the stabilizing mechanisms are absent or too weak. In the first place there are dynamic effects, which can rupture them, in particular as films approach their equilibrium thickness (see §5). But even if equilibrium is reached and a film is stabilized by repulsive forces between its surfaces, it is only metastable. The spontaneous fluctuations in thickness or density close to the equilibrium state can induce rupture. These fluctuations are all the more critical because the film is thin yet of large extent. Films on the outside of the foam (as in the bath tub or the head on a glass of beer) are even more fragile, due to the presence of external perturbations such as grease, dust, and evaporation of liquid. Film rupture causes a foam to lose gas and its volume decreases until it completely disappears. As we will describe in §5.3, it is sometimes desirable to get rid of a foam, or prevent one from forming, and this stimulated rupture is possible with antifoaming agents.

1.2 Elementary topological processes

1.2.1 Topological changes in a dry foam

A *topological change* refers to any process which modifies the number of faces (or neighbours) of the bubbles in a foam. For example, as the size of the bubbles changes during coarsening, or as a foam is made to flow by applying a strain to it, the bubbles change shape. Some Plateau borders grow longer while others shorten and vanish. In order to continue to respect Plateau's laws (see p. 30), some topological changes must therefore take place. All rearrangements other than coalescence or division are topologically equivalent to a combination of two elementary processes called T1s and T2s [84]. T1s (§1.2.2) are caused by both coarsening and flow, while T2s (§1.2.3) are due to coarsening alone.

1.2.2 T1 topological changes

When a Plateau border in a 3D foam shrinks to the point of disappearance, more than four Plateau borders find themselves in contact at a single vertex. Since this configuration is unstable, the bubbles rearrange themselves, in a transformation called a T1, towards a new configuration of lower

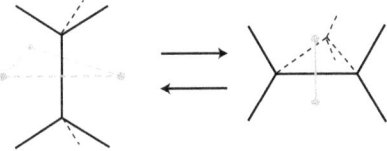

FIG. 3.3: T1 topological changes in 3D. The black lines represent the Plateau borders (solid lines for those in the plane of the figure, and dashed lines for those behind it); the grey ones illustrate the links between nearest neighbours before (dashed line) and after (continuous line) the T1.

energy, resulting in a change of each bubble's neighbours. Two possible T1 transformations are illustrated in FIG. 3.3: from left to right the vertical Plateau border disappears and a new triangular face is created; three bubbles move apart and two move closer together. From right to left the reverse transformation occurs. There is a third type of T1, which can be viewed as being composed of two of the T1s illustrated, in which a quadrilateral film separating two bubbles retracts not to a point but to a Plateau border, and then a new quadrilateral face is created, perpendicular to the first one [65]. In all cases, a T1 can be characterized by the size and directions of the links which appear and disappear (FIG. 3.3).

In 2D, a T1 affects four bubbles (FIG. 3.4): two bubbles move apart and two move closer together, a link between two bubbles disappears (they are no longer nearest neighbours), and another is created (see exercise 7.3, chap. 2). The reverse transformation is identical and there is therefore only a single type of T1 in 2D.

A T1 is a topological process in which the movement of six connected Plateau borders (four films in 2D) induces neighbour-swapping. It results in a non-equilibrium state, and is followed by a relaxation, the duration of which is controlled by dissipative processes (see §3.2.3, chap. 4).

The whole process (a T1 and its associated relaxation) induces motion of bubbles beyond the ones which swap neighbours. Superimposing the position of the bubbles

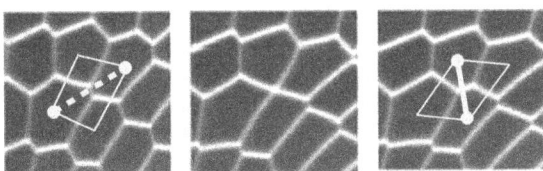

FIG. 3.4: T1 topological changes in 2D. The links (or first nearest neighbours) are represented by the thick lines (dashed for the link which disappears; continuous for the link which appears). Images courtesy of C. Raufaste and P. Marmottant [35], reproduced by kind permission of The European Physical Journal.

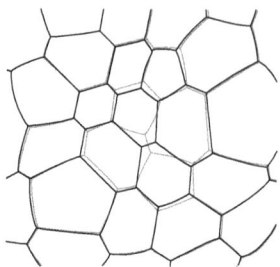

FIG. 3.5: Range of a T1 in 2D. The geometry of a foam before and after the T1 are superimposed. Only a few bubbles around the "flipped" film are affected by the transformation. The foam skeleton shown here is from FIG. 3.6 (see shaded section) [26].

before and after a T1 (FIG. 3.5) illustrates its *range*, which remains poorly understood. The effect of liquid fraction is also unclear: in a wet foam, we can still define links which appear and disappear, but they do not necessarily do so simultaneously, or in any correlated fashion, so that even the definition of a T1 is ambiguous.

Spontaneous and induced T1s

We have up until now considered the T1s that occur spontaneously in a foam, as a result of coarsening or flow. In these cases, the structure adopted by the foam immediately before the T1 is an equilibrium structure (subject to the deformation being slow), with one or several very short edges. Due to the deformation, or to changes in bubble volumes, the equilibrium associated with the initial topology is no longer stable: an edge shrinks to zero and a T1 therefore occurs.

In a similar way, it is possible to force a T1 to occur by applying a local force on a vertex (FIG. 3.6). In particular, this is possible in a 2D magnetic foam (see §6.1.4): two vertices can be brought together by focusing the magnetic field with a metal pin until a four-fold vertex is produced. When the pin is withdrawn, two equilibrium states are accessible to the foam: the initial one, and a second one obtained by changing the topology. This second state does not necessarily exist, cf. exercise 7.3, chap. 2.

The energy barrier between two states separated by a T1 (i.e. the difference in energy between the initial structure and the one with a four-fold vertex) is sufficiently high in this "induced" case that a T1 never occurs due to thermal fluctuations. In effect, this barrier is of the order of γ times the surface area of a face in 3D, which corresponds to about 10^{-5} J for a soap foam in which the bubbles are of the order of centimetres. If the bubbles are tiny, of the order of microns, the barrier can be much lower, perhaps 10^{-14} J. Langmuir foams have the smallest energy barrier, of the order of pN $\times$ μm, which is 10^{-18} J, yet this is still orders of magnitude greater than thermal fluctuations ($k_B T \approx 4 \times 10^{-21}$ J, where k_B is Boltzmann's constant).

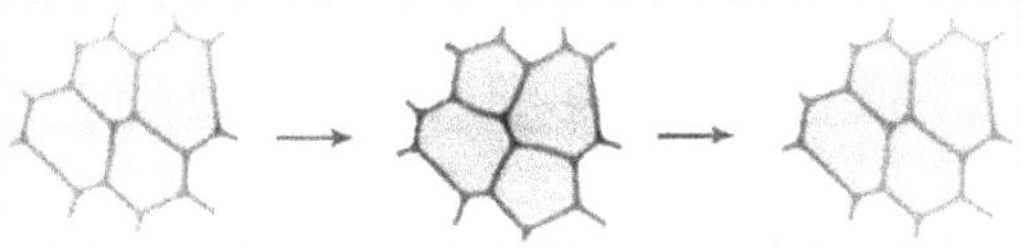

FIG. 3.6: Induced T1 and the reverse process induced in a 2D magnetic foam (see §6.1.4). On the left: initial state; in the middle: state obtained after the T1, with lower energy; on the right: return to the initial state after reversing the T1. Image courtesy of F. Elias, C. Flament, F. Graner [26], reproduced by kind permission of Taylor & Francis Ltd.

1.2.3 Topological changes that change the number of bubbles The second elementary process, called a T2, is the disappearance of a bubble: in 3D this is usually a bubble with four faces and in 2D it is a bubble with three sides (FIG. 3.7, left to right). In a foam, a T2 can only occur if the bubble volumes are changing: it is thus a process typical of coarsening. The reverse process is nucleation (FIG. 3.7, right to left), the appearance of an additional bubble, which occurs when a foam is made by bubbling.

A third topological process is coalescence (FIG. 3.8, left to right), in which a film between two bubbles disappears. Topologically, coalescence can be formally broken down into a T2 and T1s, although physically these processes are totally unconnected. The reverse process is division, where one bubble splits into two (FIG. 3.8, right to left).

As with nucleation, this occurs during foam formation. T2s and nucleation are both, a priori, isotropic, while coalescence and division have a well-defined direction defined by the links which appear and disappear [35].

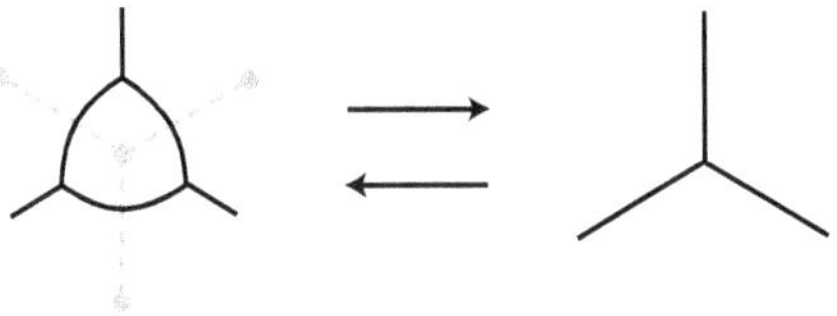

FIG. 3.7: From left to right: T2 topological change in 2D. From right to left: nucleation of a bubble in 2D. The links which disappear are represented by dotted lines. No links appear.

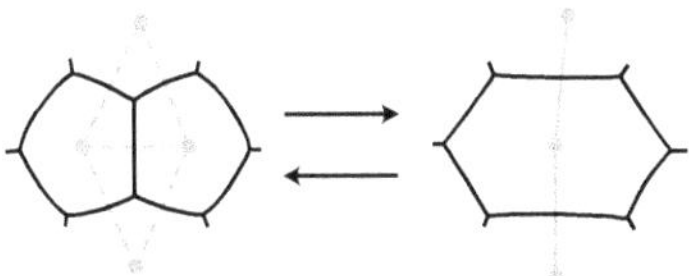

FIG. 3.8: From left to right, coalescence of two bubbles in 2D. The film separating the two bubbles ruptures. From right to left, division of one bubble into two. The links which disappear and appear are represented by dotted and continuous lines, respectively.

Finally, from the topological point of view, all these processes leave Euler's formula invariant, as expected, i.e. they do not change the sum $N - N_{\text{faces}} + N_{\text{edges}} - N_{\text{vertices}}$ in 3D or the sum $N - N_{\text{edges}} + N_{\text{vertices}}$ in 2D (see §3.1.1, chap. 2). In 2D, they also leave the topological charge of the foam unchanged, creating as much new charge as they destroy.

2 Birth of a foam

In this section, we explain why it is possible to form a stable foam. We first consider the effect that surfactants have on the static and dynamic properties of an interface and on a liquid film. We then discuss the relationship between the foaming capacity of a solution and the properties of the resulting foam's interfaces and films. While certain correlations become clear, numerous questions remain open.

2.1 Foamability: introduction to the role of surfactants

Not all liquids foam in the same way; each possesses a unique foaming capacity and therefore a different foamability (as defined on p. 75). When we try to make a solution foam, two extremes are apparent. If the foamability is poor, a small number of bubbles form and then disappear within a few seconds, like the foam under a waterfall. On the other hand, if the foamability is very good, the foam lasts for at least a few minutes: it appears stable, the bubbles don't burst instantly, and all the gas used to produce the foam is contained within it. This is the case for a solution of washing-up liquid or for shaving foam, for example. Between these two extremes, there are many intermediate possibilities: for example bubbles form and produce a few centimetres of strongly heterogeneous, unstable foam on the surface of the liquid which soon disperses, as on a fizzy drink or lager.

In order to obtain a stable foam, it is evidently necessary to add to the liquid several drops of "soap" or washing-up liquid, which contain surfactants (see p. 18). In this chapter we will describe in detail what it is that the presence of surfactants alters at the scale of a liquid interface, and then at the scale of a film. Among all the effects that are due to surfactants, we will determine those that are responsible for the existence of the foam, including the adsorption time, elastic modulus in compression, and the interactions between the interfaces.

2.2 Interfacial properties and foamability

2.2.1 Static surface tension Surfactants adsorb to the gas/liquid interfaces and thus modify the surface tension γ (§2.1.1, chap. 2). It is therefore important to know how γ depends on the surfactant concentrations in the bulk, c, and at the surface, Γ (FIG. 3.9).

By analogy with bulk systems, the problem can be tackled in terms of equations of state, which determine the relationships between different thermodynamic quantities at equilibrium. We first express the thermodynamic balance for the interface using the Gibbs–Duhem equation (in which volume is replaced by area and bulk pressure by surface pressure) to show that *at constant temperature* [1, 13]

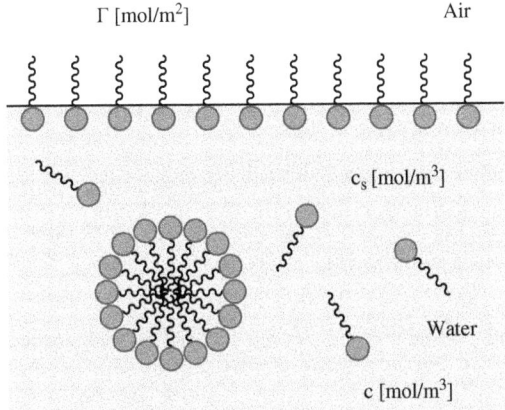

FIG. 3.9: Distribution of surfactants at the surface and in the bulk at equilibrium: c is the surfactant concentration in the bulk far from the interface, Γ the surface concentration, and c_s the concentration just below the interface. Note the micelles, the interfacial layer, and the free surfactants.

$$d\gamma = -\sum \Gamma_i d\mu_i, \qquad (3.1)$$

where $d\gamma$ and $d\mu_i$ indicate small variations in γ and in μ_i, the chemical potential of component i. In an ideal solution (without interactions), the chemical potential is $\mu = \mu_0 + k_B T \ln c$ (k_B is Boltzmann's constant), and then for a single species in solution we have

$$d\gamma = -\Gamma k_B T d(\ln c). \qquad (3.2)$$

This relationship (Gibbs' equation) describes the adsorption isotherm at the surface, and is the basis of all physical models of interfaces.

Irving Langmuir proposed a dynamical model (see eq. (3.7)) which predicts the surface concentration Γ at equilibrium:

$$\frac{\Gamma/\Gamma_\infty}{1 - \Gamma/\Gamma_\infty} = \frac{c}{c_a} \qquad (3.3)$$

where Γ_∞ is the saturated surface concentration and c_a is Szykowski's concentration, of order 10^{-1} kg.m^{-3}, discussed in §2.2.2. This allows us to calculate $dc/d\Gamma$ and then use Gibbs' equation (3.2) to obtain a value for the surface pressure Π_l:

$$\Pi_l = \gamma_0 - \gamma = k_B T \Gamma_\infty \ln\left(\frac{\Gamma_\infty}{\Gamma_\infty - \Gamma}\right) \qquad (3.4)$$

where γ_0 is the surface tension of the pure liquid. This can be rewritten in terms of the concentration c using eq. (3.3):

$$\Pi_l = \gamma_0 - \gamma = k_B T \Gamma_\infty \ln\left(1 + \frac{c}{c_a}\right). \qquad (3.5)$$

Eqs. (3.3–3.5) show how the surface tension (or, equivalently, the surface pressure), bulk concentration, and surface concentration are related. Experimentally, if we know two of these quantities (often the bulk concentration and the surface tension) it is possible to deduce the third. Although quite simple, Langmuir's equations of state are accurate to within a few percent.

There are other, more complex, equations of state for an interface [41] that include the interactions between molecules at the surface. In effect, the adsorbed molecules on the surface can interact in numerous ways, for example electrostatically, by steric repulsion, or by excluded volume (see exercise 8.2).

2.2.2 Dynamics of adsorption If we introduce surfactants into a pure solution we observe a decrease in surface tension until equilibrium is reached, but this is not instantaneous (FIG. 3.10). Likewise, if we rapidly increase the size of an interface between a gas and a surfactant solution, the tension increases, then it relaxes back towards its initial value. The temporal evolution of the surface tension sometimes differs greatly from one solution to another, even if the equilibrium values are close. In particular, it varies with bulk concentration, as illustrated in FIG. 3.10. Several processes are important in the temporal evolution of $\gamma(t)$: a molecule must diffuse to the surface (traversing a region where the concentration is lower than in the bulk) and it must then adsorb to the interface, overcoming the associated energy barrier. Either of these processes may control the dynamics.

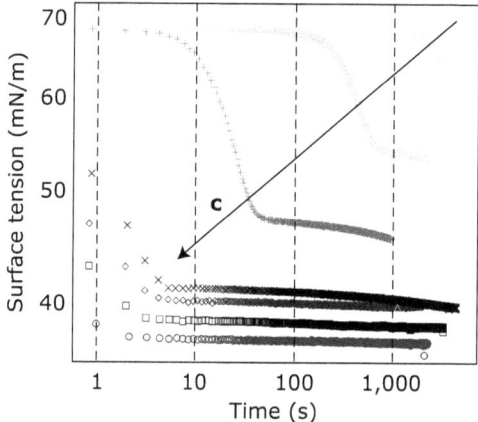

FIG. 3.10: At time $t = 0$ an interface is created rapidly in a cationic solution of surfactants. Surface tension is measured by the pendant-drop method (see §1.1, chap. 5) as a function of time for different bulk concentrations c (from $c = cmc/100$ to $c = cmc$, where cmc stands for critical micelle concentration, defined on p. 146). As the concentration is increased, as indicated by the arrow, there is a gradual shift from the diffusion-limited to the adsorption–desorption limited case and the adsorption times are reduced until equilibrium is reached. Reproduced from [70] by permission of The Royal Society of Chemistry.

In either case we assume that the equilibrium equation of state linking Γ to γ is known (Langmuir's equation (3.4) or similar, depending on the system), and so it is sufficient to determine the expression for $\Gamma(t)$ to find $\gamma(t)$.

Diffusion-limited case During the dynamic process of populating an interface, there is more adsorption than desorption. In the diffusion-limited case, a region close to the interface is depleted of surfactant molecules compared to the bulk, and the local surfactant concentration, c_s, is smaller than c; the thickness ζ of this region is typically of the order of Γ/c. To zeroth order, disregarding c_s, the typical time-scale for adsorption can be determined quantitatively as $t_{\text{diff}} = \Gamma^2/(D_v c^2)$, where D_v is the diffusion coefficient of the surfactants in the bulk liquid, which is just the time required to diffuse a distance ζ. It therefore appears to be very sensitive to the bulk liquid concentration. For example, for common surfactants of low molecular mass (see §6.1.1), Γ is of the order of mg.m^{-2} and $D_v = 10^{-10}$ m^2.s^{-1}. With $c = 10^{-2}$ g.L^{-1} we find $t_{\text{diff}} = 20$ s; with $c = 1$ g.L^{-1} we find $t_{\text{diff}} = 0.002$ s.

A quantitative description of this process is achieved by solving the diffusion equation for the bulk liquid. We assume that the process of adsorption/desorption is very rapid compared to typical diffusion times, and that Γ is therefore in equilibrium with the subsurface layer. That is, the concentration below the interface, c_s, is given by the equation of state at the surface (eq. (3.3)). One thus obtains the surface concentration $\Gamma(t)$ as a function of time [81]:

$$\Gamma(t) = \Gamma_0 + \sqrt{\frac{D_v}{\pi}} \left[2c\sqrt{t} - \int_0^t \frac{c_s(t)}{\sqrt{t-\tau}} d\tau \right]. \tag{3.6}$$

This equation is not explicit, as $c_s(t)$ itself depends on Γ. However, it simplifies to $\Gamma(t) \sim \sqrt{t}$ at short times (while $c_s \ll c$), and at long times there is an asymptotic solution showing that the difference from the equilibrium value decreases as $1/\sqrt{t}$.

Case limited by the adsorption–desorption barrier In this second case, observed for example at high concentrations where diffusion is rapid, the situation is different: the concentration in the bulk, c, is uniform, but Γ is no longer in equilibrium with the bulk. Energy barriers to adsorption are generally caused by electrostatic or steric forces. Conversely, leaving the interface also has an energy cost. For example, certain molecules such as proteins modify their spatial configurations when they are adsorbed, as a way of optimizing their amphiphilic character, making it more difficult for them to return to the bulk.

Langmuir assumed that the adsorption sites are localized at the surface, that the surfactants can adsorb and desorb, and that the interactions between molecules are negligible. With these assumptions, the flux of matter towards the interface is

$$\frac{d\Gamma(t)}{dt} = ac(\Gamma_\infty - \Gamma(t)) - b\Gamma(t). \tag{3.7}$$

The saturated surface concentration Γ_∞ is the concentration that would be found if there was only adsorption, without equilibrium with desorption. In eq. (3.7), the first term on the right corresponds to adsorption and the second to desorption, with a and b the adsorption and desorption coefficients. Their ratio $c_a = b/a$ is Szykowski's concentration (eq. (3.3)), which expresses the competition between adsorption and desorption. Eq. (3.7) predicts the equilibrium value of Γ (eq. (3.3)), and also how the system converges to it; the difference from the equilibrium value varies as $\exp(-t/\tau_b)$, with $\tau_b = 1/(ac+b)$.

Case of complex solutions With a micellar solution ($c > cmc$; see p. 146), it is necessary to consider several characteristic times: the diffusion of individual surfactant molecules, the diffusion of micelles, the adsorption of surfactants or micelles, and all the effects of micellization and de-micellization (which are rapid—of the order of ms—for ionic surfactants, but rather slower—of the order of seconds—for non-ionic ones) [12]. Likewise, for molecules more complex than common surfactants (see §6.1), it may be necessary to take into account the conformational rearrangements at the interface (protein unfolding, for example), or the formation of interfacial links between molecules. However, these effects only become important over long periods of time, as equilibrium is approached.

In addition to the variations in concentration perpendicular to the surface and the evolution of surface tension as a function of time, a gradient of γ can also appear along the surface. The interface reacts to these spatial gradients by creating an interfacial viscoelasticity via the surfactants.

2.2.3 Interfacial viscoelasticity An interface has a two-dimensional viscoelasticity which characterizes the response of the surface to shear and extensional stresses. Like bulk complex fluids (§2, chap. 4), these interfaces respond linearly to small deformations, but their response to large deformations is flow-dependent.

The presence of an adsorbed layer of surfactants at the surface, in contact with the surfactant reservoir in the bulk, gives rise to an interfacial viscoelasticity with properties that are distinct from those of an uncovered interface. These viscoelastic properties depend not only on the concentration and the solubility of the surfactant in water, but also on the time-scale of the mechanical perturbation.

We introduce here the general concepts of interfacial rheology which apply to the liquid/gas interfaces, and in particular to those interfaces covered by surfactants [21]. In §1.1, chap. 5, we describe several methods that quantify the rheology of interfaces.

We consider an element of length dl in a macroscopic interface. By transposing to 2D the 3D formalism (eq. (4.14)), we obtain an expression for the surface force which generalizes the surface tension (FIG. 2.4a):

$$df_i = \sum_j \sigma_{ij}^s \hat{n}_j dl, \tag{3.8}$$

where $\hat{\mathbf{n}}$ is a unit vector in the tangent plane of the interface, normal to the element dl. The indices i and j denote the components of vectorial and tensorial quantities. The total surface stress tensor σ_{ij}^s includes an isotropic contribution due to surface tension (γ plays the opposite role to 3D pressure), to which the components of $\bar{\sigma}_{ij}^s$ are added, normalized to a force by a length, containing the viscous and possibly the elastic terms:

$$\sigma_{ij}^s = \gamma \delta_{ij} + \bar{\sigma}_{ij}^s(\varepsilon, \dot{\varepsilon}). \tag{3.9}$$

Here γ is the surface tension at equilibrium, corresponding to the bulk concentration of surfactant far from the interface.[1] The viscous and elastic terms are connected to, respectively, the rate of deformation $\dot{\varepsilon}$ and the deformation ε of the interface by constitutive laws.

Response to an expansion or compression Variations in surface area are one type of deformation that can be applied to a liquid interface by either expanding or compressing it.

When a purely viscous interface of area A is expanded at a rate $A^{-1}dA/dt$ it is subject to surface stresses (with the notation given in FIG. 3.11a):

$$\sigma_{xx}^s = \sigma_{yy}^s = \gamma + \eta_d \frac{1}{A}\frac{dA}{dt}; \quad \sigma_{xy}^s = 0. \tag{3.10}$$

The coefficient η_d, linking the stress to the rate of expansion, is the surface dilatational viscosity, expressed in kg.s^{-1}.

In expansion and compression, in addition to a viscous response, the surfactants also induce a surface dilatational elasticity. When the interface is stretched locally by an amount δA, the surface concentration of surfactants Γ is modified, which changes the local value of surface tension. This is taken into account in the dynamic term $\bar{\sigma}^s$.

In the case where the total quantity of adsorbed surfactants is *not modified* (if they are insoluble for example), the variation in concentration $\delta \Gamma / \Gamma$ induced by the deformation is $\delta \Gamma / \Gamma = -\delta A / A$. The elastic response is thus linked to the *Gibbs–Marangoni elastic modulus*, E_{GM}, defined in terms of the values at equilibrium (eq. (3.4)):

$$E_{GM} = -\frac{d\gamma}{d\ln \Gamma}\bigg|_{eq}. \tag{3.11}$$

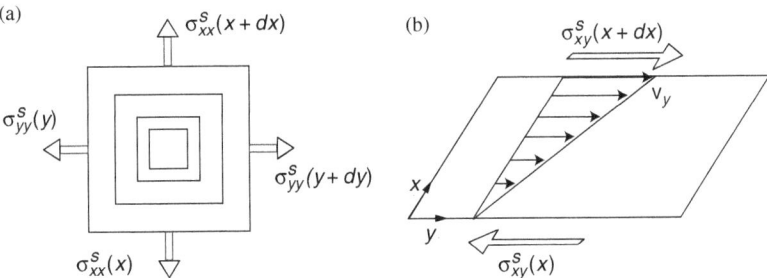

FIG. 3.11: Notation for stress on an interface. (a) Expansion and (b) shear of an interface.

---

[1] We have chosen this convention here, but the isotropic stress terms can be included in either γ or $\bar{\sigma}^s$ without loss of generality.

It is measured in a Langmuir trough, from the isotherm of the insoluble layer in compression. The interfacial stress in this case is

$$\sigma_{xx}^s = \sigma_{yy}^s = \gamma + E_{GM}\frac{\delta A}{A}; \quad \sigma_{xy}^s = 0. \tag{3.12}$$

On the other hand, with *soluble* surfactants the response to dilation becomes complicated and depends on the mechanical excitation frequency. For an oscillation of the surface of the sample at frequency ω, the complex surface dilatational elastic modulus, E_s^*, can be split into two parts:

$$E_s^*(\omega) = E_s'(\omega) + iE_s''(\omega). \tag{3.13}$$

The real part (elastic modulus) describes the part of the stress that is in-phase with the strain. The imaginary part (the loss modulus) describes the stress that is out-of-phase by $\pi/2$ with respect to the strain; the latter modulus is linked to the surface viscosity by $E_s'' = \omega \eta_d$.

Note that calling this modulus *elastic* recognizes that this process induces an in-phase response. Nevertheless, it is not associated with a *solid elastic* property of the monolayer. The exact parallel in 3D is the elastic compression or expansion of a fluid, as in sound propagation for example.

Response to shear We now consider the shear response, in which the shape of an interface changes but its surface area remains constant. For a simple shear flow imposed on a flat interface (FIG. 3.11b), the surface shear rate is

$$\dot{\varepsilon}_{xy}^s = \frac{\partial v_y}{\partial x}, \tag{3.14}$$

where v_y is the speed of the interface in the y direction, which varies with x. The surface shear stress is thus linked to the strain rate $\dot{\varepsilon}_{xy}^s$ by

$$\sigma_{xy}^s = \eta_s \dot{\varepsilon}_{xy}^s; \quad \sigma_{xx}^s = \sigma_{yy}^s = \gamma. \tag{3.15}$$

η_s is the surface shear viscosity and depends strongly on the surfactants (their concentration and molecular type, as discussed in §6.1). In the Newtonian case, this viscosity is independent of the shear rate, which is usually the case for surfactants of low molecular mass. Nevertheless, more complex behaviour is found when interfaces are covered by proteins or polymers: then viscosity can depend on shear rate, and this may be associated with the appearance of plasticity (we shall return to these ideas in §2.1, chap. 4), and even fracture.

Surface elasticity due to shear may also be observed in a surfactant monolayer. Use of the term "elastic" is perfectly justified here, as it concerns behaviour specific to solids, the in-phase response to shear. In fact, like a 3D phase, a 2D phase can be solid, liquid, or gas (see §5, chap. 2), and a 2D solid resists shear.

As in the case of expansion and compression (eq. (3.13)), we introduce a 2D complex modulus, G_s^*, for the response to shear, defined as for a 3D fluid (eq. (4.9)):

$$G_s^*(\omega) = G_s'(\omega) + iG_s''(\omega), \tag{3.16}$$

where the in-phase response to the deformation is given by G_s', and the out-of-phase stress by $G_s'' = \omega \eta_s$.

Values of the viscoelastic moduli The existence and frequency-dependence of elastic and viscous moduli result from two contributions at the surfactant scale. One, which is intrinsic to the interface, is due to hydrodynamic interactions and to friction between the adsorbed molecules, while the other comes from the exchange of surfactants between the interface and the underlying liquid.

In fact, the response to shear (in which the shape changes but the surface area remains constant) is dominated by intermolecular interactions at the interface. On the other hand, the response to expansion and compression is strongly coupled to the degree of surfactant exchange between the surface and the liquid (FIG. 3.12), which is controlled by the diffusion of surfactants in the bulk liquid and adsorption–desorption processes at the interface. Nevertheless, this response also contains an intrinsic contribution linked to mechanisms occurring at the interface and so during expansion and compression it is difficult to distinguish the different contributions.

However, in the response to expansion and compression two limiting cases are easily established. If the oscillation frequency ω is very low, or the surfactants very soluble, there is sufficient time for surfactant transfer to occur and for the surface concentration to adjust. So there is no variation in surface tension and no stress response to a compression, hence $E'_s = E''_s = 0$. In the limit of high ω, or very soluble surfactants, the layer no longer has time to react because surface and bulk are unable to exchange surfactants sufficiently rapidly, hence $E'_s = E_{GM}$ and E''_s vanishes.

Between these limits the simplest model, developed by Lucassen and Van der Tempel [50], assumes that surfactant exchange between the surface and the bulk is controlled by diffusion. Then

$$E'_s = E_{GM} \frac{1+\Omega}{1+2\Omega+2\Omega^2} \qquad (3.17)$$

and

$$E''_s = E_{GM} \frac{\Omega}{1+2\Omega+2\Omega^2}, \qquad (3.18)$$

where

$$\Omega = \sqrt{\frac{D_v}{2\omega} \frac{dc}{d\Gamma}} \qquad (3.19)$$

combines the diffusion coefficient, D_v, of surfactants in the bulk, with the surface and bulk concentrations. For common surfactants this model is adequate, implying that

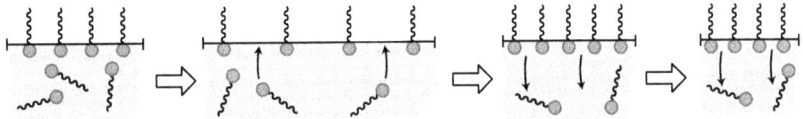

FIG. 3.12: Sketch of an interface subjected to an expansion/compression oscillation. The black arrows show how the surfactants can move between the interface and the bulk. Their response to the oscillation determines the viscoelasticity of the interface.

intrinsic effects play only a minor part in expansion and compression. In practice, there is a maximum in elasticity as the concentration is varied, which often occurs just below the *cmc* (see p. 146). This indicates that the presence of micelles complicates the exchange between surface and bulk. Refinements to this model take into account surface diffusion and energy barriers to adsorption or desorption.

Quantitatively, shear moduli are always much lower than compression moduli. Moreover, the shear elastic modulus of soluble surfactants is usually zero (it is only non-zero if there is a solid layer at the surface, for example a rigid network or a complex interconnected structure). The dilatational elastic modulus typically varies between a fraction of a $mN.m^{-1}$ to hundreds of $mN.m^{-1}$.

The shear viscosity, η_s, has a typical value of 10^{-8} to 10^{-7} $kg.s^{-1}$ for surfaces covered with surfactants of low molar mass, and can reach 10^{-5} $kg.s^{-1}$ for surfaces covered in proteins, for example. In the former case, such interfaces are referred to as *mobile*, while in the latter case they are *immobile*. This distinction is made quantitative later, in terms of the Boussinesq number (eq. (3.62)), which recognizes the contribution of bubble size too. On the other hand, η_d is typically greater than η_s by one or two orders of magnitude. Different models for the surface dilatational viscosity of a film in steady extension are given in §6.2.

2.2.4 Interfacial phenomena and foamability

As far as surface tension is concerned, a simple rule of thumb is that the lower the value of γ, the less costly in energy is the creation of an interface. Producing a foam requires a significant increase in the area, S_{int}, of gas/liquid interface. That the interfacial energy E is proportional to γ (eq. (2.15)) suggests that a low surface tension should enhance foamability.

However, it is the adsorption dynamics that matter, rather than the value at equilibrium. Even when the final value of γ is low, if this is obtained after a long and slow adsorption, the solution will foam poorly.

To create a stable foam, newly formed bubbles need to be covered with surfactant as quickly as possible. A high surface concentration is required as soon as the bubbles touch one another and create films. It is necessary to compare a typical adsorption time t_{ad} (i.e. the time-scale over which the surface tension decreases) with a characteristic bubble formation time t_{exp}, which depends on the method used to produce the foam. If t_{exp} is long it is not necessary for the surfactants to be adsorbed rapidly, and it is possible to produce a foam even with a low surfactant concentration. On the other hand, if t_{exp} is short, only systems that are able to lower the surface tension very rapidly will foam.

In order to better understand this correlation between adsorption dynamics and the way in which the foam is made, it is possible to measure the surface tension of an interface with continuously increasing area. By repeating the experiment at increasing rates of expansion, we can determine the rate at which the surfactants become less efficient at decreasing the surface tension [51].

So it is necessary for the surfactants to be able to adsorb rapidly at the surface to form a dense layer there. Yet this does not explain how and why the presence of this layer is indispensable to foam generation.

We saw in §2.2.3 that the adsorption of surfactants to an interface has a major effect on its viscoelastic properties (both in extension and under shear), and so surface rheology plays a role in foam production. The viscoelasticity of the interface in expansion and compression, characterized by E'_s and E''_s, is widely recognized as an important interfacial parameter for foamability and foam stability.

Yet there are no quantitative criteria that predict whether or not a foam will form. Several effects contribute to the link between interfacial viscoelasticity and foamability:

- For a given applied stress, a larger interfacial dilatational elastic modulus E'_s results in a smaller strain. That is, the film stretches less and is therefore less likely to rupture.
- Elasticity also acts to *restore* the interface. If an expansion is not uniform, or if there is a fluctuation in surfactant density, a depleted region is created (FIG. 3.13); a surface tension gradient appears, which is greater when E'_s is larger. This induces an elastic restoring force in the plane of the interface which serves to *counteract* the fluctuation by bringing back surfactants and by limiting the stretch of the interface. The modulus E'_s relates the size of the fluctuation or local stretch of the interface to the surface tension gradient, and consequently to the stretching or restoring force.
- In addition to transport at the interface, this mechanism of minimizing the interfacial gradients also induces a coupling between surface tension gradients and the flow of liquid below the interface, known as the *Marangoni effect*. This flow in the subsurface layer enables surfactant to be transported more rapidly to "oppose" interfacial fluctuations.
- Even more significantly, at the scale of a foam film (FIG. 3.13), the Marangoni effect not only draws surfactants to the surface but also draws liquid into the film. As a surface tension gradient is often associated with a fluctuation in thickness (as in FIG. 3.13), there is then a secondary benefit: if the interfaces are elastic, the restoring force will simultaneously correct both the fluctuations in concentration and thickness via the Marangoni effect, thereby reducing the risk of film rupture.

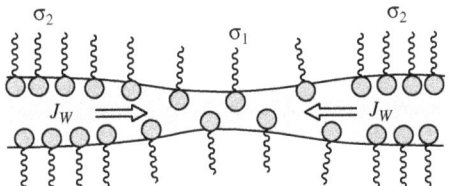

FIG. 3.13: Sketch of a film with elastic interfaces, where a local stretch causes a gradient in the surface tension coupled with a reduction in thickness. The combination of flows induced at the surface and in the bulk enable differences in thickness and concentration to be restored. In particular, the flux of liquid J_W in the film brings surfactant back to the interface and re-inflates the film.

- A low value of the modulus $|E_s^*|$ facilitates the stretching of a surface; however, if it stretches too quickly the adsorption dynamics of the surfactants may become significant, as discussed previously. At the other extreme, if E_s' is high, then the response of the interface could be solid-like, leading to the possibility of fracture.
- The elasticity of the interfaces during expansion and compression, as for surface viscosity during shearing, modifies the process of drainage in the films and Plateau borders. In fact, flows at the surface, governed by interfacial parameters, are coupled to bulk flows (see §4).
- Interfacial elasticity also appears to control the appearance of bell-shaped liquid drops (see §5) in films which form rapidly. These *dimples* have a destabilizing effect on a film.

Given all these competing effects, how can elasticity be tuned to optimize foamability? It is often found that an intermediate value of the dilatational elasticity—of the order of tens of mN.m^{-1}—is optimal, offering a balance between the elasticity-controlled effects described above [46]. For common surfactants, this often corresponds to a concentration close to the *cmc*.

To summarize, it is clear that foamability is enhanced by a viscoelastic interface which is rapidly covered by surfactants. However, it is not always possible to explain the existence of a foam by these two criteria alone. As we will now describe, the interactions between the interfaces at the scale of a film play a crucial role.

2.3 Properties of liquid films and foamability

2.3.1 Interactions between two interfaces

During foaming, surfactants adsorb to the surface of each bubble. When two bubbles meet, what forces are acting between the two interfaces? This question proves crucial to our understanding of the stability of the thin liquid film which separates the bubbles of a foam.

We consider first a liquid film consisting of two infinite interfaces separated by a thickness h of liquid. We use the disjoining pressure Π_d (§4.1.3, chap. 2) to characterize the forces per unit area between these two interfaces, which result from the following attractive and repulsive contributions [4, 58, 74].

First, the *London–van der Waals* forces, also known as dispersion forces, arise because, on a microscopic level, two molecules which have electrostatic dipoles, whether permanent or oscillating, interact: the correlation between their dipoles induces an interaction potential $V(r)$ which varies as $1/r^6$, where r is the distance between them. Macroscopically, after integration over all molecules of a thin film, this always results in an attraction between both interfaces:

$$\Pi_{\text{vdW}} = -\frac{A_h}{6\pi h^3}, \qquad (3.20)$$

where A_h is the Hamaker constant, which is positive for a thin film in air. Since interfaces tend to get closer, the film tends to thin: van der Waals forces always make a film unstable. This fundamental fragility must then be balanced by other contributions, hence the need for surfactants.

A second contribution is the *electrostatic* interaction, which is often significant because the interfaces are generally electrically charged, due to the ionic character of many surfactants (see §6.1). These charged surfaces repel each other, but this repulsion is screened by the presence between them of an ionic solution, containing co-ions with the same charge as the surface and counter-ions with the opposite charge. The effective range of the electrostatic potential ψ of a charged interface (see FIG. 3.14), taking into account the role of the co-ions, is given by the Debye length,

$$\lambda_D = \sqrt{\frac{\varepsilon k_B T}{8\pi n e^2}}. \tag{3.21}$$

Here ε is the relative permittivity of the fluid, n is the charge density, and e the elementary charge. λ_D is also the typical size of the ionic cloud induced by a charged surface. Therefore, when two charged surfaces approach each other the interaction becomes significant when the separation h falls below $2\lambda_D$.

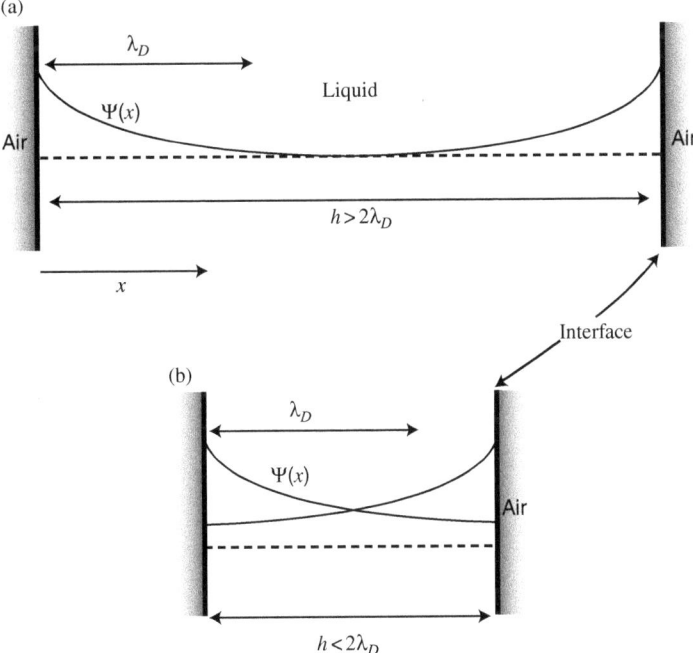

FIG. 3.14: Electrostatic potential $\psi(x)$ of two charged thin film interfaces a distance h apart. λ_D is the effective range of the electrostatic potential ψ. **(a)** There is no repulsive interaction as long as $h > 2\lambda_D$. At the centre of the film, we find the electrostatic potential in the bulk (dashed line). **(b)** On the other hand, for $h < 2\lambda_D$, the electrostatic potential in the film is nowhere equal to that in the bulk. The concentration of counter-ions is greater than in the bulk, and this excess of counter-ions is responsible for the repulsive force between the thin film interfaces.

Third, at small separations, the ionic clouds overlap to generate a repulsion. This interaction is therefore entropic in origin, as suggested by the presence of $k_B T$. With the usual assumption of a small overlap and a constant potential, the electrostatic disjoining pressure is approximately

$$\Pi_{el} \sim \exp(-h/\lambda_D). \tag{3.22}$$

Adding Π_{vdW} and Π_{el} (eqs. (3.20) and (3.22)) results in the DLVO model, named for Derjaguin, Landau, Vervey, and Overbeck, which is sufficient to describe the stability of many colloidal systems [28]. However, in foams there are often further contributions.

At very small separation the surfactant layers adsorbed on the interfaces interact via *steric* repulsion, since the surfactant layers cannot inter-penetrate. For films with surfactants of low molecular mass this is associated with steric hindrance between water molecules and the hydrophilic head of the surfactants. FIG. 3.15 illustrates the van der Waals, electrostatic, and steric contributions to the total disjoining pressure in a film:

$$\Pi_d = \Pi_{vdW} + \Pi_{el} + \Pi_{ste}. \tag{3.23}$$

These interactions between bubbles only act over short distances, generally less than 100 nm. In FIG. 3.15, we have represented the electrostatic interaction as sufficiently strong compared to dipolar attractions that a maximum appears in the curve. Similarly, at short distances steric interactions prevail over dipolar interactions and Π_d becomes positive again. This type of non-monotonic curve is usual for common surfactants and yields one or two possible stable thicknesses for a given pressure (FIG. 3.20) [4].

Fourth, the *supramolecular* interaction is also significant for surfactant solutions and foams. In foaming liquids, there are often organized structures in the bulk, on a much larger scale than that of the molecules, for example micelles or bilayers formed

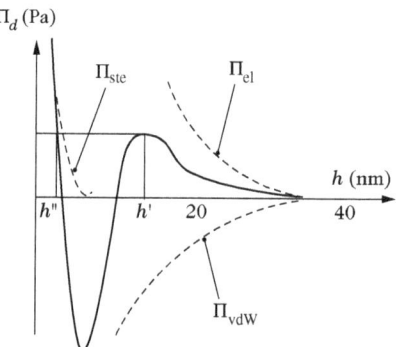

FIG. 3.15: Disjoining pressure between two interfaces as a function of interface separation h. The different contributions to the disjoining pressure—van der Waals, electrostatic, steric—are represented by the dashed lines, and the continuous line represents their sum.

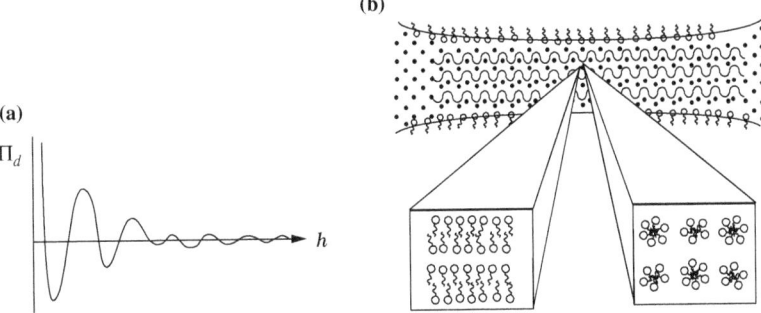

FIG. 3.16: Stratification. (a) The disjoining pressure Π_d oscillates as a function of the thickness of the film h. (b) These oscillations result from the confinement and self-organization of supramolecular structures in the film, for example packings of bilayers or micelles.

by surfactants at high concentration. The confinement of these structures in the films induces an additional disjoining pressure, which is long-range and depends on distance (FIG. 3.16). Due to the confinement, we may observe stratification in a film, i.e. organization into parallel layers. The concentration profile therefore oscillates in a direction perpendicular to the film.

Steric effects due to confinement and non-inter-penetration can appear at a thickness greater than 100 nm. For proteins, adsorbed polymers, and solid particles adsorbed at an interface, steric repulsion is sometimes observed at thicknesses of several hundreds of nanometres. Then the films are not uniform in thickness; instead they appear bumpy with a congealed gel between the surfaces (FIG. 3.17).

The fact that the disjoining pressure is non-monotonic has an important consequence: thermodynamically, the parts of the curve where $d\Pi_d/dh$ is positive are unstable and corresponds to inaccessible film thicknesses.

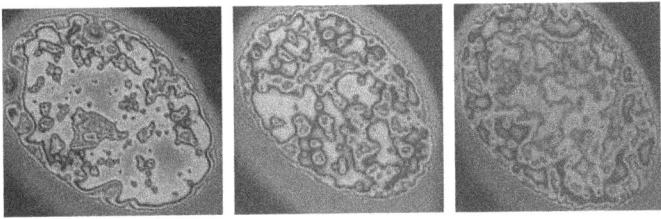

FIG. 3.17: Inhomogeneous films stabilized by milk protein (casein, see p. 147), obtained using the film balance technique (§1.1.4, chap. 5). The bulk concentration of the protein, and the stability of the film, increases from left to right. Interference images (§1.1.3, chap. 5) show that the thickness is of the order of several hundreds of nanometres and that it is non-uniform. The surface area of the film in the image plane is 3 mm². Photographs courtesy of A. Saint-Jalmes. See PLATE 6.

96 *Birth, life, and death*

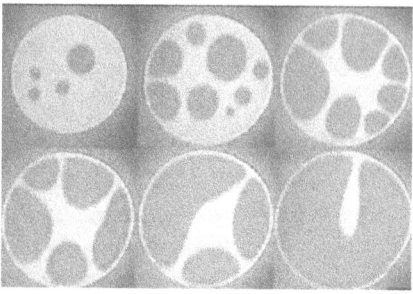

FIG. 3.18: Thickness discontinuity. View of the top of a single film showing a discrete jump in thickness from a thicker film (appearing clear) to a thinner film (appearing darker). The images correspond to different times, and the pressure on the film is constant. The dimensions and techniques are the same as in FIG. 3.17. Photographs courtesy of A. Saint-Jalmes. See also PLATE 7.

For the usual case described in FIG. 3.15, if, by departing from $\Pi_d = 0$, we increase the disjoining pressure (or if we push the two interfaces together), h first decreases continuously to a critical value h' (at the local maximum), then jumps *directly*—at constant pressure—to a smaller value on the second stable branch at h''.

This discontinuity in film thickness can be seen using interference (FIG. 3.18; see §1.1.3, chap. 5): if a film with uniform initial thickness greater than h' is subjected to a sufficiently large pressure, then over time some darker domains appear, corresponding to reduced thicknesses. These progressively join together and, at equilibrium, the film ends up with a lower uniform thickness. Between these two values, a whole range is inaccessible. It is the same for stratified films: for a curve of oscillating disjoining pressure like that in FIG. 3.16, film thinning occurs via several discrete jumps in thickness.

Fifth, a flow of liquid in a film can also induce a *dynamic* pressure, which stabilizes the thickest films. For this contribution to play a role the flows must be fast, both along the films and along the adjacent Plateau borders. This requires large radii of curvature (see §4.4.3), which is the case for centimetre-sized bubbles or for high liquid fractions.

2.3.2 Conditions for local stability and foamability Returning to a foam as a whole, we ask how the film thickness is determined by the disjoining pressure, and thus by which of the above contributions. In order to find out, we need to again consider the equilibrium of pressures between a flat film and the adjacent Plateau border (FIG. 3.19), introduced in §4.1.3, chap. 2.

The Plateau border curvature is such that the liquid in a film is sucked into the adjacent Plateau borders. This is known as *capillary suction* and is always present since Plateau borders surround every film.

In order for a flat film to exist at equilibrium, it is necessary that, in addition to the liquid pressure p, there is a *positive* disjoining pressure in the film, which balances the gas pressures and resists the suction from the Plateau borders. At equilibrium, the

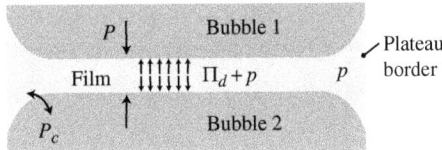

FIG. 3.19: Local equilibrium of pressures: in order to balance the different gas pressures and to oppose capillary suction into the Plateau borders (and so enable a stable film to exist), repulsive forces must act between the interfaces, inducing a disjoining pressure Π_d.

disjoining pressure is equal to the capillary pressure P_c (eq. (2.30)), $\Pi_d = P - p = P_c$. This relationship is fundamental to film stability (and to a foam's existence!).

The capillary pressure P_c depends on the properties of a foam and is determined by experimental conditions. It is inversely proportional to the radius of curvature of the Plateau borders r (eq. (2.30)), which itself depends both on bubble diameter d and the liquid fraction ϕ_l. Very different ranges of capillary pressure may therefore be found: either low (a few Pa) for wet foams and/or large bubbles, or high (of the order of several hundreds, even thousands, of Pa) for small bubbles and/or dry foam, corresponding to small radii of curvature of the Plateau border, $r < 0.1$ mm.

Consider a foam with uniform and constant bubble size and liquid fraction, which fixes the capillary pressure and the capillary suction. The equilibrium of pressures implies that this foam can only exist if there are repulsive forces in the films which give rise to a disjoining pressure Π_d with $\Pi_d = P_c$, allowing an equilibrium film thickness h to be found. To evaluate the limit of stability of a foam it is therefore necessary to compare the possible capillary pressures in the foam with the possible disjoining pressures in a film. Foamability is enhanced when repulsive interactions are strong and long-range.

Returning to FIG. 3.15, which corresponds to common surfactants of low molecular mass, there are ranges of thickness where the electrostatic and steric forces are greater than the dipolar forces. At low capillary pressures ($P_c = P_{c1}$, of the order of several hundreds of pascals), a film thickness of several tens of nanometres can be stable (FIG. 3.20) [4]. Such a film, stabilized by the equilibrium between electrostatic and dipolar forces, is called a *common black film* (CBF), with a typical thickness in the range 10 to 80 nm.

For much greater capillary pressures ($P_c = P_{c2}$), there is a single equilibrium state fixed by the equilibrium between steric and dipolar forces; it corresponds to a nanometric liquid film. Although extremely fragile, such a film, known as a *Newton black film* (NBF), is nevertheless observable in surfactant-stabilized foams, with a thickness in the range 2 to 4 nm depending on the surfactant chain lengths. In principle, by carefully decreasing the pressure from $P_c = P_{c2}$ to P_{c1}, it is possible that a NBF could still be observed, at thickness h'_{eq1} (FIG. 3.20). These films are called *black* (see §1.1.3, chap. 5) because at these thicknesses they no longer reflect light (only destructive interference occurs).

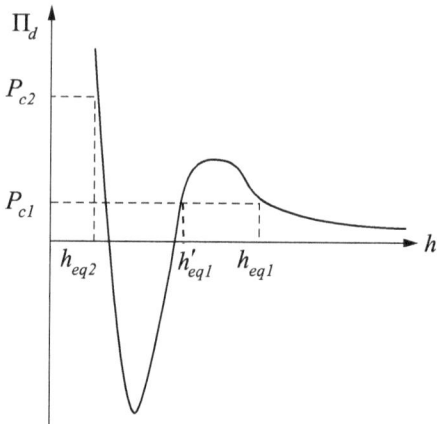

FIG. 3.20: Equilibrium thicknesses h_{eq1}, h'_{eq1}, and h_{eq2} for two values of capillary pressure P_{c1} and P_{c2}. For large thickness, a *common black film* is stable while at the nanometre scale a *Newton black film* is found.

2.4 Summary of the microscopic origins of foamability

We have seen that macroscopic foamability is associated with certain microscopic properties peculiar to each interface, as well as to the forces which appear between two interfaces. Can we now improve and predict the foamability of a solution?

2.4.1 Improving foamability There is a correlation between isolated film stability and foam stability. A good foaming solution must be able to *rapidly adsorb* large quantities of surfactants at the surface, and the surfactants must produce *a positive and significant disjoining pressure*, that is, strong and long-range repulsion, for instance with electrostatic charge. There is also an optimal value of the *surface dilatational elasticity* (neither too large, nor too small) to restore the surface after fluctuations—notably via the Marangoni effect—and limit the dynamic instabilities of the film interfaces.

We have seen that by increasing the surfactant concentration c, the adsorption kinetics increase, the equilibrium value of γ decreases, and the viscoelastic moduli and the range of the repulsive forces both increase. Consequently, all the effects which tend to stabilize the foam are increased: by increasing c we obtain a better foaming solution and a more stable foam. Nevertheless, foamability cannot be improved indefinitely by varying only the concentration: when the concentration reaches the *cmc*, the surface is saturated and the surface concentration no longer depends on the bulk concentration; thus it is pointless to go much beyond the *cmc* (unless it is to increase stratification effects in the films). In addition, for a weakly surface-active molecule (which doesn't adsorb much at the interface), increasing the concentration does not populate the interfaces, but instead produces structures in the bulk.

2.4.2 Predicting foamability Are measurements at the liquid interfaces (for adsorption dynamics) and on single films (to test their stability) sufficient to understand and predict the foamability of a solution? The answer is usually yes, at least for simple surfactant systems of low molecular mass, but there are cases where measurements on the substructures don't reflect the actual properties of the 3D foam, in particular for complex chemical formulations: surfactant mixtures, surfactant–polymer mixtures, proteins, etc. In certain cases, measurements of the adsorption dynamics, $\gamma(t)$, on a free interface are poorly correlated with foamability. Thus certain systems foam at concentrations for which the $\gamma(t)$ measurements at the scale of a single gas/liquid interface suggest that only very little material adsorbs, and that it does so slowly (suggesting a poor foamability). In this case, a measurement on a single interface does not reflect the properties of the foam: it doesn't take into account, for example, rapid, forced adsorption phenomena, which occur when bubbles come into contact during the process of foam generation.

Herein lies the main problem: due to experimental limitations, we can only study certain foam substructures at time-scales, frequencies, amplitudes, or length-scales different from those that occur during foaming and which are therefore not directly relevant. Likewise, we can often only measure properties in the linear regime, which is not always representative of foam formation. Finally, in the majority of measuring techniques on isolated substructures, we often suppress collective effects, associated with confinement or with dynamics for example. Thus, the information obtained on the intermediate scales, although always useful, often remains incomplete and the conclusions that one draws are often inadequate.

Consequently, there still remains much to learn in order to be able to respond with certainty in every case to the question "why does that foam?".

3 Coarsening

A bubble with few neighbours has a higher pressure than its neighbours. If the gas that it contains can diffuse through its films, the bubble shrinks and then loses neighbours, and consequently empties more and more quickly (§3.1). In this way, bubbles disappear, one after another, and so the average size of those which remain increases. Finally there is only one large bubble remaining, which contains all the gas. This coarsening mechanism and the resulting foam structure are generic (§3.2). However, the rate at which coarsening occurs depends on the particular characteristics of the system being considered; for a foam, it depends on the liquid fraction, the average bubble size, and the physical chemistry of the gas and the liquid (§3.3). We begin with 2D (§3.1.1), which is much simpler than 3D (§3.1.2).

3.1 Growth rate of a bubble in a dry foam

The following is specific to the growth of a bubble in a dry foam. A large part of the literature on this subject in fact boils down to the study of film curvature, described in §3.3.2, chap. 2.

3.1.1 Dry 2D foam: von Neumann coarsening

We consider an ideal 2D foam (see p. 26), that is, a dry foam with thin films. If the gas diffuses from a bubble i to its neighbour j solely through the film ij, the flux from one bubble to the other is proportional to $\ell_{ij}(P_i - P_j)$: it is the product of the driving force, here $P_i - P_j$, and the size of the region of gas exchange, here the length ℓ_{ij} of the film. If the flux is positive, the amount of gas contained in bubble i decreases, as does the bubble area A_i (the gas density remains roughly constant). More precisely, the change in area of bubble i, dA_i/dt, results from the sum of the fluxes to each neighbour j (FIG. 3.21):

$$\frac{dA_i}{dt} = -a_1 \sum_{j=1}^{n}(P_i - P_j)\,\ell_{ij}, \tag{3.24}$$

where the coefficient a_1, which is the same for all films in a given foam, will be interpreted in §3.3.

Surprisingly, the right-hand side of eq. (3.24) appears in eq. (2.24), relating pressure to topology and curvature. By combining them we obtain:

$$\frac{dA_i}{dt} = -\frac{2a_1\lambda}{e}\sum_{j=1}^{n_i}\kappa_{ij}\ell_{ij} = -\frac{2a_1\lambda\pi}{3e}(6 - n_i). \tag{3.25}$$

This remarkable relationship between topology, geometry, forces, and the evolution of bubble i, which is at least correct within the limit of the *ideal foam* model (see §2.2.1, chap. 2), is readily confirmed experimentally (FIG. 3.22a). It can be written in the following abbreviated form, due to von Neumann [79], and even in terms of the geometric charge defined in eq. (2.23):

$$\frac{dA_i}{dt} = -D_{\text{eff}}q_i = \frac{\pi}{3}D_{\text{eff}}(n_i - 6). \tag{3.26}$$

The *effective diffusion constant* D_{eff} is positive, and is expressed in m^2.s^{-1}. It is the same as in 3D and is proportional to the rate of gas diffusion D_f in the foaming liquid, but also includes several other factors (see §3.3).

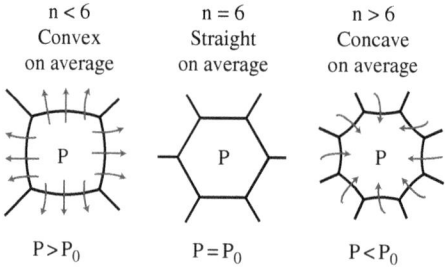

FIG. 3.21: A bubble with 5 sides or less has convex sides, so its pressure is greater than the average pressure of its neighbours, P_0, and it shrinks. The curvature of the sides of a bubble with 6 neighbours is zero on average, and it is therefore at the same pressure as the average of its neighbours. A bubble with 7 sides or more has concave sides, its pressure is lower than its neighbours, and it grows.

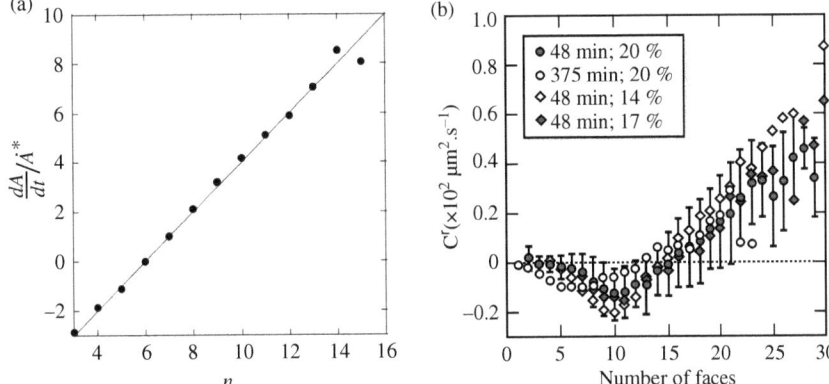

FIG. 3.22: Growth rate of a bubble as a function of its number of neighbours. (a) Bubble area variation vs. number of sides in a dry 2D foam (from [18]). The individual bubble growth rates are averaged over the duration of the experiment, and fitted by a straight line that passes through 0 at $n = 6$, as predicted by eq. (3.26)). The slope is $\dot{A}^* = 140$ mm^2.h^{-1}, giving $D_{\text{eff}} = 133$ mm^2.h^{-1}. (b) Relative bubble volume variation vs. number of faces in three wet 3D foams, with liquid fractions between 14 and 20% [44]. The graph is directly comparable with FIG. 2.17, with the opposite sign, in agreement with eq. (3.30). The anomalous negative slope at low f is probably linked to the liquid fraction (see §3.3.3). Unlike in FIG. 2.17, the data come from dynamic experiments (X-ray tomography, see §1.2.4, chap. 5) which enables the prefactor D_{eff} to be determined.

Consequently, bubbles with positive geometric charge (3, 4, or 5 sides, see §3.3.2, chap. 2), which have a higher pressure than the average of their neighbours (balanced by the length of the shared edges), lose their gas to bubbles which have more than 6 sides (FIG. 3.21). The geometric details (size and shape of the bubbles) do not play a role, only the topology is important. This result, which had been predicted theoretically before being confirmed experimentally [72], has simplified considerably the study of coarsening in 2D.

If the foam has a free surface through which it can exchange gas with the environment, the peripheral bubbles are rather bulbous (FIG. 2.12), and, since the foam pressure is greater than ambient pressure, overall the foam gradually loses gas.

On the other hand, for a foam of fixed total area $A_{\text{foam}} = \sum_{i=1}^{N} A_i$ (a foam in a box, or even a numerical simulation of a foam using periodic boundary conditions), any gas lost from one bubble is redistributed among it neighbours, and the average of dA_i/dt over the foam is zero:

$$\frac{1}{N} \sum_{i=1}^{N} \frac{dA_i}{dt} = \left\langle \frac{dA_i}{dt} \right\rangle = 0. \tag{3.27}$$

Eqs. (3.26) and (3.27) are compatible with the fact that the average value of n in the foam is 6 (see p. 32). However, $d\langle A_i\rangle/dt$ is not zero: the average size increases, as discussed on p. 105.

3.1.2 Dry 3D foam: Mullins coarsening

The von Neumann relationship in 2D, relating the curvature of a bubble's sides to its number of neighbours (eq. (2.24)), cannot be generalized to 3D in an exact way (cf. eq. (2.27)). The same is therefore true for the growth rate: in 2D it is an exact function of the number of neighbours (eq. (3.25)), but this is not the case in 3D.

In 3D, the flux of gas across a film is proportional to the pressure difference across it, this time multiplied by its surface area S. More precisely, dV_i/dt is the sum of the fluxes to each neighbour j. With a_1 the same constant as in eq. (3.24), we obtain

$$\frac{dV_i}{dt} = -a_1 \sum_{j=1}^{n}(P_i - P_j) S_{ij} = -2\gamma a_1 \sum_{j=1}^{n} H_{ij} S_{ij}, \tag{3.28}$$

where we have introduced the mean curvature H_{ij} of the film ij using the Young–Laplace law (eq. (2.4)). The term $H_{ij}S_{ij}$ consists of a curvature (m^{-1}) multiplied by a surface area (m^2), so it is a length. In order to compare experiments on different foams it is natural to divide eq. (3.28) by a length, usually $V_i^{1/3}$. We consequently define the *relative growth rate* C^r of bubble i as

$$C_i^r = \frac{1}{V_i^{1/3}}\frac{dV_i}{dt} = \frac{3}{2}\frac{d}{dt}\left(V_i^{2/3}\right). \tag{3.29}$$

As in 2D (§3.1.1), the proportionality constant of eq. (3.28) is an effective diffusion coefficient, denoted D_{eff}, expressed in m^2.s^{-1}. We thus write

$$C_i^r = -D_{\text{eff}} \sum_{j=1}^{n} \frac{H_{ij}S_{ij}}{V_i^{1/3}} = -D_{\text{eff}}\, q_i, \tag{3.30}$$

in terms of the geometric charge q_i (eq. (2.26)).

Bubble growth and number of faces As in 2D (eq. (3.26)), the change in volume of a bubble depends only on its shape, and in determining the growth rate we need not concern ourselves with its neighbours. Nevertheless the geometric characteristics which influence the growth rate are not as simple as in 2D.

We note first that C_i^r (eq. (3.30)) is only the relative growth rate; the actual growth rate, $dV_i/dt = C_i^r V_i^{1/3}$, depends explicitly on the bubble's volume V_i. Comparing bubbles of the same volume, FIG. 3.22 and eq. (3.30) together show that almost all bubbles with a small number of faces, typically $f \leq 11$, shrink over time, whereas those bubbles with a large number of faces, for example $f \geq 16$, get larger more quickly as f increases. Bubbles which have between 12 and 15 faces are almost stationary, neither losing nor gaining much gas. The same result can be seen in the (theoretical) curve of geometric charge (FIG. 2.17) and in numerical simulations [75]. However, in contrast to the 2D case (eq. (3.26)), the equation linking the growth rate of a bubble with its number of faces is not exact. If a bubble has faces which all have roughly the same number of edges and the same shape, we observe that it grows faster (or shrinks more slowly) than a more asymmetric bubble with the same number of faces (FIG. 2.17).

In particular, the sign of the growth rate of a bubble with $f \approx 13$ depends on its exact shape (FIG. 3.22). It is even possible that it might show no change in volume ($C^r = 0$), as for the idealized bubble without curvature (with 13.4 faces, see p. 44).

Recall that, in a dry foam, the growth rate of a bubble satisfies

$$\frac{dV_i}{dt} = -D_{\text{eff}} V_i^{1/3} q_i. \tag{3.31}$$

It is thus linked to the bubble diameter via $V_i^{1/3}$, to the different physical parameters at the local scale via D_{eff} (see §3.3), and to the bubble shape via the geometric charge q. The latter can be expressed exactly as a function of the total length of the edges of the bubble and its mean width (see eq. (2.28)), and is well-correlated to its number of faces. In the following, $q(f)$ will denote the average geometric charge of all bubbles with the same number of faces in a disordered foam. Note also that in a disordered foam, the number of faces of a bubble is closely correlated with its volume (see p. 38), that is, large bubbles have more faces than small ones statistically, and it is the largest bubbles that grow and the smallest bubbles which disappear (cf. FIG. 5.7b).

Evolution of the total foam volume For a foam of fixed volume (with no contact with the outside, for example confined in a box), $V_{\text{foam}} = \sum_{i=1}^{N} V_i$ is constant. Any gas that leaves one bubble goes into another, and the gas flux must be zero on average at any instant in time:

$$\frac{1}{N} \sum_{i=1}^{N} \frac{dV_i}{dt} = \left\langle \frac{dV_i}{dt} \right\rangle = 0. \tag{3.32}$$

However, in contrast to the 2D case (eq. (3.27)), this is not an automatic consequence of eq. (3.30). In fact, eq. (3.30) determines $dV^{2/3}/dt$, not dV/dt (see eq. (3.29)), and besides, the average geometric charge on the foam in 3D is not zero, $\langle q_i \rangle \neq 0$.

In practice, the individual growth rate of each bubble (eq. (3.30)) and global gas conservation (eq. (3.32)) are compatible if, and only if, the proportion of bubbles which are growing, multiplied by their individual growth rates, equals exactly the proportion of those that are shrinking, multiplied by their individual shrinkage rates. More precisely, this imposes a constraint on the distribution of the geometric charge, and thus on the numbers of faces of the bubbles and the correlations between neighbours. Mullins [56] thus confirmed that his expression (eq. (2.27)), for an experimental distribution of f, leads to gas conservation (eq. (3.32)).

On the other hand, if the foam has a free surface, as in a glass of beer, the average f is smaller, the peripheral bubbles bulge, the foam has a pressure greater than atmospheric pressure, and overall it gradually loses gas.

3.2 Evolution of bubble distributions in a dry foam

Mullins coarsening affects the bubble size and the bubble size distribution in a dry foam because of the occurrence of T2 topological processes (p. 81). The 2D case (von Neumann), which is similar, is not described here. We assume here that D_{eff} is constant in time and uniform throughout the foam (its variations will be discussed in §3.3).

3.2.1 Inevitable disappearance of bubbles

We follow a bubble i which has few neighbours, for example $f_i < 12$. It tends to lose gas, so its volume decreases. We first assume that it can do so without changing shape, and therefore without changing its geometric charge. Eqs. (3.29) and (3.30) would then imply that

$$V_i(t)^{2/3} = V_i(0)^{2/3} - \frac{2 D_{\text{eff}} q_i}{3} t. \tag{3.33}$$

An individual characteristic shrinkage time appears in this equation:

$$t_i = \frac{3 V_i^{2/3}}{2 D_{\text{eff}} q_i}. \tag{3.34}$$

In practice, when the volume of the bubble decreases, it gradually loses neighbours (FIG. 2.15) through T1 topological changes (FIG. 3.3). Eqs. (3.33) and (3.34) then imply that $V_i(t)^{2/3} = V_i(0)^{2/3}(1 - t/t_i)$. Each time a neighbour is lost, q_i increases and the shrinkage time t_i decreases. Consequently, between two successive T1s, $V_i^{2/3}$ decreases like t, and V_i like $t^{3/2}$, but at each T1, the prefactor of $t^{3/2}$ increases. So the bubble shrinks even more quickly in a vicious circle. Eventually, the bubble disappears in a T2 (see §1.2.3), usually in the form of a small tetrahedron (see FIG. 5.7b).

We now consider a bubble j which was a neighbour of this small bubble i. It had f_j neighbours, but since its neighbour i disappeared, it finds itself with one less neighbour (f_j decreases by 1). If f_j was large, larger than 15 say, and bubble j was previously growing, it continues to grow, but now more slowly. If f_j was 13 or 14, it starts to shrink. If it was already shrinking, for example because f_j was less than 12, its shrinking accelerates.

To summarize, every bubble eventually loses neighbours, decreases in volume more and more quickly, then disappears. For a bubble of volume V with a small number f of faces which has begun to shrink, we estimate the time it takes to disappear from t_i (eq. (3.34)), up to an unknown factor of order one, taking into account the changes in f which occur before the final T2:

$$t_{\text{disapp}} \approx \frac{V^{2/3}}{D_{\text{eff}} q(f)}. \tag{3.35}$$

3.2.2 From the individual to the group

We now think about the foam as a whole. At each moment in time, some bubbles are shrinking, whilst others are growing. From time to time, a bubble disappears (T2 process), the number of bubbles N decreases by one, and the average bubble volume $\langle V \rangle = V_{\text{foam}}/N$ increases discontinuously. Since new bubbles are not created, N is always decreasing (step-by-step, at each

T2), and at the end of the process there remains just one large bubble containing all the gas.

If the foam is contained in a box, the coarsening proceeds at constant total volume $V_{\text{foam}} = \sum_{i=1}^{N} V_i$ (eq. (3.32)). *Between* two successive T2s, $\langle dV_i/dt \rangle = 0$ (eq. (3.32)) and $\langle V \rangle$ is constant, i.e. $d\langle V_i \rangle / dt = 0$. *At the instant* at which a T2 occurs, we still have $\langle dV_i/dt \rangle = 0$ (eq. (3.32)), but at the same time, because N varies, we also have

$$\frac{d}{dt} \sum_{i=1}^{N} \frac{V_i}{N} = \frac{d\langle V \rangle}{dt} > 0. \qquad (3.36)$$

There is no simple expression for the rate of disappearance of bubbles (T2 rate) but we can estimate it based on the lifetime of all the bubbles with $f < 13$ (which, since they disappear most rapidly, are the most important contribution). The disappearance-rate of bubbles with f faces is the number of bubbles of this type, $N(f)$, divided by their characteristic lifetime given by eq. (3.35). We denote by $\langle . \rangle_f$ the average over all bubbles with the same f, and, by smoothing over time the discontinuities due to T2s, obtain

$$\frac{dN}{dt} \approx -D_{\text{eff}} \sum_{f=4}^{13} N(f) q(f) \left\langle V^{-2/3} \right\rangle_f. \qquad (3.37)$$

Observe that the temporal evolution of N is not universal: it depends on the distribution of the number of faces.

3.2.3 Asymptotic regime
From the moment a foam is created, it begins to coarsen and the average bubble size increases in time. We describe here the growth of dry foams in 3D. The results also apply to wet foams (§3.3). They are analogous to the coarsening of bubbly liquids (Ostwald ripening, see §1.1.3), and the 2D case is comparable, although easier to study as all the bubbles are visible (FIG. 3.23).

Evolution of the average size One experimental observation [57] (interpreted below using eq. (3.39)) is that foams which have had time to coarsen reach an asymptotic regime in which the average bubble size increases in a simple way. That is, at long times, N varies as $t^{-3/2}$, the average bubble volume $\langle V \rangle$ varies as $t^{3/2}$, the average surface area $\langle S \rangle$ varies as t, and the average diameter $\langle d \rangle$ varies as $t^{1/2}$ (FIG. 3.27). The quantities $\langle V \rangle^{-1/3} d\langle V \rangle/dt$, $d\langle S \rangle/dt$ and $\langle d \rangle d\langle d \rangle/dt$ become constant. More precisely, it is observed that

$$\langle d(t) \rangle^2 = \langle d(t_o) \rangle^2 + K(t - t_o). \qquad (3.38)$$

This averaged expression recalls the one for the growth of an individual bubble (eq. (3.33)), but the actual link is not obvious since the growth of $\langle d \rangle$ is the result of some bubbles growing and others shrinking. The averaged coarsening constant K depends not only on D_{eff}, and thus on the physico-chemical characteristics of the foam (§3.3), but also on the bubble size distribution (via the parameter b in eq. (3.43) below). In 2D, eq. (3.38) again holds: the average bubble radius grows as $t^{1/2}$ and the average bubble area as t (see exercise 8.1).

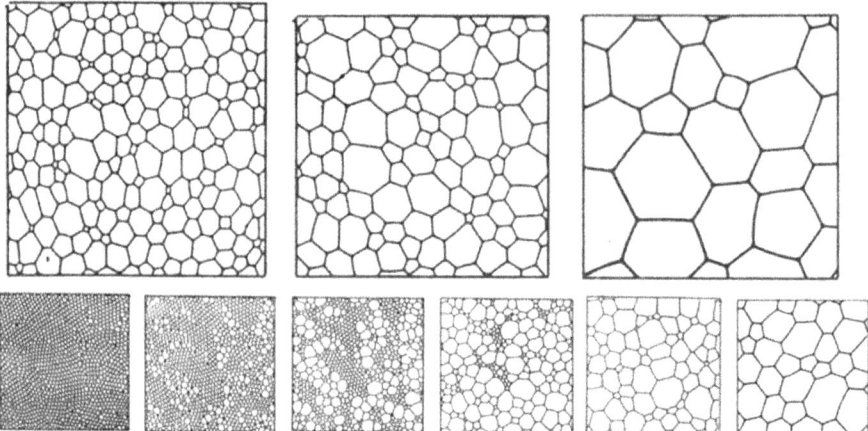

FIG. 3.23: Self-similar growth. Images of coarsening were obtained by leaving a 2D foam on a photocopier. At the top: three successive images (1.95 then 21.5 and 166.15 hours) of a small part of an initially disordered dry foam. At the bottom: with an initially ordered foam, the transition to self-similar growth takes longer, but the final state is similar to the one in the top row. Reproduced by kind permission of J. Glazier [33], © 1987 The American Physical Society.

Evolution of distributions It is more difficult to study the change in the detailed distribution of bubble sizes than the growth of the average size, both experimentally (§1, chap. 5) and by simulation (§2, chap. 5). This is because it is necessary to observe thousands of individual bubbles over a long time. In fact, the foam needs to coarsen for a long enough time that it passes through a transient and reaches the limiting regime; this requires that enough bubbles remain (despite coarsening) to be able to characterize this limiting regime with good statistics.

So-called *self-similar growth* is defined as temporal evolution in which the dimensionless statistical distributions remain stationary. That is, after the transient [45, 75], in this limiting regime the foam structure at two different times has (geometric and topological) statistical characteristics which differ only in their length-scale, i.e. the average bubble size (eq. (3.38)). Two successive photos of the foam, if the scale is not specified, appear alike (FIG. 3.23). This self-similarity is demonstrated by several statistical quantities, including the proportion $N(f)/N$ of bubbles with f faces, the relative volumes at a given time $V_i/\langle V(t)\rangle$, the correlations between the relative volume of a bubble and its number of neighbours f_i, or even the correlations between the relative volumes of two neighbouring bubbles $V_i/\langle V(t)\rangle$ and $V_j/\langle V(t)\rangle$ (FIG. 3.24).

The self-similar growth regime appears *universal* [45, 75]. The size distributions and the number of neighbours don't depend on the initial state of the foam (FIGS. 3.23 and 3.24). The transient is the time that the foam takes to completely lose any memory of its initial configuration, through the disordering effects of T1s and T2s. This is why the duration of the transient varies, and can be long if the foam is initially far from the

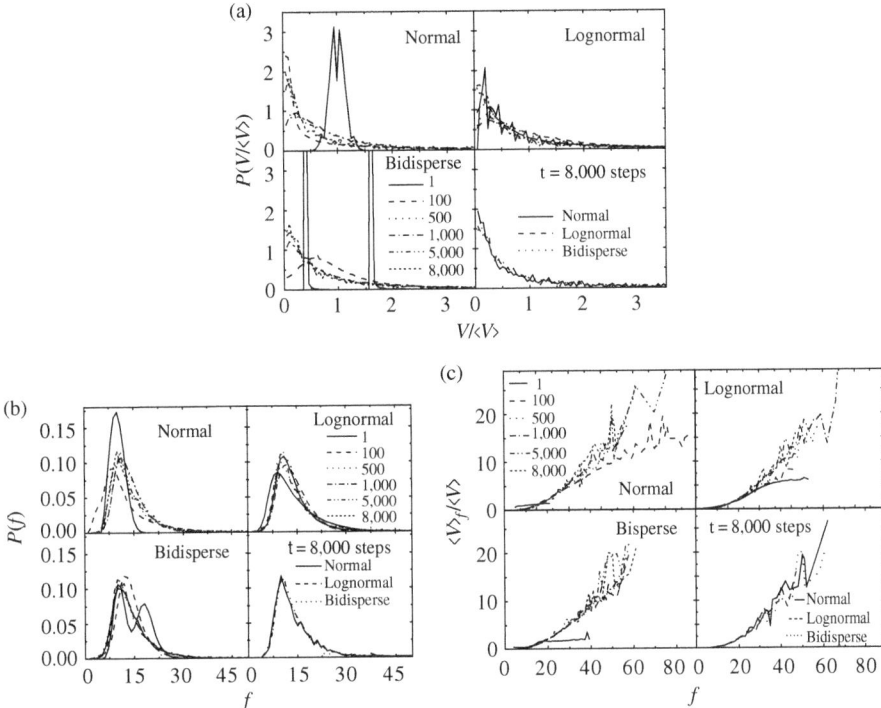

FIG. 3.24: Self-similar growth in 3D, simulated with the Potts model (§2.1.1, chap. 5) [75]. **(a)** Volume distribution; **(b)** distribution of number of neighbours; **(c)** volume as a function of number of neighbours. Simulations are shown for three different initial distributions: normal (Gaussian), log-normal, and bidisperse (mixture of two bubble-volumes). The duration of the transient is different in each case. After 8,000 time steps, all three have left the transient and show the same distributions, which remain constant in time.

limiting regime. In extreme cases, for example in a very ordered foam, the transient can be longer than the duration of the experiment, and the self-similar growth regime is not observed.

The link between the evolution of bubble size and of distributions Mullins, on the basis of dimensional analysis, suggested that in the self-similar growth regime the average bubble volume should increase as $t^{3/2}$ [57]. However, the converse is not true: observing this exponent does not prove that, and in fact it often occurs before [29, 45, 75], the self-similar growth regime has been reached. The growth of the average volume with $t^{3/2}$ (eq. (3.38)) is observed much more often than the self-similar regime, and historically was recorded much earlier. More generally, in a self-similar growth regime, any growth law compatible with the Young–Laplace law yields a growth exponent of $t^{1/2}$ for the bubble size [32], both in 2D and 3D.

The link between the self-similar growth regime and the exponent can be established exactly in 2D (see exercise 8.1). In 3D, thanks to the approximate equation (3.37), we can describe the self-similar growth regime by introducing the average volume $\langle V \rangle = V_{\text{foam}}/N$, with V_{foam} constant. We obtain

$$\frac{dN}{dt} \approx -D_{\text{eff}} \frac{N^{5/3}}{V_{\text{foam}}^{2/3}} \sum_{f=4}^{13} \frac{N(f)}{N} q(f) \left\langle \left(\frac{V}{\langle V(t) \rangle} \right)^{-2/3} \right\rangle_f, \qquad (3.39)$$

from which we deduce that

$$\text{3D:} \qquad \frac{dN}{dt} \propto -N^{5/3} \qquad (3.40)$$

with a prefactor that is constant in the self-similar growth regime, since $N(f)/N$ and $\langle (V/\langle V(t) \rangle)^{-2/3} \rangle_f$ are then independent of time. Using eq. (3.40), we deduce that $N \propto t^{-3/2}$, and consequently the average volume increases as $t^{3/2}$. Thus the change in the average volume governs all the time-scales of the system. Likewise, the topological changes become less and less frequent during coarsening, the frequency varying with $t^{2/3}$ (FIG. 3.25) [9].

The value of the prefactor is another piece of information that can be obtained from eq. (3.39). The given form is already very approximate, and if we also take into account the fact that q and $V/\langle V \rangle$ are of order one (see FIG. 2.17), we obtain a prefactor of the order of $D_{\text{eff}}/V_{\text{foam}}^{2/3}$. Then an approximate expression for the number of bubbles is

$$N(t) \approx N(0) \left(1 + b\, D_{\text{eff}} \langle V \rangle_0^{-2/3} t \right)^{-3/2} \qquad (3.41)$$

where b is a constant of order one that depends on the bubble size distribution and $\langle V \rangle_0 = \langle V(0) \rangle$ is the average initial bubble volume. The bubble size then varies as

$$d(t) \approx d_0 \left(1 + b\, D_{\text{eff}} \langle V \rangle_0^{-2/3} t \right)^{1/2}. \qquad (3.42)$$

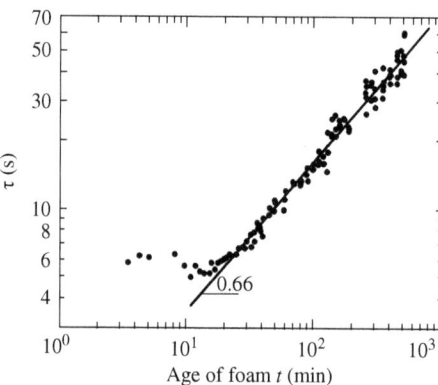

FIG. 3.25: Average time interval between T1 topological changes, τ, in a coarsening foam measured by multispeckle DWS (see §1.2.5, chap. 5). The sample is a Gillette shaving foam with $\phi_l = 7.5\%$. After 20 min, $\tau \sim t^{0.66}$ [9].

Finally, we can define a characteristic coarsening time t_c such that $d(t) = d_0\sqrt{1 + t/t_c}$ by

$$t_c \approx \frac{\langle V \rangle_0^{2/3}}{b\, D_{\text{eff}}}. \tag{3.43}$$

t_c depends on the effective diffusion constant D_{eff}, and is related to the constant K in eq. (3.38) by $K t_c \approx d_0^2$.

3.3 Effects of different parameters

3.3.1 Time-scale and effective diffusion coefficient

We explore the significance of the effective diffusion coefficient D_{eff}, defined by eqs. (3.26) in 2D and (3.30) in 3D, which fixes the time-scales associated with coarsening (eq. (3.43)). Following Princen [63], we reason in the following way.

A film has one interface in contact with a bubble of pressure P_1 and the other in contact with a bubble of pressure P_2 (FIG. 3.26). On the liquid side of the interface, the gas concentration is proportional to the pressure in the neighbouring bubble: $c_1 = He\, P_1$, $c_2 = He\, P_2$, where the Henry coefficient He reflects the solubility of the gas in the liquid. Within the liquid, the gas diffuses between these two values c_1 and c_2. Once a steady profile is established, the gas concentration therefore varies linearly with z, with a gradient determined by the film thickness h:

$$\frac{\partial c(z)}{\partial z} = -\frac{He\,(P_1 - P_2)}{h}. \tag{3.44}$$

The flux of gas (per unit surface area of film) is determined by the actual diffusion coefficient D_f of the gas molecules within the liquid. The molar flux is $D_f\, He\,(P_1 - P_2)/h$ and so, with V_m the volume of a mole of gas at ambient temperature and pressure, the volumetric flux is $V_m\, D_f\, He\,(P_1 - P_2)/h$. Eqs. (3.29) and (3.30) express the volumetric flux per unit surface area as $D_{\text{eff}}\, H_{12} = D_{\text{eff}}\,(P_1 - P_2)/2\gamma$ (using eq. (2.4)). Comparing coefficients gives, for a dry foam,

$$D_{\text{eff}} = D_f\, \frac{2 He\, \gamma\, V_m}{h}. \tag{3.45}$$

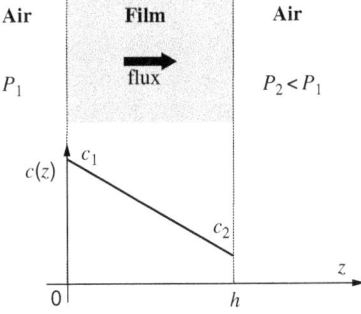

FIG. 3.26: Profile of concentration of dissolved gas across a film (eq. (3.44)).

The effective diffusion coefficient for coarsening is thus proportional to the actual diffusion coefficient of the gas molecules. The units of eq. (3.45) are correct: γ/h is a pressure, while HeV_m is the inverse of a pressure. A dimensionless geometric correction factor for the liquid fraction must be taken into account (see §3.3.3), so that

$$D_{\text{eff}} = D_f \frac{2He\gamma V_m}{h} a(\phi_l). \tag{3.46}$$

We will now comment on each term of this equation.

3.3.2 Effect of chemical composition The data in FIG. 3.27 compare well to an expression with an exponent $1/2$ (eq. (3.42)), implying a characteristic time t_c and therefore a value of D_{eff} (see eq. (3.43)) which depends linearly on the solubility He and on the diffusivity D_f, in agreement with expression (3.46).

It is generally assumed that the diffusion coefficient D_f of the gas in the liquid films is equal to that in the bulk liquid, and therefore independent of h, and inversely proportional to the viscosity of the liquid. Since D_f varies like the inverse square-root of the molecular mass of the gas, in practice it does not depend much on the gas. In pure water, D_f lies between 10^{-6} and 10^{-5} cm^2.s^{-1} for common gases. The variation of D_f is responsible for only a factor of about 2 in the variation of the coarsening rates shown in FIG. 3.27.

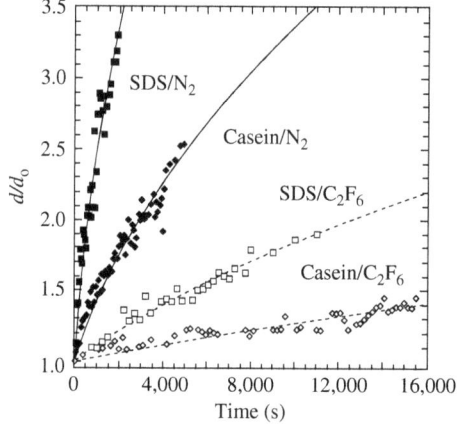

FIG. 3.27: Effect of different surfactants and gases on coarsening. The liquid fraction is kept roughly constant at $\phi_l = 0.15$ by periodically inverting the sample to avoid drainage. The average bubble diameter is measured as a function of time by diffuse light transmission (see §1.2.5, chap. 5). An expression of the form $d/d_0 \sim (1+t/t_c)^{1/2}$ (eq. (3.42)), where d_0 is the average diameter at $t=0$ (identical for the 4 foams), is fitted to each set of data, allowing us to deduce a characteristic time $t_c = 190$, 980, 4200, and 19100 s, respectively for the pairs SDS/N$_2$, casein/N$_2$, SDS/C$_2$F$_6$, and casein/C$_2$F$_6$. The coarsening dynamics therefore depend strongly on the physical chemistry of the components (gas and surfactants) [66].

The Henry constant He depends strongly on the gas and the liquid. Thus, in pure water, $He = 3.4 \times 10^{-4}$ mol.m^{-3}.Pa^{-1} for CO_2, whereas it is 6×10^{-6} mol.m^{-3}.Pa^{-1} for N_2 and only 6×10^{-7} mol.m^{-3}.Pa^{-1} for a fluoro-carbon gas such as C_2F_6. The variation of He is responsible for a factor of about 10 in the difference of the coarsening rates between N_2 and C_2F_6 in FIG. 3.27 (a CO_2 foam would coarsen too quickly to plot). An even more extreme effect of gas solubility is discussed in §3.3.4.

Surfactants affect eq. (3.46) directly via the surface tension γ. They also control film thickness h via the disjoining pressure (see p. 48), and therefore affect the denominator: the thicker the films, the slower the coarsening. For instance, in FIG. 3.27, when going from SDS to casein, the surface tension changes from 36 to 43 mN.m^{-1}, and is responsible for only a small change in the coarsening rate. As shown in FIG. 3.17, casein results in inhomogeneous films a few hundreds of nanometres thick, which is significantly different from the roughly 50 nm thick film for SDS and is thus responsible for most of the change in the coarsening rate.

Experimentally, to test the effect of the physico-chemical conditions on the dynamics of 3D foam coarsening requires that the liquid fraction ϕ_l is kept constant, which is difficult (see §4). One possibility (provided the foam is not too wet and the bubble diameter is less than a millimetre) is to periodically change the vertical orientation of the foam sample, thereby changing the direction of drainage and keeping the liquid fraction constant at the centre of the sample.

3.3.3 Effect of geometrical parameters

We now fix the chemical components of the foam and investigate how the dynamics depend on the initial bubble diameter d_0 and on the liquid fraction ϕ_l.

As shown in eq. (3.43), the coarsening time t_c depends strongly on the initial bubble diameter: $t_c \sim V^{2/3} \sim d^2$. For bubbles of diameter 100 μm, t_c is several tens of seconds, but for bubbles of 1 cm in diameter, the coarsening time is multiplied by 10^4! Clearly there is practically no coarsening in such a foam.

Experimentally, for the reasons mentioned in §3.3.2, there are few measurements of the coarsening rate at constant ϕ_l, either for different values of d_0 (the initial bubble diameter) or of ϕ_l. These two dependencies are coupled (see eq. (3.46)): D_{eff} depends on ϕ_l via the function $a(\phi_l)$, but also because h depends on ϕ_l through the value of d_0.

Film surface area The function $a(\phi_l)$ is the proportion of the surface of the bubbles covered by thin films rather than Plateau borders. It is through this effective surface that gas diffuses from bubble to bubble. Diffusion across Plateau borders is much slower because they are several orders of magnitude thicker. Two different expressions for $a(\phi_l)$ are used [36, 38], neither of which takes into account polydispersity. First, $a(\phi_l) = 1 - (\phi_l/\phi_l^*)^{1/2}$ (model 1 in FIG. 3.28). This comes from eq. (2.38), which gives the size r of a Plateau border as a function of the liquid fraction: the area of a film in a dry foam is of the order of $S_0 = d^2$, which in a wet foam becomes $S(\phi_l) = d^2 - dr = d^2(1 - (\phi_l/\phi_l^*)^{1/2})$. Then $a(\phi_l) = S(\phi_l)/S_0$ and the expression follows [38]. This approach, which is based on orders of magnitude, does not enable

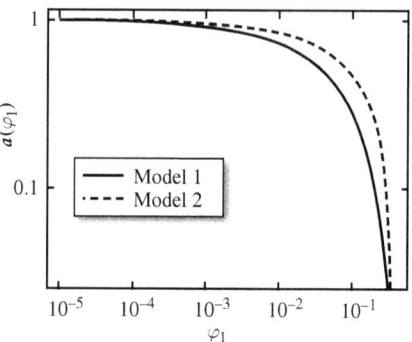

FIG. 3.28: Two models for the evolution of the effective surface area for gas transfer as the liquid fraction varies. Model 1 has $a(\phi_l) = 1 - (\phi_l/\phi_l^*)^{1/2}$, with $\phi_l^* = 0.36$ [38]. Model 2 has $a(\phi_l) = \left[1 - (\phi_l/\phi_l^*)^{1/2}\right]^2$, with $\phi_l^* = 0.43$ [36].

a prediction of ϕ_l^*, but explains the form of the dependence on ϕ_l. More recently, numerical simulations have suggested a second expression, $a(\phi_l) = \left[1 - (\phi_l/\phi_l^*)^{1/2}\right]^2$ (model 2 in FIG. 3.28) [36].

Potts model simulations [29] (§2.1.1, chap. 5) indicate that the rate of gas diffusion is the weighted sum of the contributions through films and Plateau borders. Thus the rate at which an individual bubble grows is intermediate between the dry and wet limits. Similarly, the exponent for the increase of the average bubble size in the self-similar growth regime interpolates between $t^{1/2}$ (eq. (3.38) and $t^{1/3}$, as $1/2 - \phi_l^{0.2}/6$ [29].

Film thickness The thickness h of a film depends on d and on ϕ_l, since they control the capillary suction $P_c \approx \gamma/r$ of the Plateau borders (see §2.3.2). At equilibrium, the disjoining pressure Π_d, which governs the thickness of the films, is equal to the capillary pressure, P_c. Using eq. (2.38) we find $P_c = \Pi_d \approx \gamma/(d\sqrt{\phi_l})$.

However, a typical curve of $\Pi_d(h)$, as in FIG. 3.15, shows that if Π_d is high (of the order of several hundreds of Pascals) a variation of Π_d (induced by ϕ_l) makes little difference to h. This is the case for small bubbles ($d \leq 1$ mm) or dry foams ($\phi_l \sim$ a few percent), where $h(\phi_l)$ is almost constant, and in these cases the dependence of D_{eff} on liquid fraction is entirely due to the function $a(\phi_l)$. In fact, a coarsening experiment at constant liquid fraction ϕ_l, such as in FIG. 3.27, allows us to infer (using eq. (3.46), where the only unknown is h) that the thickness of the film is less than 100 nm (the range of common black films) and independent of ϕ_l.

On the other hand, if the disjoining pressure is weak, of the order of several tens of Pascals, the thickness h depends strongly on Π_d, and consequently on ϕ_l. In this case, which corresponds to wet foams or large bubbles, the dependence of D_{eff} on ϕ_l is then more complex than just the function $a(\phi_l)$ and includes a non-trivial contribution $h(\phi_l)$.

3.3.4 Can coarsening be controlled?
In summary, which parameters have the greatest effect on the dynamics of coarsening? Is it more effective to vary physical parameters or the chemistry of the solution?

As we have seen, t_c is strongly dependent on bubble size (see eq. (3.43)) and on liquid fraction (see eqs. (3.43) and (3.46) and FIG. 3.28). Changing the gas also changes the dynamics, essentially by altering the solubility He (eq. (3.46)). Adding to the usual gas (e.g. air or CO_2) tiny amounts of a much more insoluble gas slows down the coarsening rate, which is dominated by the slowest gas. It even changes the bubble size distribution and its time evolution, which loses its self-similarity. Taken to extremes, adding traces of the very insoluble gas C_6F_{14} dramatically changes even the final state. In fact, the quantity of such an insoluble gas in each bubble remains constant, and if there is enough of it then there is a coexistence volume at which the Laplace pressure and the partial pressure of the trapped gas balance, and the coarsening process stops [30, 83, 85].

Surfactants of low molecular mass appear to have a much weaker influence: in fact, surface tension and film thickness only vary over a small range. Nevertheless, there are particular systems in which surface concentration and surface tension depend on bubble size. If the molecules adsorbed at the surface are *irreversibly* attached, then when a bubble becomes smaller the surface concentration increases and we observe a strong reduction in the coarsening rate. This is the case, in particular, for solid particles or certain proteins [17]. In fact, the surface tension γ can continue to decrease and coarsening can be stopped entirely [8]. This behaviour is linked to the elasticity of the surface in compression (see §2.2.3).

Finally, while a foam is coarsening, gravity-driven drainage may also occur, redistributing the liquid in the foam (see §4). It is therefore clear that the coarsening rate, which is sensitive to ϕ_l, will be modified by drainage. Conversely, coarsening changes the average bubble diameter, a parameter upon which drainage is strongly dependent. The two processes are therefore strongly coupled. We will see (§4) that it is possible to define a characteristic drainage time, so that comparing typical coarsening and drainage times for a given foam enables us to determine whether or not we need to take this coupling into account.

4 Drainage

The aim of this section is to present a description and model of liquid drainage through a foam. We will show that it is comparable to the problem of liquid flow in porous media, but with two significant differences. First, in a foam, the liquid flows through a network of pores whose diameter depends on flow itself: the bubbles can move apart to allow liquid to pass and then move back. Second, the interfaces are fluid, and are partially entrained by the flow. From this point of view, drainage is a good example of coupling between different length-scales in a foam: to model this macroscopic phenomenon, we will see that it is necessary to take into account the properties of the interfacial monolayer that forms the pore surfaces and the characteristic size of the pores themselves.

4.1 What is drainage?

Drainage is the passage of liquid through a foam. It is easily observed in daily life, in beer foam for example. Gravity-driven drainage is a mechanism by which an aqueous foam evolves: as noted in §1.1, under the effect of gravity the liquid contained in the foam flows downwards and the gas bubbles are displaced upwards.

4.2 Free drainage

The evolution of the liquid fraction of a foam under gravity towards the equilibrium profile (§4.3, chap. 2) is known as *free drainage*. FIG. 3.1 shows a foam sample of height H_0 and of uniform liquid fraction ϕ_{l0} at $t = 0$ in a container with a closed base. Over time, the downward flow of the liquid dries the foam and the liquid accumulates at the base of the container. The interface between the foam and the drained liquid rises over time (height $L(t)$ on FIG. 3.1). At a local scale, by watching the surface of the foam through the wall of a transparent container, it is possible to see the liquid network gradually emptying and the bubbles moving closer together.

The liquid fraction at a given vertical position z in the foam can be estimated by the Plateau border size (see §1.2.3, chap. 5) or measured more precisely by foam conductivity (see §1.2.6, chap. 5 and experiment 7.4), which gives the average liquid fraction in a slice of foam of thickness dz. The profiles obtained are shown at successive times in FIG. 3.29. The behaviour is complex, but two important stages of the drainage process are visible, defined by the time τ at which the drainage front reaches the bottom of the foam:

(i) For $t < \tau$, the liquid fraction is only modified in the upper part of the foam and varies linearly with z between a small value and its original value, ϕ_{l0}. The lower limit of this region of dry foam is known as the *drainage front*. It moves downwards at constant velocity as the foam drains, and the width of the front increases (the slope $d\phi_l/dz$ is reduced). The lower part of the foam undergoes steady drainage, with constant liquid fraction. That is, the same amount of liquid leaves as enters this region and the velocity of the liquid there is constant.

(ii) For $t > \tau$, the drainage is characterized by a slow relaxation towards the equilibrium profile described in §4.3, chap. 2. To explain this slowing down, we note that gravity is not the only force acting on the liquid. In a drier foam, the radius of curvature of the Plateau borders is smaller, and the Young–Laplace law indicates that the capillary pressure is higher (see §4.1.2, chap. 2), so that the liquid fraction gradient gives rise to a capillary pressure gradient. Dry regions imbibe liquid, independent of the effects of gravity. The capillary pressure gradient (directed from the bottom upwards) increases during drainage, and reduces the drainage rate until the two are in balance, thus leaving the foam in static equilibrium. At equilibrium, in contrast to the initial state, the part of a foam in direct contact with the drained liquid is very wet. At the interface with the liquid, the liquid fraction is close to 0.36, which corresponds to the void fraction between randomly close-packed spheres (see FIG. 4.17).

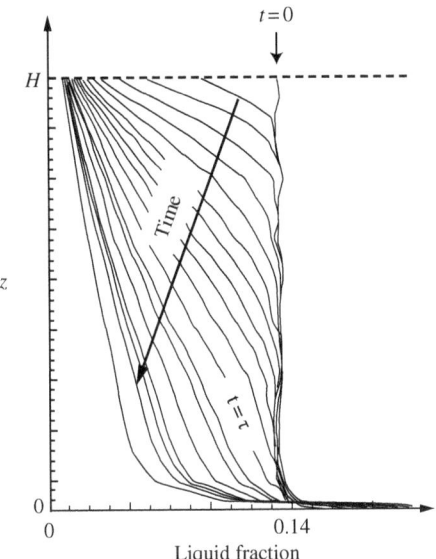

FIG. 3.29: Free drainage experiment: vertical profile of liquid fraction as a function of time measured by electrical conductivity (see p. 238) in a foam of height 40 cm. At $t = 0$, the liquid fraction is uniform and constant, $\phi_{l0} = 0.14$. Drainage is evident in the downward propagation of a front (shown by the arrow) that reaches the base of the foam at $t = \tau$. The bottom of the foam gets wetter and ϕ_l increases above its initial value. The equilibrium profile is reached eventually. Graph courtesy of A. Saint-Jalmes.

No liquid leaves the foam if the liquid fraction at the base is less than the value $\phi_l^* = 0.36$. This wetting phase can take a relatively long time and if the total quantity of liquid in the foam is too low (a dry foam, or a small volume of foam), it is even possible that no liquid drains at all! Nonetheless, even if no liquid leaves, drainage may still occur and give rise to liquid fraction gradients (see exercise 8.3).

4.3 Forced drainage

Free drainage, although conceptually straightforward, is not the best experiment through which to understand drainage, since there are regions of very different liquid fraction present simultaneously (from very dry to very wet). Moreover, the experiment can take a long time, so that other ageing effects are often at work, complicating the interpretation of the results. Finally, the initial profile of liquid fraction has a significant influence on the subsequent dynamics, so that to generate reproducible results requires that foams must be generated with an initial liquid fraction that is uniform across the full height of the foam.

Instead, a *forced drainage experiment* can be performed, which consists of pouring the foaming solution at constant flow rate into a foam that has already been allowed to

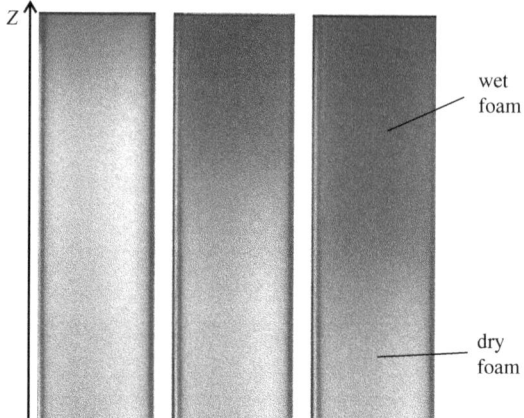

FIG. 3.30: Forced drainage experiment: liquid is introduced at the top of an initially dry foam of height 10 cm in a vertical cylinder. Three photos taken at successive times show, by diffuse light transmission (§1.2.5, chap. 5), the progress of a wetting front, enabling the liquid velocity in the foam to be measured and its permeability to be determined. Photograph courtesy of O. Pitois.

drain freely. A horizontal wetting front develops, separating wet foam (at the top) from dry foam (FIG. 3.30). The front moves downwards at a constant speed u_{front} which is easy to measure. Above the front, ϕ_l is constant and depends on the imposed flow rate, and the liquid therefore flows only under the force of gravity. The experiment shows that u_{front} depends on the bubble size and the surfactant. We will make this more precise later (§4.7.1), but for the moment it is worth noting that the influence of the surfactant is explained by the partial mobility of the interfaces which confine the flow, that is, the Plateau border walls. These interfaces may be entrained by the liquid and then the surfactant plays an important role.

4.4 Modelling flows in solid porous media

Modelling of foam drainage is inspired by the field of flow in solid porous media, and in the following we present some of the significant results in that field.

4.4.1 Flow at the scale of a pore—Poiseuille's law
We consider the flow of liquid (with viscosity η and density ρ_l) through a cylindrical pore of radius r_p and length ℓ, oriented at an angle ψ to the vertical (FIG. 3.31). $\vec{e}_\psi$ denotes the unit vector along the pore axis, oriented upwards. The velocity field in the liquid is approximately of the form $\vec{u} = u(r)\vec{e}_\psi$, where r is the radial coordinate. The force acting on the liquid contained in the pore due to the pressure p is $\pi r_p^2(p(0) - p(\ell))$. Per unit volume, this force is equal and opposite to the pressure gradient along the axis of the pore, $-dp/d\ell$. The force of gravity (per unit volume) along the pore is $\rho_l \vec{g} \cdot \vec{e}_\psi = -\rho_l g \cos \psi$. In a foam we can safely assume that the Reynolds number Re, based on the characteristic speed of the liquid u and from r_p, is small compared to one, $Re = \rho_l u r_p/\eta \ll 1$. In

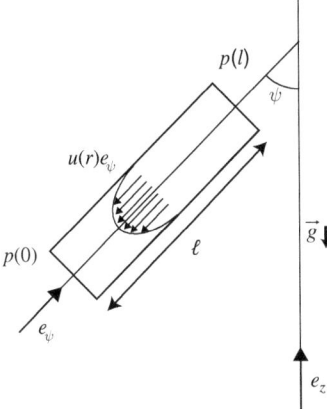

FIG. 3.31: A cylindrical pore of length ℓ, oriented at an angle ψ to the vertical, through which liquid flows under the action of pressure and gravity forces. The velocity field is $\vec{u} = u(r)\vec{e}_\psi$.

this case, the flow is governed by the equilibrium between the driving forces (pressure and gravity) and viscous forces, which oppose the flow, giving

$$\eta \frac{1}{r}\frac{d}{dr}\left(r\frac{du}{dr}\right) = \frac{dp}{d\ell} + \rho_l g \cos\psi. \quad (3.47)$$

Assuming that the velocity is zero at the wall (no-slip boundary conditions), the solution is

$$u(r) = -\frac{r_p^2}{4\eta}\left(1 - \frac{r^2}{r_p^2}\right)\left(\frac{dp}{d\ell} + \rho_l g \cos\psi\right). \quad (3.48)$$

The direction of flow is determined by the sign and magnitude of the pressure gradient: if it is positive, it enhances the effect of gravity; if it is negative, it opposes it and can even negate it. The flux of liquid through the pore is obtained by integration, giving the average velocity of liquid in the pore:

$$\bar{u} = \int u\, ds = -\frac{s}{\eta}K_p\left(\frac{dp}{d\ell} + \rho_l g \cos\psi\right), \quad (3.49)$$

where the cross-section $s = \pi r_p^2$ of the pore and a geometric coefficient $K_p = 1/(8\pi)$ have been introduced. This relationship indicates that the average liquid velocity is directly proportional to the driving force, which is known as *Poiseuille's law*.

4.4.2 Flow at the scale of the porous medium—Darcy's law

Henry Darcy showed in 1856 that the relationship between velocity and driving force at the scale of the pore (eq. (3.49)) remains true if we average the velocity and the pressure over a collection of pores within a representative volume of the porous medium. This result is known as *Darcy's law*:

$$u_m \vec{e} = \frac{\alpha}{\eta} \left(\rho_l \vec{g} - \vec{\nabla} p \right), \qquad (3.50)$$

where α is the permeability of the medium. The minus sign arises because the liquid moves from regions of high pressure to low pressure. Here u_m is the flux, defined as the liquid flow rate divided by the surface area of the sample transverse to the direction of flow, which is denoted $\vec{e}$. u_m depends on the average liquid velocity $\langle \bar{u} \rangle$ and the volume fraction of the medium occupied by liquid, ϕ_l, according to

$$u_m = \langle \bar{u} \rangle \phi_l, \qquad (3.51)$$

where the average represented by the angle brackets is over all the pores in the representative volume element.

Note that we assume that the pores of the medium are completely saturated with liquid, so the porosity plays the same role as the liquid fraction, justifying our choice of notation.

4.4.3 Permeability of a porous medium

Whilst simple to use, eq. (3.50) still requires knowledge of the permeability coefficient α. The latter expresses the ease with which a liquid flows through a porous medium. Intuitively, we expect that α increases with ϕ_l and hence with the radius of the pores. To estimate it, we consider a model consisting of a random network of identical cylindrical pores. We will calculate the flow of liquid in the vertical direction ($\vec{e}_z$) under the action of a macroscopic pressure gradient $\vec{\nabla} p = dp/dz \, \vec{e}_z$ and the force of gravity $\rho_l \vec{g} = -\rho_l g \vec{e}_z$. In a pore with orientation ψ, the microscopic pressure gradient is $dp/d\ell = (dp/dz) \cos \psi$ and the average vertical velocity is

$$\bar{u}_z^\psi = \bar{u} \vec{e}_\psi \cdot \vec{e}_z = \bar{u} \cos \psi = -\frac{s}{\eta} K_p \left(\frac{dp}{dz} + \rho_l g \right) \cos^2 \psi. \qquad (3.52)$$

The average vertical velocity of liquid in the medium is the average of $\bar{u}_z^\psi$ over all orientations:

$$\langle \bar{u}_z \rangle = \int_0^{\pi/2} \bar{u}_z^\psi p(\psi) d\psi. \qquad (3.53)$$

Here $p(\psi)d\psi$ is the probability that ψ falls between ψ and $\psi + d\psi$. To determine this probability, we calculate the relationship between the surface area at the ends of those pores inclined at an angle between ψ and $\psi + d\psi$, equal to $2\pi l^2 \sin \psi d\psi$, and the surface area of the hemisphere including the ends of all the pores. Thus $p(\psi)d\psi = 2\pi l^2 \sin \psi d\psi / 2\pi l^2 = \sin \psi d\psi$, and the average vertical velocity is

$$\langle \bar{u}_z \rangle = -\frac{s}{\eta} \left(\frac{dp}{dz} + \rho_l g \right) K_p \int_0^{\pi/2} \cos^2 \psi \sin \psi d\psi = -\frac{s}{\eta} \left(\frac{dp}{dz} + \rho_l g \right) \frac{K_p}{3}. \qquad (3.54)$$

To relate this to the flux u_m, we use eqs. (3.50) and (3.51) to give

$$\langle \bar{u}_z \rangle = \frac{u_m}{\phi_l} = -\frac{\alpha}{\eta \phi_l}\left(\frac{dp}{dz} + \rho_l g\right). \tag{3.55}$$

Comparing with eq. (3.54) shows that the permeability is

$$\alpha = \frac{sK_p \phi_l}{3}, \tag{3.56}$$

linking the macroscopic permeability of the medium to the geometric and hydraulic characteristics of its pores. For the cylindrical pores considered here, the permeability is

$$\alpha = \phi_l \frac{r_p^2}{24}. \tag{3.57}$$

More elaborate models combine several pore sizes in order to form a more realistic network. It is also possible to expand the cylindrical pore model by using the concept of *hydraulic radius*, m, the ratio of the volume of a pore to its surface area. For example, for cylindrical pores of radius r_p, $m = r_p/2$. We generally express m as a function of the specific surface area of the porous medium, A_s, defined as the ratio of the surface area of the pores to the total volume:

$$m = \frac{\phi_l}{A_s}. \tag{3.58}$$

The expression for α retains the form given in eq. (3.57) even though the pore geometry may change, so we can write

$$\alpha = \phi_l \frac{m^2}{C_K} = \frac{\phi_l^3}{C_K A_s^2}. \tag{3.59}$$

This relationship, known as the *Carman–Kozeny model*, introduces a constant C_K which represents the geometry of the network. For cylindrical pores we find $C_K = 6$, from eq. (3.57). In a more complex network this constant is determined by fitting eq. (3.59) to experimental data. For a packing of spheres with diameter d, the specific surface area is $A_s = (1 - \phi_l)(6/d)$ and C_K is close to 5; then the permeability of a medium consisting of spherical grains (a model often used for soil) is

$$\alpha = \frac{\phi_l^3 d^2}{180(1 - \phi_l)^2}. \tag{3.60}$$

This relationship provides a reference value for studying the permeability of many different porous media.

4.5 Modelling the permeability of a liquid foam

4.5.1 Foam: a special porous medium

Foam can be viewed as a porous medium: the liquid forms the interstitial phase, and the bubbles play a similar role to grains in a solid medium. There are nevertheless important differences between a foam and a solid porous medium:

Birth, life, and death

1. The liquid network in a foam is special, in that its geometry is fixed by Plateau's laws (cf. chap. 2). In particular, eq. (2.38) links the dimensions of a Plateau border, i.e. its radius of curvature r and length ℓ, to the liquid fraction (or porosity) ϕ_l defined at the macroscopic scale.
2. The fluid nature of the interfaces which make up the liquid network allows the foam to expand or contract. Within this network, the Young–Laplace law directly links the pressure p of the liquid, which causes the expansion, to the radius of curvature r (eq. (2.30)):

$$p = P - \frac{\gamma}{r} \approx P_{\text{atm}} - \frac{\gamma}{r}, \qquad (3.61)$$

where P is the bubble pressure. In the context of deformable porous media, this relationship between liquid pressure and the space available for flow is the simplest of its kind.

3. The mobility of the liquid/gas interfaces poses a more significant challenge for models of foam drainage. The flow of a liquid in contact with a solid wall is governed by a no-slip condition (zero velocity at the wall), but at the surface of a bubble there is instead a condition on the stress, since the flow in a Plateau border exerts a tangential viscous stress on the interface, pulling it in the direction of flow. The surfactant molecules adsorbed on an interface change the rheological behaviour of the interface (see §2.2.3), changing the resistance to flow; the surface shear viscosity, analogous to the bulk shear viscosity, appears to play the most important role in this resistance [47].

We now take account of these different effects to estimate the permeability of a liquid foam. We first consider the flow in a single Plateau border, and then modify the models presented in §4.4.3 for solid porous media.

4.5.2 Flow in a Plateau border

The permeability coefficient for a Plateau border must take into account not only its shape but also the way in which the liquid/gas interface is entrained by the flow. For the problem of drainage, the effect of interfacial mobility was considered in the 1960s by Leonard and Lemlich [47], who considered an infinitely long Plateau border. Then the flow can be considered to be in the axial direction and the surface concentration of surfactant is constant. The bulk viscous stresses must be in equilibrium with the viscous shear stresses at the interface, a coupling between bulk and surface flow that is described by the dimensionless Boussinesq number

$$Bo = \frac{\eta_s}{\eta r}, \qquad (3.62)$$

where η_s is the surface shear viscosity (§2.2.3) and η is the bulk viscosity (see shaded section below). Note that the value of Bo depends on the radius of curvature r, and therefore on the geometry of the foam, not just the chemical composition. Leonard and Lemlich make the additional hypothesis that the liquid velocity is zero along the three lines where a Plateau border meets the films.

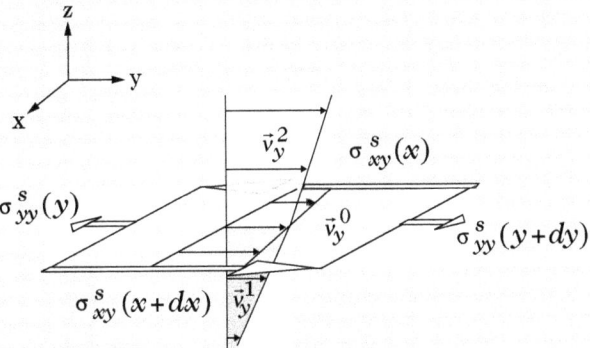

FIG. 3.32: Element of an interface lying in the (x, y) plane subjected to a surface stress.

Equilibrium of viscous stresses at the Plateau border walls

To better understand the distortion of the gas/liquid interface that constitutes the wall of a Plateau border, we consider the example of steady parallel flow of two fluids 1 and 2 which flow in the y direction with velocity vectors $\vec{v}_y^1$ and $\vec{v}_y^2$. Any given element of the interface between the fluids, of area $dxdy$, is subjected to the action of forces on its edges, and must be in equilibrium. For this two-dimensional analysis, we introduce forces per unit length, $\sigma_{xy}^s(x,y)$ and $\sigma_{yy}^s(x,y)$, shown in FIG. 3.32. The equilibrium of forces is:

$$\left[\sigma_{xy}^s(x+dx, y) - \sigma_{xy}^s(x, y)\right] dy + \left[\sigma_{yy}^s(x, y+dy) - \sigma_{yy}^s(x, y)\right] dx$$
$$- \sigma_{yz}^1\big|_0 \, dxdy + \sigma_{yz}^2\big|_0 \, dxdy = 0, \quad (3.63)$$

where $\sigma_{yz}^1|_0$ and $\sigma_{yz}^2|_0$ are the viscous stresses that the bulk exerts on the interface at $z = 0$ for fluids 1 and 2 respectively. Rewriting the finite differences in eq. (3.63) as gradients of the surface stress and dividing by the area of the element, we obtain

$$\frac{d\sigma_{xy}^s}{dx} + \frac{d\sigma_{yy}^s}{dy} - \sigma_{yz}^1\big|_0 + \sigma_{yz}^2\big|_0 = 0. \quad (3.64)$$

The bulk shear stresses evaluated at the interface are

$$\sigma_{yz}^1\big|_0 = \eta_1 \left.\frac{dv_y^1}{dz}\right|_0 \quad \text{and} \quad \sigma_{yz}^2\big|_0 = \eta_2 \left.\frac{dv_y^2}{dz}\right|_0. \quad (3.65)$$

If the interface is ideal then the surface stress $\sigma_{xy}^s(x)$ is proportional to the interfacial velocity gradient dv_y^0/dx (eq. (3.14)), so

$$\sigma_{xy}^s(x) = \eta_s \frac{dv_y^0}{dx}. \quad (3.66)$$

122 Birth, life, and death

Note that the surface stress σ^s_{yy} is none other than the surface tension γ (see eq. (3.15)). If an elongation is coupled to the shear, a correction appears (eqs. (3.10) and (3.12)) and induces a Marangoni effect (see §2.2.4), which we do not take into account here. We then have

$$\eta_s \frac{d^2 v^0_y}{dx^2} + \eta_2 \left.\frac{dv^2_y}{dz}\right|_0 - \eta_1 \left.\frac{dv^1_y}{dz}\right|_0 = 0. \qquad (3.67)$$

If fluid 1 is water and fluid 2 is air, we set $\eta_1 = \eta$ and neglect η_2. We introduce a characteristic length corresponding to the radius of curvature of the Plateau borders, r, and a characteristic velocity $r^2|\frac{dp}{d\ell}|/\eta$ to write this equation in dimensionless form using the Boussinesq number (eq. 3.62):

$$Bo \frac{d^2 \tilde{v}^0_y}{d\tilde{x}^2} = \left.\frac{d\tilde{v}^1_y}{d\tilde{z}}\right|_0. \qquad (3.68)$$

With these hypotheses, solving the equations of fluid motion enables us to obtain liquid velocity profiles [16, 42], shown in FIG. 3.33, and to determine the average velocity and the Plateau border permeability coefficient, K_p (eq. (3.49)), shown in FIG. 3.34, for different values of Bo. These figures highlight the influence of Bo on the flow through the Plateau border:

- For $Bo \ll 1$, the profile is close to a *plug* flow (i.e. it is flat over most of the cross-section), and velocity gradients are only apparent in the corners of the Plateau border. This corresponds to *high mobility* of the wall in response to the bulk flow. It is accompanied by an increase in the average velocity of the liquid and in the value of K_p.
- For $Bo \gg 1$, on the other hand, the interface is almost *immobile*. It is barely entrained and resists flow in such a way as to generate high velocity gradients across the entire cross-section. This type of flow is close to a *Poiseuille* flow (with zero velocity at the wall).

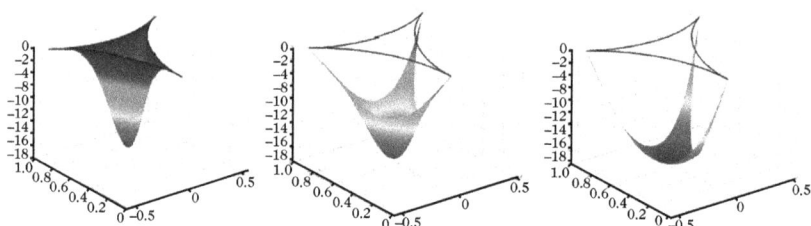

FIG. 3.33: Velocity profiles in a Plateau border for three values of the Boussinesq number, $Bo = 10$, 1, and 0.1. The corners of the Plateau border are fixed. The greater Bo, the more the interface resists the flow. Simulations by W. Drenckhan. See PLATE 8.

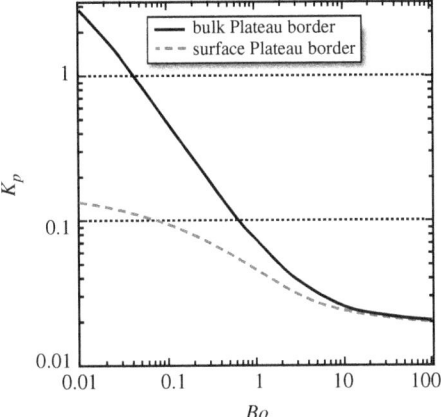

FIG. 3.34: Evolution of the permeability coefficient K_p with Boussinesq number for a bulk Plateau border (solid line) and a surface Plateau border (dashed line). After [42].

Some authors prefer to use the *interfacial mobility* $M = 1/Bo$ [66, 73], which is small for immobile surfaces and large for mobile ones.

More quantitatively, the minimum value of K_p, found when $Bo \gg 1$, is 0.02 (FIG. 3.34). This corresponds to pure Poiseuille flow in the Plateau border and is less than for a pore with a circular cross-section, $1/8\pi \simeq 0.04$. This difference in K_p indicates that a greater proportion of liquid is in contact with the surface and consequently that the flow velocity decreases. For smaller Bo, K_p can be approximated by the analytic expression [42]

$$K_p = \frac{1}{6}\left[\frac{3}{25} + \sqrt{\frac{2}{Bo}}\arctan\sqrt{\frac{1}{8\,Bo}} - \arctan\left(\frac{1}{2\pi\,Bo}\right)\right]. \quad (3.69)$$

In FIG. 3.34, we also show K_p for a Plateau border in contact with a solid wall (a surface Plateau border, see §5.1, chap. 2). Here again, K_p increases as Bo decreases, but less than in the bulk case because the zero velocity condition is imposed along the entire area of contact with the solid wall, irrespective of Bo. From these curves it is possible to estimate the influence that the surface Plateau borders have on the permeability of a foam (see exercise 8.4): to ensure that any measurement accurately represents the bulk permeability, the liquid flow in the bulk Plateau borders must represent a significant fraction of the total flow. We find that for a foam with about 30 bubbles across the diameter of the container, this proportion reaches 80% for $Bo = 10$ and 95% for $Bo = 0.1$.

4.5.3 Plateau border-dominated case Firstly, we will focus on the case of dry foams ($\phi_l \leq 0.02$), in which the liquid network is composed of elongated Plateau borders with radius of curvature r and length ℓ. Eq. (2.38) allows us to describe the link between r and the bubble radius R_V:

$$\frac{r}{2R_V} = g_1(\phi_l) = \frac{\delta_b}{2}\phi_l^{1/2}, \tag{3.70}$$

where $\delta_b = 1.73$ is a geometric constant. We consider a precise structure, the Kelvin foam (see p. 50), to estimate the relative length of the Plateau borders ℓ:

$$\frac{\ell}{2R_V} = 0.36. \tag{3.71}$$

It is then possible to confirm that for $\phi_l = 0.01$, for example, the Plateau borders are indeed long and thin, since $\ell \approx 4r$.

The geometric analogy with the network of cylindrical pores described in §4.4 suggests the use of the expression for the permeability α of a solid porous medium (eq. (3.50)) as a function of ϕ_l, the permeability coefficient K_p, and the pore cross-section s (eq. (3.56)). For a Plateau border, we have derived an expression for K_p above and we note that its cross-section is

$$s = \left(\sqrt{3} - \frac{\pi}{2}\right) r^2 \approx \delta_a \delta_b^2 R_V^2 \phi_l, \tag{3.72}$$

with $\delta_a = \sqrt{3} - \pi/2 \approx 0.161$. Then

$$\alpha = \frac{sK_p\phi_l}{3} = \frac{\delta_a \delta_b^2 K_p}{3} R_V^2 \phi_l^2 = C_p R_V^2 \phi_l^2, \tag{3.73}$$

which defines the constant $C_p = \delta_a \delta_b^2 K_p/3 \approx 0.16\, K_p$. In the special case of completely immobile interfaces, for which $K_p = 0.02$, the foam permeability is

$$\alpha = 3.2 \times 10^{-3} R_V^2 \phi_l^2. \tag{3.74}$$

As we will see in §4.7, this dependence of permeability on ϕ_l^2 is compatible with experimental results. Given the simplicity of expressions (3.73) and (3.74), we often use this limiting case, based on just the Plateau borders (or channels), to estimate the permeability, even though the hypotheses of the model limit its validity to relatively dry foams.

Finally, note that when Bo decreases, K_p increases, and therefore so does the permeability of a foam. There is a limit to this increase since Bo cannot decrease indefinitely: for a foam with bubbles of one millimetre in diameter, liquid fraction $\phi_l = 0.01$, and in which the interfaces are characterized by a very low surface viscosity, $\eta_s = 10^{-8}$ kg.s^{-1}, we obtain $Bo = 10^{-1}$ if the liquid which drains has a viscosity close to that of water. Eq. (3.69) indicates that the liquid velocity will be around 15 times greater than if the interfaces were completely immobile. Of course, this estimate is based on the hypothesis that only the Plateau borders slow the liquid down, and that the vertices which connect them do not participate. This hypothesis is legitimate when the Plateau borders have low permeability and occupy a large proportion of the total volume of liquid in the foam. If these conditions are not satisfied, it is necessary to take into account the effect of the vertices [43], as we do below.

4.5.4 Network model of Plateau borders and vertices

We now derive expressions for the permeability of an assembly of Plateau borders and vertices, considering as the basic element of the liquid network a Plateau border with one quarter of a vertex at each end (since a vertex connects four half-Plateau borders) [7, 43, 48, 59, 69].

The pressure gradient necessary to impose a flow rate $Q = \delta_a r^2 \bar{u}$ through a Plateau border is

$$\frac{\delta p}{\ell} = -\frac{\mathcal{R}_p + \mathcal{R}_v/2}{\ell} Q = -\frac{\mathcal{R}_p + \mathcal{R}_v/2}{\ell} \delta_a r^2 \bar{u}, \tag{3.75}$$

where $\mathcal{R}_p$ and $\mathcal{R}_v$ are the hydraulic resistances of a Plateau border and a vertex, respectively, which link the pressure difference between the ends of these elements to the liquid flow through them (in the absence of gravity). We determine an expression for the permeability by returning to the calculation in §4.4.3: eq. (3.49) is replaced by eq. (3.75), and eq. (3.56) becomes

$$\alpha = \frac{\phi_l}{3} \frac{\eta \ell}{(\mathcal{R}_p + \mathcal{R}_v/2)\delta_a r^2}. \tag{3.76}$$

We introduce the dimensionless parameters

$$\tilde{\mathcal{R}}_p = \frac{r^4 \mathcal{R}_p}{\eta \ell_{pb}} = \frac{1}{\delta_a^2 K_p} \tag{3.77}$$

and

$$\tilde{\mathcal{R}}_v = \frac{r^3 \mathcal{R}_v}{\eta} \tag{3.78}$$

where ℓ_{pb} is the Plateau border length, which differs from ℓ because it is necessary (when the liquid fraction is greater than $\phi_l \simeq 0.02$) to account for the decrease in length of the Plateau borders as the vertices swell. In order to describe this change, we generally resort to a function $\xi(\phi_l)$, the form of which will be discussed after eq. (3.81), which improves upon eq. (3.71):

$$\frac{\ell_{pb}}{2R_V} = 0.36 - \xi(\phi_l)\frac{r}{2R_V}. \tag{3.79}$$

In the same way, at high liquid fraction we obtain a more appropriate expression for r by altering the exponent in eq. (2.37):

$$\frac{r}{2R_V} = g_2(\phi_l) = 0.62 \, \phi_l^{0.45}. \tag{3.80}$$

Then the foam permeability is

$$\alpha = \frac{0.48 \, R_V^2 \phi_l g_2(\phi_l)}{\delta_a} \left[\frac{\tilde{\mathcal{R}}_v}{2} + \frac{1}{\delta_a^2 K_p}\left(\frac{0.36}{g_2(\phi_l)} - \xi(\phi_l)\right) \right]^{-1}. \tag{3.81}$$

This expression is more general than eq. (3.73) and allows us to describe foam permeability over a wide range of liquid fractions, typically between 0.001 and 0.1, for different types of surfactant.

We have introduced two additional dimensionless parameters, $\tilde{\mathcal{R}}_v$ and $\xi(\phi_l)$. The former captures the viscous dissipation in the vertices of the network, for which there is currently no precise model; several experimental and numerical results seem to indicate that $\tilde{\mathcal{R}}_v$ is of the order of several hundreds [10, 48, 69] and can reach several thousand [43], without specifying its dependence on B_0 (cf. eq. (3.69)). The latter, ξ, accounts for the decrease in length of a Plateau border when the liquid fraction increases. In dry foams ($\phi_l \leq 0.02$) ξ is treated as a constant between 1.5 and 2.3; for larger values of ϕ_l, ℓ_{pb} slowly decreases to zero at $\phi_l^* \approx 0.36$. One choice for $\phi_l \leq 0.1$ is a quadratic polynomial, $\xi(\phi_l) = a\phi_l^2 + b\phi_l + c$ with for example $a = 52.5$, $b = -13.2$, and $c = 2.24$ [49].

In a dry foam, we can set $g_2(\phi_l) \simeq g_1(\phi_l)$ (cf. eqs. (3.70) and (3.80)), $\xi(\phi_l) \simeq 0$, and $\tilde{\mathcal{R}}_v \approx 0$ to recover the Plateau border-dominated case introduced in the previous section (eq. (3.73)). Conversely, imagine a situation in which it is the vertices which determine the flow rate (the resistance of the vertices is larger than that of the Plateau borders), even in a dry foam where the contribution of the vertices to the liquid fraction is negligible. Then we obtain the *vertex-dominated* case, with

$$\alpha \simeq \frac{5.2}{\tilde{\mathcal{R}}_v} R_V^2 \phi_l^{3/2} = C_v R_V^2 \phi_l^{3/2}, \tag{3.82}$$

which defines the constant $C_v = 5.2/\tilde{\mathcal{R}}_v$. Now the permeability varies with $\phi_l^{3/2}$, in contrast to the ϕ_l^2 dependence found in the Plateau border-dominated case (eq. (3.73)). Several experimental results show such a variation in permeability with $\phi_l^{3/2}$ over a significant range of liquid fractions, particularly for large bubbles and mobile interfaces.

4.5.5 Carman–Kozeny model The Carman–Kozeny model, as we saw in §4.4.3, expresses the permeability of a solid porous medium in terms of its specific surface area A_s (eq. (3.59)). This macroscopic parameter can be measured directly (by determining for example the isotherm of nitrogen sorption in the porous medium) and therefore offers an attractive description of the geometry of the medium at a microscopic scale. Adapting the model to liquid foams is relatively easy, as the surface area can be obtained from simulations (§2.1.2, chap. 5). The interfacial area in contact with the flowing liquid is the total surface area of the bubbles minus the area of the thin films. This is because the films do not actively contribute to drainage: even if there is a liquid flow there, the flow rate is negligible in comparison to that in the network of Plateau borders and vertices. Thus [61]

$$A_s = \frac{3}{R_V}\left(\frac{S(\phi_l)}{S_o} - \frac{S_f(\phi_l)}{S_o}\right), \tag{3.83}$$

where $S(\phi_l)$, $S_f(\phi_l)$, and S_o are the total surface area of a bubble, the surface area of this bubble that is covered by thin films (see FIG. 3.28), and the surface area of a spherical bubble of the same volume respectively. FIG. 3.35 shows the specific

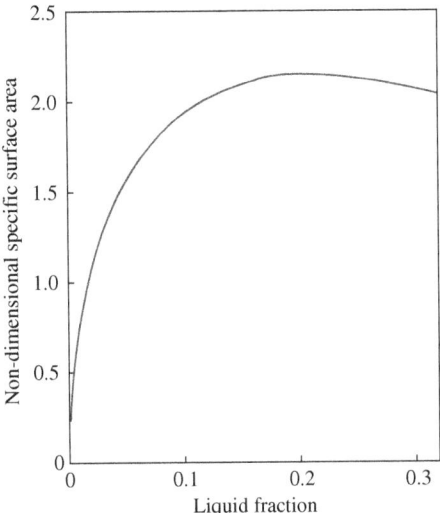

FIG. 3.35: The dimensionless specific surface area of a Kelvin foam as a function of liquid fraction [61].

surface area of a Kelvin foam as a function of the liquid fraction. Although wet foams do not necessarily have a Kelvin structure, we expect the trends to be similar: (i) when the liquid fraction increases, the interfacial area decreases (the bubbles are more and more spherical); (ii) at the same time, the surface of each bubble covered by films decreases strongly, giving rise to an increase in the specific surface area; (iii) at high liquid fraction these areas only change very slightly, unlike the total volume of the system, and so A_s reaches a maximum close to $\phi_l \simeq 0.2$, as shown in FIG. 3.35. The foam permeability is then deduced from eq. (3.59) [61]. Since this approach does not account for the mobility of the interfaces, the permeability obtained corresponds to the case of immobile interfaces (Poiseuille flow).

4.6 Drainage equations

Knowing the permeability of a foam allows us to use mass conservation to write down a corresponding *drainage equation* to describe the spatio-temporal evolution of the liquid fraction $\phi_l(\vec{x}, t)$.

Consider a fixed volume of foam with liquid fraction ϕ_l. We write ρ_l for the density and $\vec{u}$ for the local velocity of the liquid, and assume that the liquid is incompressible. The equation of continuity states that changes in liquid fraction are balanced by a flux in or out of the foam:

$$\frac{\partial \phi_l}{\partial t} + \vec{\nabla} \cdot (\phi_l \vec{u}) = 0. \qquad (3.84)$$

The liquid flows under the action of gravity $\rho_l \vec{g}$ and pressure $-\vec{\nabla} p$. Darcy's law (§4.4.2) relates the liquid flux $\vec{u}_m = \phi_l \vec{u}$ to these forces in terms of the permeability α and the viscosity η. In addition, the Young–Laplace law (eq. (2.4)) relates the liquid pressure to the curvature of the Plateau borders, $\vec{\nabla}p = -\vec{\nabla}(\gamma/r)$, so that

$$\vec{u}_m = \frac{\alpha}{\eta}\left[\rho_l\vec{g} + \vec{\nabla}\left(\frac{\gamma}{r}\right)\right]. \qquad (3.85)$$

The radius of curvature r is linked to ϕ_l by eq. (3.70), giving

$$\vec{u}_m = \frac{\alpha}{\eta}\left[\rho_l\vec{g} + \frac{\gamma}{\delta_b R_V}\vec{\nabla}\left(\phi_l^{-1/2}\right)\right]. \qquad (3.86)$$

Inserting this in the conservation equation (3.84), we obtain the general drainage equation

$$\frac{\partial \phi_l}{\partial t} + \frac{\rho_l}{\eta}\vec{\nabla}\cdot(\alpha \vec{g}) - \frac{\gamma}{2\delta_b R_V \eta}\vec{\nabla}\cdot\left(\frac{\alpha}{\phi_l^{3/2}}\vec{\nabla}\phi_l\right) = 0, \qquad (3.87)$$

describing the evolution of the liquid fraction under gravitational and capillary forces. The important parameter is the permeability α (eq. (3.81)). Eq. (3.87) is usually studied in the limiting cases of Plateau border-dominated and vertex-dominated flow (eqs. (3.73) and (3.82) respectively).

This drainage equation is three-dimensional; for simplicity we consider just the vertical drainage by projecting eq. (3.87) in the z direction. In the Plateau border-dominated case, we use eq. (3.73) to give

$$\frac{\partial \phi_l}{\partial t} - \frac{C_p}{\eta}\frac{\partial}{\partial z}\left(\rho_l g \phi_l^2 R_V^2 + \frac{\gamma}{2\delta_b}R_V \phi_l^{1/2}\frac{\partial \phi_l}{\partial z}\right) = 0. \qquad (3.88)$$

This is rendered dimensionless by introducing a characteristic length $\hat{z} = \lambda_c = \sqrt{\gamma/\rho_l g}$, time $\hat{t} = \eta \delta_b^2/\sqrt{\gamma \rho_l g}C_p$, and liquid fraction $\hat{\phi}_l = \lambda_c^2/R_V^2 \delta_b^2$, giving [34, 78, 82]

$$\frac{\partial \tilde{\phi}_l}{\partial \tilde{t}} - \frac{\partial}{\partial \tilde{z}}\left(\tilde{\phi}_l^{\,2} + \frac{\tilde{\phi}_l^{1/2}}{2}\frac{\partial \tilde{\phi}_l}{\partial \tilde{z}}\right) = 0. \qquad (3.89)$$

In the vertex-dominated case, we change the non-dimensionalization to use C_v instead of C_p in the characteristic time, and use eq.(3.82); then eq. (3.87) becomes [43]

$$\frac{\partial \tilde{\phi}_l}{\partial \tilde{t}} - \frac{\partial}{\partial \tilde{z}}\left(\tilde{\phi}_l^{\,3/2} + \frac{1}{2}\frac{\partial \tilde{\phi}_l}{\partial \tilde{z}}\right) = 0. \qquad (3.90)$$

4.7 Comparison of theoretical predictions with experiments

Most drainage experiments are free drainage experiments, but forced drainage (§4.3) results in a considerable simplification of the drainage equations. We will therefore begin by comparing the models with forced drainage experiments of the type presented in §4.1 before turning to free drainage.

4.7.1 Forced drainage With a boundary condition of a constant flow rate at the top of the foam and an initial condition of uniform liquid fraction ϕ_{l0} at $t = 0$, the solution of eqs. (3.89) or (3.90), expressed in terms of the function $\tanh^2(z)$ or $\exp(-z)$ respectively [43, 82], shows a uniform value for the liquid fraction at the top of the foam which decreases rapidly to ϕ_{l0} in a transition zone or *wetting front* (FIG. 3.36).

The front velocity, u_{front}, is the average vertical speed of the liquid in the foam, that is $u_{\text{front}} \approx \langle \bar{u}_z \rangle$, and is directly linked to the foam permeability α. The liquid fraction is uniform above the front, so in this region the liquid flows solely due to the force of gravity (there is no capillary pressure gradient). Then from eq. (3.55) we have

$$u_{\text{front}} \approx \frac{\alpha}{\phi_l} \frac{\rho_l g}{\eta}. \tag{3.91}$$

In the Plateau border-dominated case (eq. (3.73)), where the contribution of the vertices is neglected, and in the vertex-dominated case (eq. (3.82)), where the contribution of the Plateau borders is neglected, the front velocities are therefore

$$u_{\text{front},p} = C_p \frac{\rho_l g R_V^2}{\eta} \phi_l \qquad u_{\text{front},v} = C_v \frac{\rho_l g R_V^2}{\eta} \phi_l^{1/2}. \tag{3.92}$$

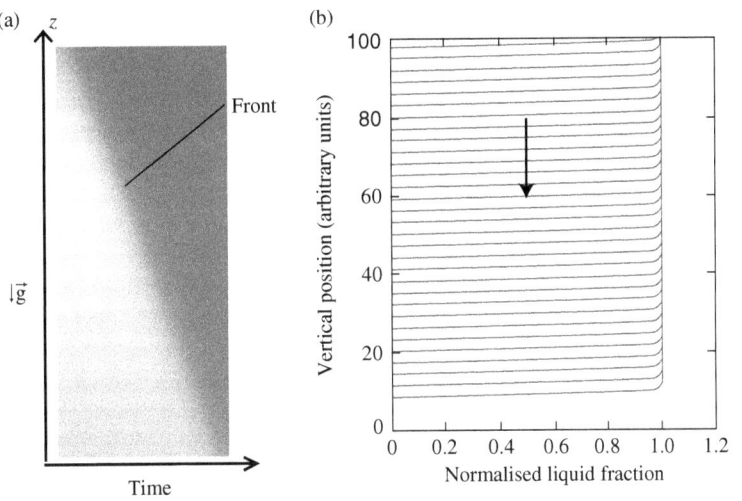

FIG. 3.36: Forced drainage. (a) The advancing wetting front in an experiment. The image is obtained by extracting a vertical line from each of a series of images (such as those in FIG. 3.30) and then placing them side by side. This *spatio-temporal* image (time is the abscissa) enables the speed of the front to be determined directly from the gradient of the line separating the light region (initially dry foam, with low liquid fraction) from the dark one (wet foam). (b) The Plateau-border dominated drainage equation (3.89) predicts that the wetting front moves at constant speed towards the bottom of the foam, as represented by the arrow. The liquid fraction is normalized by its value above the front. Numerical solution courtesy of S.J. Cox.

The exponents predicted by these two limiting cases are found experimentally, depending on the Boussinesq number [69].

There is no strict equality between u_{front} and $\langle \bar{u}_z \rangle$ because the front velocity is generally measured within the laboratory frame of reference, which does not take into account the upward displacement of bubbles due to the expansion of the foam (the sign of this correction depends on the experimental conditions). However, for the liquid fractions considered here ($\phi_l \leq 0.1$), this effect can be disregarded. Beyond a certain value of imposed flow rate, and therefore above a certain liquid fraction, we observe a phenomenon known as a *convective instability* [37], in which the bubbles themselves move around. It occurs because the yield stress of a foam (see §3.2, chap. 4) depends on both the liquid fraction and bubble size. The liquid fraction above which the instability is triggered is inversely proportional to the bubble size; it is about 0.05 for bubbles of several millimetres in diameter and 0.15 for submillimetric bubbles.

Foam permeability The liquid fraction above the front can be determined from $\phi_l = Q/(S\, u_{\text{front}})$, where Q is the imposed liquid flow rate and S the cross-sectional area of the container. Then eq. (3.91) gives information about the foam permeability α.

FIG. 3.37 shows the foam permeability for two surfactant solutions with different interfacial rheological behaviour. The figure compares theoretical predictions with experimental data from forced drainage experiments. The vertex-dominated case correctly describes the permeability of dry foams with mobile interfaces and the Plateau border-dominated case is a good approximation, and the network model even better, for foams with less mobile interfaces (note that a TTAB-dodecanol solution doesn't give perfectly immobile interfaces).[2] Nevertheless, these two limiting cases are only valid in the case of dry foams ($\phi_l \lesssim 0.05$). The network model (eq. (3.81)) is more appropriate for predicting the permeability α over a larger range of liquid fraction. FIG. 3.37 shows that this model reproduces the variation of α with ϕ_l and takes account of the difference in permeability between the two solutions. The dependence of α on bubble size is less well understood: FIG. 3.37 suggests that $\alpha \sim d^2$, but this is not always the case. Eq. (3.81) predicts the dependence on d^2 (via R_V), but this is modified by the dependence of K_p and $\tilde{\mathcal{R}}_v$ on d via Bo (eq. (3.69)). At the present time, there isn't a more satisfactory model, and the network model remains the one which most accurately describes the drainage of liquid foams.

4.7.2 Free drainage

We now return to free drainage, which, in practice, is the more usual case. Consider a foam sample of height Z and cross-sectional area S, with uniform initial liquid fraction ϕ_{l0} at $t = 0$, and constant and uniform bubble size R_V. We seek the vertical profile of liquid fraction in the foam and its temporal evolution during free drainage. The boundary conditions are that the flow rate is zero at the top of the foam and that the liquid fraction at the bottom of the foam, in contact with the liquid reservoir, is $\phi_l = \phi_l^* \approx 0.36$. The drainage curves can be calculated numerically, as in FIG. 3.38 which shows the result in the Plateau border-dominated limit.

---

[2] Other surfactants, such as proteins, further reduce the differences between the data and the Plateau border-dominated model for immobile interfaces.

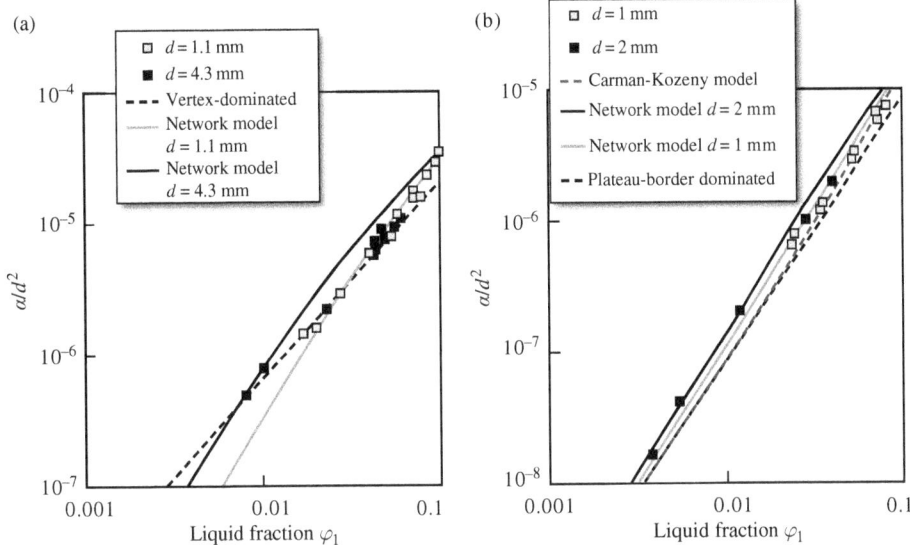

FIG. 3.37: Permeability of foams with two different mobilities, determined from forced drainage experiments. (a) TTAB foams (at 3 g.L^{-1}), which have fairly mobile interfaces, for two bubble sizes. The line for the vertex-dominated case is eq. (3.82) with $\tilde{\mathcal{R}}_v = 2{,}000$. The network model is eq. (3.81) with $\tilde{\mathcal{R}}_v = 400$, $\eta_s = 5 \times 10^{-8}$ kg.s^{-1}, and the function $\xi(\phi_l) = a\phi_l^2 + b\phi_l + c$, given in §4.5.4. (b) Foams made from TTAB (3 g.L^{-1}) and dodecanol (0.2 g.L^{-1}), giving less mobile interfaces, for two bubble sizes. The line for the Plateau border-dominated case is eq. (3.74) and the line for the Carman–Kozeny model is eq. (3.83). The network model is eq. (3.81) with $\tilde{\mathcal{R}}_v = 800$, $\eta_s = 5 \times 10^{-7}$ kg.s^{-1}, and the function $\xi(\phi_l) = a\phi_l^2 + b\phi_l + c$, given in §4.5.4. Note the difference in the scale on the vertical axes. Data from [48].

A qualitative comparison with FIG. 3.29 suggests that the foam drainage equation predicts the experimental results well. From the numerical results we can again define a characteristic time τ separating the first rapid phase, in which the liquid fraction is only modified in the upper part of the foam, from a second slower phase of relaxation towards the equilibrium profile.

The numerical results also give the shape of the liquid fraction profiles more precisely. For $t < \tau$, the liquid fraction varies almost linearly with z, from the top of the foam to the highest point that is still at the initial liquid fraction:

$$\frac{\phi_l}{\phi_{l0}} = A_p(t)(Z - z), \qquad (3.93)$$

where $A_p(t)$ is the observed gradient. The volume of drained liquid, $V(t)$, is also linear in time at first, and then tends asymptotically towards the value corresponding to the total volume of liquid initially present in the foam: $V_{l0} = S\phi_{l0}Z$. The end of the linear

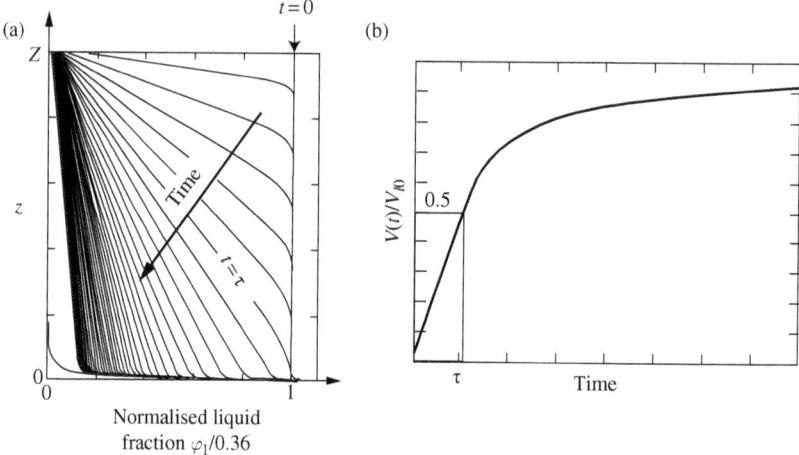

FIG. 3.38: Prediction of the Plateau border-dominated drainage equation (eq. (3.89)), showing similar characteristics to experiments (FIG. 3.29). (a) Liquid fraction profiles at successive times. At time $t = \tau$ the drainage front has reached the base of the column. The equilibrium profile, eq. (2.48), is visible in the bottom left-hand corner. (b) The volume of liquid V that has drained out of the foam, normalized by the total volume of liquid in the foam at $t = 0$, V_{l0}. The variation is linear until $t = \tau$, the instant at which $V = V_{l0}/2$. Numerical solution courtesy of S.J. Cox.

regime, at time $t = \tau$, offers a natural time-scale for characterizing foam drainage. At this time, $A_p(\tau) = 1/Z$ and the volume of drained liquid is

$$\frac{V(\tau)}{V_{l0}} = \frac{S\phi_{l0}}{V_{l0}} \int_0^Z \left(1 - \frac{\phi_l}{\phi_{l0}}\right) dz = \frac{1}{2}. \tag{3.94}$$

We determine the characteristic time τ by considering the lowest part of the foam. For $t < \tau$, the liquid fraction here is equal to the initial liquid fraction, ϕ_{l0}, so that the flow of liquid into this region is equal to the flow leaving it; thus the drainage rate here is constant and characterized by the front velocity: $u_{\text{front}} = V(\tau)/S\phi_{l0}\tau = Z/2\tau$. Using eq. (3.91) or eq. (3.92), the value of τ is therefore

$$\tau = \frac{\phi_{l0}Z}{2} \frac{\eta}{\alpha \rho_l g} = \frac{\eta Z}{2C_p \rho_l g R_V^2 \phi_{l0}}. \tag{3.95}$$

The parameter $C_p = 0.16\,K_p$ (eq. (3.73)) depends on Bo (FIG. 3.34). In the vertex-dominated limit, a similar calculation shows that the liquid fraction profiles are parabolic; eqs. (3.93) and (3.95) become

$$\phi_l/\phi_{l0} = A_v(t)(Z-z)^2 \qquad \tau = 2\eta Z/(3C_v \rho_l g R_V^2 \phi_{l0}^{1/2}), \tag{3.96}$$

where C_v is given by eq. (3.82)). Here τ corresponds to the instant at which 2/3 of the liquid has drained out of the foam (see exercise 8.5).

Numerical solution of the drainage equations also allows a more detailed comparison with experiments. For example, it confirms the experimental observation that for $t \gtrsim \tau$, the volume of drained liquid at time t differs from the volume of drained liquid in the final, equilibrium, profile (eq. (2.48)) by an amount that varies in time as t^{-1} (Plateau border-dominated) or t^{-2} (vertex-dominated), provided the system is sufficiently far from equilibrium. In that limit, when $t \gg \tau$, the approach to the equilibrium profile is very slow and is characterized by exponential dynamics, $\exp(-t/\tau)$. Finally, note that the equilibrium profile, eq. (2.48), is a solution of the drainage equation (eq. (3.89)), shown in FIGS. 2.23 and 3.38a, and that there is a similar solution in the vertex-dominated case in the long-time limit.

4.8 Summary and remarks

The Plateau border-dominated and vertex-dominated models introduced in §4.5 can be viewed as limiting cases of a network model (§4.5.4). The transition from one limiting case to the other is accomplished by changing the interfacial mobility (for example via the interfacial viscosity) via the Boussinesq number: $Bo \gg 1$ corresponds to a foam with "immobile interfaces", in which the drainage can be described by the Plateau border-dominated case, while $Bo \ll 1$ corresponds to a foam with "mobile interfaces", in which drainage can be described by the vertex-dominated case. Real foams are rarely at one extreme and we must consider both contributions (FIG. 3.37).

We finish this section with some remarks on these results and on the limit of validity of these drainage models. The theory is based on numerous hypotheses, involving relatively few parameters. In particular, it should only apply to dry foams, in which the bubble size is constant, and which are made from a simple chemical solution. The agreement between prediction and experiment is satisfactory bearing in mind the difficulty of the problem.

Nevertheless, at the present time, there remains a great deal to do in order to describe drainage from every angle. Firstly, as we saw in §4.7, existing models are not entirely satisfactory: either they depend on hypotheses which are too restrictive or they do not faithfully reproduce certain experimental aspects. Current efforts focus on a more finely tuned model of the coupling between Plateau borders, vertices, and films [20, 60, 62, 73]. A precise measurement of surface viscosity would also be advantageous.

Secondly, research into very wet foams, which are often crucial in industrial processes, is currently lacking [68]. Moreover, in many industrial applications the interstitial liquid in a foam is often a non-Newtonian fluid. There are no models that can describe the drainage of such foams, in which the liquid phase has a complex rheological behaviour.

Thirdly, bubble size, d, is one of the parameters which most influences drainage kinetics (in addition to the liquid fraction). Coarsening usually occurs at the same time as drainage, and therefore modifies the drainage kinetics. This coupling between coarsening and drainage [67, 77] is significant (FIG. 3.39), but is absent from existing models, which assume constant d.

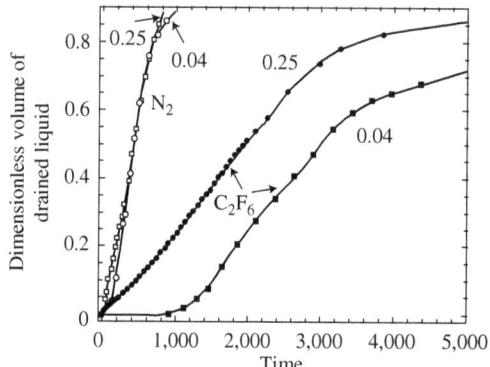

FIG. 3.39: Free drainage experiment for foams in which the bubbles contain nitrogen and C_2F_6, for two values of initial liquid fraction ($\phi_l = 0.04$ and 0.25). The foam height is 30 cm, and the initial size of bubbles is $d = 0.15$ mm. The C_2F_6, which is not very soluble in the foaming liquid, significantly slows down the ageing of the foam (see FIG. 3.27). The volume of drained liquid increases much more quickly in the case of nitrogen. This illustrates the strong coupling which exists between drainage and coarsening, the latter being responsible for the increase in average bubble size. After [66].

In extreme cases, the coarsening may completely control the drainage, and the dependence on the initial liquid fraction disappears. An initial high liquid fraction gives rise to rapid drainage, but slow coarsening (§3.3.3). Conversely, a dry foam drains slowly, but coarsens more quickly: this is then accompanied by an acceleration of drainage, as the bubble size increases, which can compensate for the reduction of the initial liquid fraction. This has been termed "self-limiting" drainage: increasing the initial liquid fraction doesn't increase the drainage rate (observe the behaviour of nitrogen-based foams in FIG. 3.39).

Finally, we note that there are other experimental parameters which have an influence on the drainage dynamics, such as the shape of the container or the uniformity of the foam [5, 66].

5 Rupture and coalescence

We have seen that a foam ages over time under the effects of drainage and coarsening. The rupture of a film between two neighbouring bubbles is another process which tends to make a foam disappear. In this section we will discuss the mechanisms by which a single film can break, the collective effects at the scale of the foam, and the stimulated rupture of a film by antifoaming agents.

5.1 Rupture at the scale of a single film

We saw in §2 that, in order to create a foam, significant repulsive forces must be present in the films so as to induce a positive disjoining pressure to counterbalance capillary suction. Without this positive disjoining pressure, the surfaces of a film do not repel each other, its thickness decreases to zero, and, since there is nothing to prevent coalescence, it breaks. This is clearly the simplest cause of film rupture.

It is also possible that repulsive interactions exist, but that they are weak (in other words, Π_d is positive for large thicknesses, but the maximum is small, see FIG. 3.20). As we saw in §2.3.2, capillary pressure P_c must be equal to Π_d for a film to exist. This allows the existence of foams in which the capillary suction from the Plateau borders is weak, such as in the wet limit and for bubbles of large diameter. However, over time, gravity-driven drainage dries the foam, reducing the cross-sectional area of the Plateau borders and thereby increasing the suction applied to the films. The pressure equilibrium condition ($\Pi_d = P_c$) becomes untenable, the repulsive forces in the films are insufficient to counterbalance the increased P_c, and the films thin and then break.

The description of stability presented in §2.3.2 applies to a quasi-static situation. Even in the case where there are strong repulsive interactions and a large disjoining pressure, there are dynamic effects (for example during film formation) or fluctuations which can cause rupture.

5.1.1 Film thinning and dynamic effects When two bubbles come into contact, a thin film forms and its area increases while its thickness decreases towards an equilibrium value h. During this dynamic process, there is a possibility of hydrodynamic instabilities causing film rupture at thicknesses greater than those at which the disjoining pressure intervenes.

The simplest models assume that the interfaces are flat, parallel, horizontal, and immobile (with no flow in the plane of the interface). This leads to the following expression for the rate of thinning ($v_{Re} = -dh/dt$), known as the *Reynolds velocity*, of a circular, horizontal film of radius R:

$$v_{Re} = \frac{2h^3(P_c - \Pi_d)}{3\eta R^2}, \tag{3.97}$$

where P_c is the capillary pressure and Π_d the disjoining pressure.

Experimental results show a significantly greater thinning velocity [40], which means that these simple hypotheses are unrealistic. Increasing the interfacial mobility and adding a coupling between the flow at the surface and within the film (as in the Plateau borders, cf. §4) partially improves the agreement.

The film between two bubbles thins over time due to gravity, but more importantly due to capillary suction into the Plateau borders. Experimentally, it is easy to observe that there are complex flows in a film. A film exchanges liquid with the adjacent Plateau borders, which accelerates film thinning, and the surface of the film becomes distorted, with regions of different thickness. This leads to *pinching* at the periphery, accompanied by an excess of liquid at the centre of the film known as a *dimple*, illustrated in FIG. 3.40. With immobile interfaces (high interfacial viscoelasticity), the dimple thins symmetrically and remains at the centre of the film. On the other hand, if the interfaces have high mobility (low viscoelasticity), the dimple may suddenly move to one part of the periphery [71] (FIG. 3.40) and then disappear by emptying

136 *Birth, life, and death*

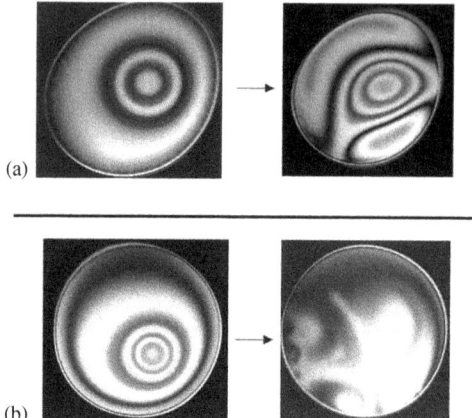

FIG. 3.40: Asymmetric drainage of a dimple in a horizontal film with high interfacial mobility, studied with the film balance technique (§1.1.4, chap. 5). **(a)** A dimple is stuck in the film (on the left), then empties into the peripheral Plateau border (on the right). This occurs very rapidly (<1 s) and induces significant flows in the film. **(b)** The dimple on the left is rapidly absorbed into the peripheral Plateau border, resulting in the large-scale flow pattern seen on the right. Photographs courtesy of A. Saint-Jalmes. See PLATE 9.

into a Plateau border. This response provides a way to estimate the magnitude of the interfacial viscoelasticity or mobility.

This exchange between a film and a Plateau border is also seen when a vertical film drains, where it is known as *marginal regeneration*. It occurs when the interfaces are very mobile, allowing rapid exchange between thick and thin regions (*regeneration*) close to the boundary of the film (the *marginal* zone) [58] (FIG. 3.41).

Even if all conditions are chosen so that an equilibrium thickness can be found, the presence of dimples and the process of marginal regeneration increase the drainage rate in the film and can even cause films to rupture.

5.1.2 The effect of fluctuations We reconsider the fluctuations which can occur in a liquid film, that were introduced in §2.2.4 in relation to the interfacial viscoelasticity. Here again, even though all conditions necessary for stability are fulfilled, a film may break because of natural (especially thermal) fluctuations in thickness or surfactant concentration, which can either trigger a transition to a Newton black film (more fragile), or cause rupture directly.

One of the main causes of film rupture is due to a non-uniform distribution of surfactants at the surface and the creation of gaps large enough to destabilize the film. A decrease in surfactant concentration is accompanied by an increase in surface tension, and by a decrease in Π_d, causing a local decrease in film thickness. Dilatational elasticity stabilizes a film against these fluctuations by causing, through the Marangoni effect, surfactant and fluid flow within the film (§2.2.4): the bulk flow induced by

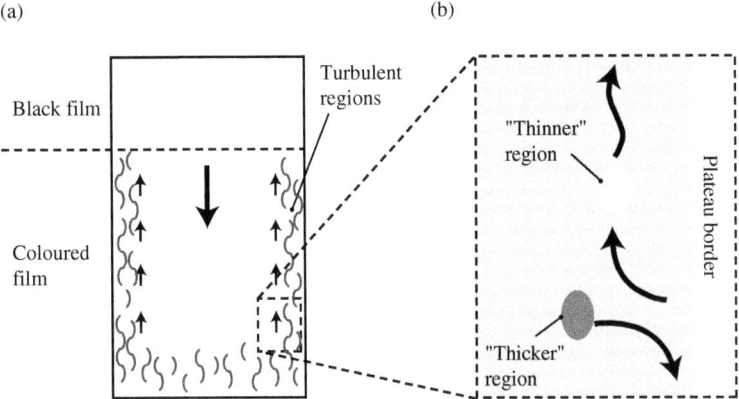

FIG. 3.41: Marginal regeneration in a film held on a vertical frame. (a) General situation: the film drains from the top, creating a region of very thin black film; lower down the flow is perturbed close to the frame. (b) Detailed situation, at the scale of the Plateau border attached to the frame: regions of thick film rejoin the Plateau border, whilst thinner areas detach and rapidly rise to the black film region. This exchange between the thick and thin regions increases the film drainage rate. After [2].

the surface concentration gradient eventually reduces the fluctuations in interfacial concentration (and the associated fluctuations in thickness).

Rupture of an initially stable film may also occur when the interface curves and the thickness fluctuate locally [80], as sketched in FIG. 3.42. These oscillations may be the result of external (thermal or mechanical) perturbations, for example due to flow or the bursting of neighbouring films. Curvature of an interface sometimes generates hydrodynamic instabilities in the form of amplified capillary waves. In addition, although the surface tension tends to smooth the interfaces, short-range attractive forces between interfaces (§2.3.1) can dominate and increase the curvature and pinching of the film. For example, if the local thickness of a film is less than that of a common black film, it becomes locally unstable (§2.3.2). Film rupture is thus observed at the point where the distance between the two interfaces is smallest: due to the pinching of the film a hole is formed.

Analysing the stability of a film of average thickness h_0 is possible if we assume an initially sinusoidal fluctuation in the wavelength λ_0 (see FIG. 3.42) and follow its

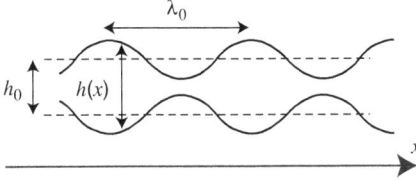

FIG. 3.42: Oscillations of the interfaces of a film generated by mechanical or thermal perturbations sometimes cause film rupture.

temporal evolution. Ivanov [40] and Vrij *et al.* [80] have shown that if the interaction between the interfaces is dominated by van der Waals forces, then all wavelengths greater than $\lambda_c \sim (4\pi^3 \gamma h_0^4/A_h)^{1/2}$ are unstable, where A_h is the Hamaker constant (see eq. (3.20)). Only films of radius R less than this critical length are stable, which gives the film thickness h_c at which rupture occurs spontaneously in terms of the size of a film: $h_c = \left[A_h R^2/(4\pi^3\gamma)\right]^{1/4}$. This expression is based on the hypotheses that the film is at equilibrium and that the average film thickness is constant in time.

5.1.3 Opening of a hole

When a film breaks, a hole appears and grows. Can we understand the dynamics of this growth? As soon as a hole is formed, surface tension tends to enlarge it. For a Newtonian fluid, the displaced liquid accumulates at the rim, which takes a toroidal form (FIG. 3.43). In a cylindrical coordinate system (r,θ), with origin at the centre of a hole of radius a, Newton's second law applied to the part of the rim lying between θ and $\theta + d\theta$, with mass dM and velocity $w\vec{e}_r$, resolved parallel to $\vec{e}_r$, gives

$$\frac{d(dMw)}{dt} = 2\gamma a d\theta. \tag{3.98}$$

In fact, the film is at rest before it is drawn into the rim and the only force acting at the rim is the surface tension of the film. Moreover, it is reasonable to assume that the liquid is entirely collected in the rim, which gives $dM = d\theta\, \rho_l a^2 h/2$. Finally, by assuming that the velocity $w = da/dt$ is constant, Culick [11] showed that

$$w = \sqrt{\frac{2\gamma}{\rho_l h}}. \tag{3.99}$$

This predicts that in a film one micron thick the velocity at which a hole grows is of the order of 10 m.s^{-1}. A circular film with a radius of 1 mm therefore disappears in 10^{-4} s! Assuming equilibrium between capillary and kinetic energies (inertial regime) leads to an expression for the velocity that has the same dependence on the physical parameters (ρ, γ, and h) but with a different numerical coefficient.

The experiments of McEntee and Mysels [53] validate Culick's prediction (eq. (3.99)) for films with thickness between 0.1 and 10 μm. Note that energy is not conserved in this model: the lost capillary energy is twice the gain in kinetic energy. The failure of the model to conserve energy is attributed to the presence of an inelastic shock between the rim, which is displaced at velocity w, and the liquid contained in the film at rest; this is a visco-inertial regime. This macroscopic argument is somewhat

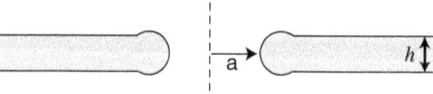

FIG. 3.43: The opening of a hole in a liquid film, represented here as a cross-section through the film. For a Newtonian fluid the liquid gathers at the rim in a toroidal shape.

limited: it doesn't take into account either the surfactants, the precise shape of the liquid at the rim, or the stability of the film.

If a film is very thin and/or the liquid is extremely viscous, then inertia is negligible. Equilibrating viscous forces (dissipation) with capillary forces gives

$$w = \frac{2\gamma}{\eta}. \tag{3.100}$$

By comparing this expression for the velocity with the other limit (eq. (3.99)), we find that the critical thickness h_c of a film separating this viscous regime from the visco-inertial one, corresponding to a Reynolds number of order one, is

$$h_c = \frac{\eta^2}{2\gamma\rho_l}. \tag{3.101}$$

For a typical foaming solution this thickness is of the order of tens of nanometres, close to a common black film.

For films formed from non-Newtonian fluids (for example long-chain polymers without surfactants), the rim does not thicken and the radius of the hole grows exponentially in time [14].

Note also that it is not only holes that form in films, but also thin regions (§2.3.1). They occur during drainage, following a jump in the local film thickness, and their area increases in time, as shown in FIG. 3.18.

Finally, the liquid rim may be unstable, and at certain velocities and thicknesses may form a string of drops (FIG. 3.44), illustrating another dynamic effect which may compromise film stability.

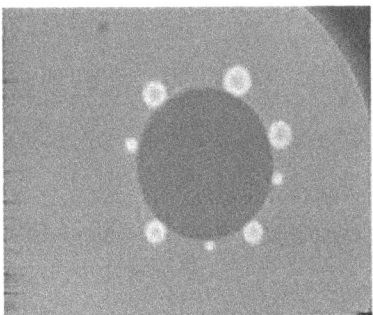

FIG. 3.44: The thickness of a stratified film (FIG. 3.18) decreases discontinuously. When a thin region develops and grows, the liquid rim may become unstable, forming several large droplets. The film shown consists of a mixture of surfactants and polyelectrolytes (see §6.1), studied with the film balance technique (§1.1.4, chap. 5); its diameter is 0.2 mm and its thickness is 20 nm. Photograph courtesy of A.Saint-Jalmes. See PLATE 10.

5.2 Rupture at the scale of a foam

The link between film rupture and the collapse of a foam is obvious: a foam disappears via a succession of film ruptures. The bubble films at the surface of a foam are most susceptible to rupture: they are the thinnest (due to drainage), the most curved (bubbles with few neighbours are closer to a sphere), they evaporate the most quickly (because they are in contact with ambient air), they are subject to external perturbations (notably dust particles which, at the scale of a nanometric film, resemble enormous meteorites), and, lastly, they are the first to encounter environmental variations (temperature for example).

Once the bubbles at the top have burst, the bubbles beneath come into contact with the atmosphere and become in turn more susceptible to bursting. Thus a foam often collapses from the top. However, if the walls of the container are dirty, for example if particles or oil are present, a foam may start to rupture here. In addition, a film within a foam is also sensitive to mechanical perturbations caused by its environment: the liquid flow caused by drainage, ejection of droplets by the destabilization of the rim of a hole in a nearby film, even sound waves emitted during film rupture (the intermittent "crackling" of a dry foam that you hear in your bath). For all these reasons, the probability of rupture of an isolated film is not the same as that of a film in a foam, which seems to depend on its neighbours.

Observation of two-dimensional foams and recording the sound made by a rupture event at the surface of a 3D foam allow the identification of individual rupture events. They tend to occur in *avalanches*, that is, each rupture is not independent of the previous event; instead, rupture occurs in correlated bursts [55, 76]. In fact, if these events were independent, they would follow Poisson's law.[3] FIG. 3.45 shows the distribution of the number of events n in a dry foam, indicating that, at least in this case, the experimental results are poorly described by Poisson's law.

The mechanisms by which a foam collapses are still poorly understood, for example the correlation between stability of an isolated film and the stability of the same film within a foam [54], or even the origin of the avalanches of rupture events. Similarly, it is not yet clearly understood how foam collapse depends on the bubble size or on interfacial properties. The best approach has yet to be established: is the most significant parameter a critical film thickness, a critical disjoining pressure, or a critical liquid fraction?

5.3 Defoamers and antifoams

There are many situations where foam is considered a nuisance or a drawback. A number of industries are directly affected by the problem of a too-stable foam. Foam is sometimes a necessary part of a manufacturing process: its presence is therefore temporarily desirable, but it is necessary to destroy it when it has served its purpose.

---

[3] Recall that in this probability distribution, if n is the number of events occurring in a period of time and $\langle n \rangle$ denotes the average number of events in this period, the probability $p(n)$ of finding n events is $p(n) = e^{-\langle n \rangle} \frac{\langle n \rangle^n}{n!}$.

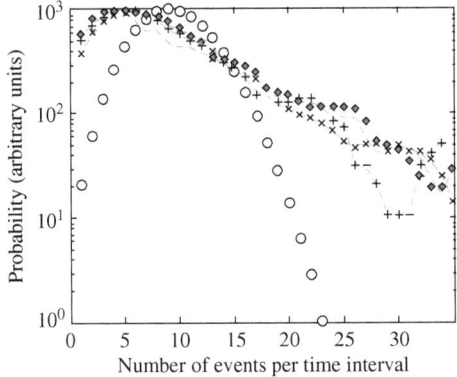

FIG. 3.45: Temporal distribution of rupture events in a dry 3D foam at different ages. The distributions were recorded acoustically over time intervals of equal duration, and have been normalized so that their maxima are equal. The white circles represent the best possible fit of the data to a Poisson distribution [55].

We present here some of the strategies put in place either to prevent the formation of a foam, or to destroy an existing, stable foam [31]. There is a subtle difference between the two: in the first case the *antifoaming* agent must be present in the solution, while in the second, the *defoamer* must be dispersed in the foam. Nonetheless, the way in which they work, and the products used, turn out to be very similar. These products are grouped into three categories: liquids (generally in the form of droplets), solids (especially ones that are finely ground), and liquid/solid mixtures (the most efficient).

5.3.1 Liquid antifoams A liquid antifoam consists of an immiscible liquid (e.g. oil in aqueous solution) that is added to the foaming solution in the form of droplets. In order to be effective, each droplet must reach the solution/gas interface, represented schematically in FIG. 3.46. As a droplet emerges from the solution, an antifoam/gas interface (with surface energy γ_{ga}) appears, and a solution/antifoam interface and a solution/gas interface (with surface energies γ_{sa} and γ_{sg}, respectively) disappear.

The difference in surface energy (per unit area) between the final state and the initial state is $\Delta U = \gamma_{ga} - \gamma_{sa} - \gamma_{sg}$. For emergence of the droplet at the solution/gas

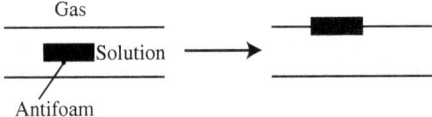

FIG. 3.46: The emergence of an antifoam droplet at the solution/gas interface. An antifoam/gas interface appears, and a solution/antifoam interface and a solution/gas interface both disappear.

interface to be energetically favourable requires that $\Delta U < 0$. The *entry coefficient* is generally defined as

$$E_\gamma = -\gamma_{ga} + \gamma_{sa} + \gamma_{sg}, \qquad (3.102)$$

and the condition for emergence is $E_\gamma > 0$. To estimate the sign of E_γ requires the values of surface energy for the three fluids, obtained when they are mutually saturated and the surfactants have had time to adsorb. If the antifoam droplet does emerge at the interface it may be energetically favourable for it to spread out. This is determined by the *spreading coefficient*, defined as the difference in surface energy (per unit area) between the non-spread and spread configurations, $S_\gamma = \gamma_{sg} - \gamma_{sa} - \gamma_{ga}$, estimated in the same way. The drop spreads out if $S_\gamma > 0$. We consider two cases, both of which can cause a film separating two bubbles to rupture:

(a) **Case $E_\gamma > 0$ and $S_\gamma > 0$** In this case an antifoam droplet emerges and spreads across the solution/gas interface. There are at least two reasons for this to contribute to the destabilization of the liquid film.

On the one hand, the solution/gas interface is replaced by the antifoam/gas interface, which does not behave a priori like a surfactant. Consequently, this new interface does not contribute to the stabilization of the film, so it is less stable than before. On the other hand, the film is destabilized by the spreading out of the antifoam at the solution/gas interface, in a process called Marangoni spreading (FIG. 3.47). The spreading of the antifoam droplet causes bulk flow in the film, by viscous drag of the neighbouring liquid layers, leading to its thinning then rupture [27]. The driving force for the spreading out of a lens-shaped droplet on a surface is the imbalance in the surface tensions along the triple line. The net force per unit length (energy per unit area) is none other than the spreading coefficient S_γ. For example, by applying conservation of volume to a silicone oil droplet of initial radius a that spreads up to some maximum size, and assuming that a thickness δ_m of liquid is entrained, it has been shown that [64]

$$\delta_m = a \left(\frac{\eta_a^2}{S_\gamma \rho_a d_m} \right)^{1/3}, \qquad (3.103)$$

where η_a and ρ_a are the viscosity and the density of the antifoam, and d_m is the thickness of the antifoam lens when it is fully spread (typically one molecule thick). The equation predicts that the antifoam is more efficient the larger the droplet; this is certainly true for small droplets, but in reality there is an upper limit (of the order of a few microns [64]) to the optimal size of antifoam droplets.

(b) **Case $E_\gamma > 0$ and $S_\gamma < 0$** This situation is very different from the previous one, because the antifoam lens arrives at the solution/gas interface, but does not spread. An antifoam effect is possible only if the lens can bridge the two interfaces of the film (FIG. 3.48).

If it can form, the stability of an antifoam bridge depends on the value of the equilibrium contact angle θ between the antifoam and the solution. If it is greater

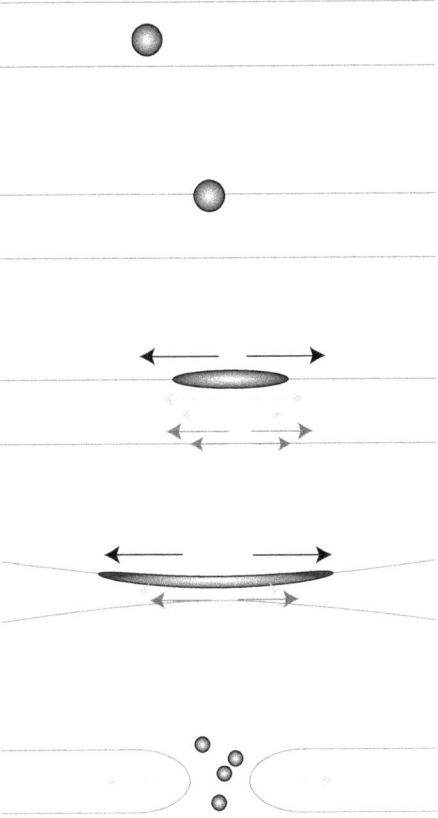

FIG. 3.47: Emergence and spreading of an antifoam droplet at the surface of a film causing it to thin and the liquid to move. The film becomes so thin that it breaks, releasing the antifoam droplets.

than $\pi/2$ then the curvature of the interface close to the droplet imposes a greater pressure than is found in more distant parts of the film. This pressure gradient causes liquid to move and the film thins until the two triple lines on each side of the lens meet. Then rupture occurs and the antifoam is dispersed into several smaller droplets. This reduction in the average diameter of antifoam droplets has been observed during several cycles of use, and is correlated with a loss in efficiency of the antifoam, since if the droplet becomes smaller than a critical value then bridging no longer occurs and the antifoam no longer works.

5.3.2 Solid antifoams Small (micron-sized) solid particles with hydrophobic surfaces (FIG. 3.50, right), for example carbon or pyrogenic silica, naturally play a role in antifoaming, as follows. The way in which a film is ruptured by a solid particle is similar to the situation $E_\gamma > 0$ and $S_\gamma < 0$ described above. The particle must emerge

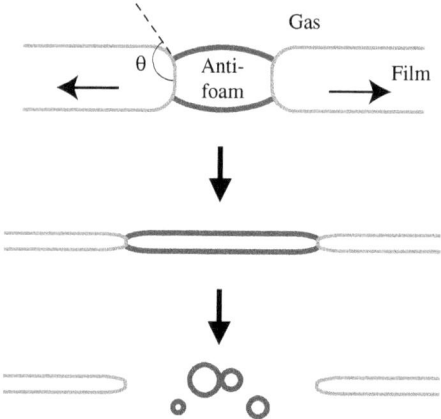

FIG. 3.48: Bridging and rupture of a soap film by an antifoam droplet.

from the film, then bridge the liquid film in such a way as to cause it to thin. The liquid/solid contact angle is the parameter which determines whether or not the particle can adopt a mechanically stable equilibrium position at the interface. For particles of this size, capillary forces f_c determine the equilibrium position, and a spherical particle assumes its equilibrium position when $f_c = 0$. This force is equal to the surface tension of the liquid/gas interface integrated along the triple line forming the liquid/gas/solid interfaces (FIG. 3.49):

$$\vec{f_c} = -\pi\gamma a \cos\psi \cos(\psi + \theta)\vec{e}_z, \qquad (3.104)$$

where a is the particle diameter.

Equilibrium occurs when $\psi = \pi/2 - \theta$. When the sphere bridges the two interfaces of a film, the liquid must move so that the interfaces may remain flat. As in the case of an antifoam droplet described above, it is the pressure generated by the curvature of the interface which gives rise to film thinning.

If the contact angle θ is less than $\pi/2$, the interfaces reach equilibrium before the film thickness decreases to zero ($\psi > 0$). On the other hand, if θ is greater than $\pi/2$, the equilibrium position of the interface corresponds to an angle of $\psi < 0$. This is incompatible with the presence of the second interface, and the film breaks when the two triple lines meet.

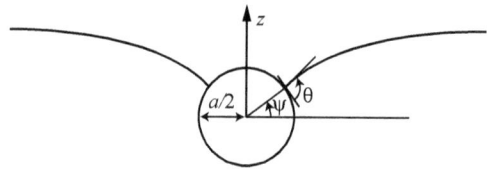

FIG. 3.49: Solid sphere at a liquid/gas interface.

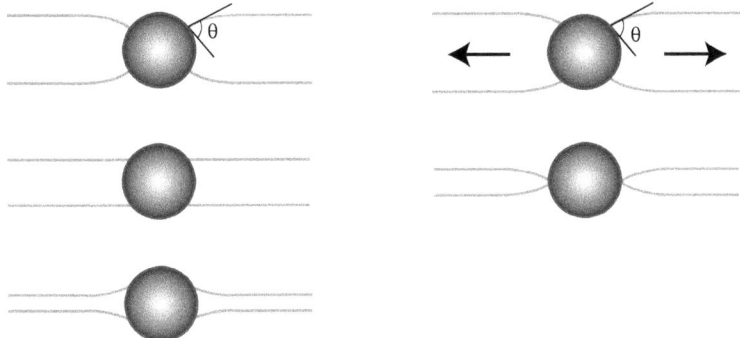

FIG. 3.50: Behaviour of a solid particle in a soap film as a function of the solid/liquid contact angle. $\theta < \pi/2$ (left): the interfaces remain attached to "their" half of the sphere, and thinning of the film is prevented by the appearance of negative curvature around the sphere. $\theta > \pi/2$ (right): the interfaces are unable to simultaneously find a stable position on the sphere, causing film rupture.

Solid particles thus have an antifoam effect if the solid/liquid contact angle is greater than $\pi/2$, i.e. if they are hydrophobic (FIG. 3.50, right). The process of rupture is even more effective if the particles are large, because they emerge more rapidly.

5.3.3 Super defoamer It is generally much more efficient to use mixed antifoaming agents, containing both oil and hydrophobic solid particles, rather than one or the other. One of the most plausible explanations for this is related to the stability of the pinched film between the antifoam droplet and the gas phase. In fact, the first step in all of the mechanisms for rupture presented here is the emergence of the antifoam at the solution/gas interface. During this motion the liquid trapped between the two interfaces is expelled, a flow which occurs more slowly as the amount of the liquid in the film that is entrained becomes smaller. Rupture can only happen when the thickness of this region is equal to some critical distance (of the order of a hundred nanometres). If there are solid particles on the surface of the antifoam droplet, their size fixes the thickness at which the film separating the antifoam/gas phases breaks. This scenario has been tested experimentally. The results have shown that films trapped between a droplet and the gas phase exist for less time in the presence of solid particles, and that this reduction can be correlated to the "roughness" of the interfaces.

6 Appendices

6.1 Stabilizing agents

Foams are stabilized by amphiphilic molecules, or surfactants (see p. 18), which are diverse in nature: detergent molecules, proteins, polymers, and even solid particles. In this section, we describe several large families of molecules that are found in the most common detergents, cosmetics, or other products of daily life, classified as a function of their molecular mass.

6.1.1 Surfactant molecules of low molecular mass
Small surfactants with a molecular mass $M_w < 1{,}000$ g.mol^{-1} move rapidly in solution through Brownian motion; they adsorb (and desorb) more quickly on an interface than surfactants of high molecular mass. Most of the surfactants used in industry fall into this category.

Chemical nature Depending on the charge of the polar head, we refer to surfactants as anionic (negative charge), cationic (positive charge), non-ionic (no net charge), or amphoteric (positive or negative charge depending on the pH of the solution). We list below some of the surfactant families of low molecular mass which are currently used. This list illustrates in particular the large chemical differences which exist between surfactants, and the variety of compounds used. Anionic surfactants are frequently found in detergent products, washing-up liquids, etc., while non-ionic surfactants are present in shampoos and other personal cleaning products. Cationic surfactants are often chosen for their corrosive and bactericidal effects. Nevertheless, commercial products generally have complex formulations, combining several surfactants (sometimes with opposite charges), charged polymers, viscosifiers, etc. (see p. 148) with the aim of meeting several objectives simultaneously.

Anionic surfactants Alkyl sulphates (like sodium dodecyl sulphate, or SDS, with the formula $C_{12}H_{25}OSO_3^-$, Na^+), alkyl ether sulphates and alkyl benzene sulphates are the standard ones. There are also alpha-olefin sulfonates and sulfoccinates (like sodium bis(2-ethylhexyl) sulfosuccinate, or AOT: $(C_8H_{17}-O-COCH_2)_2SO_3^-$, Na^+).

Cationic surfactants These include alkyl trimethyl ammonium bromides (C_nTAB: $C_nH_{2n+1}N(CH_3)_3^+$, Br^-) and alkyl trimethyl ammonium chlorides (like DTAC: $C_{12}H_{25}N(CH_3)^+$, Cl^-) and cetylpyridinium chloride (CPCl: $C_{16}H_{33}N(CH_2)_5^+$, Cl^-).

Non-ionic surfactants These include polyoxyethylene alkyl ethers ($C_n EO_m$), alkyl glucosides C_iG_j, *spans* and *tweens* (the latter two are also used for emulsions), Triton X-100 ($C_{14}H_{22}O(C_2H_4O)_n$), and polyglycerol ethers.

Amphoteric surfactants These are sodium carboxylates, $\mathbf{R}CO_2Na$ (like natural soaps, in which $\mathbf{R}$ represents a carbon chain), betaines (like cocoylamidopropyl betaine, CAPB), and tetradecyl dimethylamine oxide ($C_{14}H_{29}NO(CH_3)_2$).

Structure These surfactant molecules are distributed between the interfaces, where they lower the surface energy, and the solution. At low concentrations they are isolated when in the solution, but above a *critical micelle concentration* (*cmc*, see FIG. 2.4b), the surface concentration and the surface tension no longer change and the surfactants then aggregate in the solution. In these aggregates the hydrophobic tails group together and the polar heads form a shell around them, thus protecting them from contact with the solution (FIG. 3.9). As the surfactant concentration increases, different three-dimensional structures appear: (i) *micelles* are spheres, with a radius of the order of the length d_m of a molecule; (ii) *cylindrical micelles* are flexible filaments of circular cross-section of radius d_m; (iii) *bilayers* are membranes of thickness $2d_m$, which can reach very large lateral dimensions compared to d_m, and which sometimes form multilayers. These structures, in particular cylindrical micelles which intermix like polymers, modify the viscosity of the solution. The presence of salt also affects the structure of aggregates, and the addition of salt is a way to increase the number of cylindrical micelles in a solution (of shampoo, for example) and therefore the viscosity.

At very high concentrations it is possible to obtain phases in which the surfactants are arranged in crystalline structures.

The presence of micelles in a solution enables the dissolution of molecules that are insoluble in the aqueous phase: the grouped lipid tails in a micelle behave like small oil droplets in which some molecules can dissolve. Thus, certain molecules, called co-surfactants, for example fatty alcohols such as dodecanol ($C_{12}H_{25}OH$), are not surface-active by themselves (or only very weakly) and are also poorly soluble; but when added to an ionic surfactant, for example, they move into the micelles and then to the interfaces between surfactants. These co-surfactants improve foamability and foam stability by significantly increasing the interfacial viscoelastic moduli, partly due to their poor solubility (see §2.2.4).

6.1.2 Surfactant molecules of large molecular mass

Synthetic polymers Synthetic surfactant polymers are polymers that have been designed or modified to become amphiphilic. Among the homopolymers (that show the same repeated pattern), polyvinyl alcohol (PVA) and modified polysaccharides are common. Copolymer blocks consist of hydrophilic polymers grafted onto hydrophobic ones. Various assemblies of this type are known commercially as *Pluronic*. Diverse configurations are possible at the interfaces, depending on the number of blocks, their relative sizes, and the interfacial concentration. Diblocks, composed of one hydrophilic block and one hydrophobic block, show a *mushroom* configuration of hydrophobic chains at the surface at low densities, but above a certain density the chains extend perpendicular to the surface, in a *brush* configuration. By introducing a hydrophobic block between two hydrophilic blocks, it is possible to make a *triblock* which adsorbs on the interface only by its hydrophobic part.

Natural polymers These are often edible proteins. Examples of *globular* proteins include bovine serum albumin (BSA), β-lactoglobulin (BLG) from milk, or ovalbumin from egg. Caseins (β and κ), which are also found in milk, are compact when in solution but unfold at the interface.

These large surfactant molecules adsorb on an interface only very slowly (see §2.2.2). This means that a highly concentrated solution is needed to make a foam. Once in place, these molecules desorb with difficulty and can even remain at the interfaces permanently. They also rearrange themselves at the interface, giving rise to modifications of surface tension or viscoelastic moduli over time-scales of several hours. In addition, proteins often form interfacial gels over long times, with large elastic modulus.

Solid particles Aqueous dispersions of nano- or micrometric solid particles can also be made to foam. If their wettability allows it, they adsorb at the interfaces and stabilize them efficiently. It is mostly emulsions (*Pickering* emulsions) which are produced in this way at the moment, but solid particles are being increasingly used to stabilize foams [8].

6.1.3 Surfactant/polyelectrolyte complexes Many surfactant mixtures are possible, and often necessary to optimize the properties of an industrial foam. Witness, for example, the long list of ingredients in shampoo, shaving foam, washing powder, and other household cleaning products. Moreover, in spite of their varied uses, these products often have numerous ingredients in common!

These mixtures often act both as surfactants (of small size) and polymers (which are usually charged and therefore known as *polyelectrolytes*). FIG. 3.51 shows the four typical configurations that are found in the bulk and at the surface.

(a) In the case in which the components have the same charge, the surfactant generally invades the surface first and prevents the adsorption of the polyelectrolyte by repulsion. The interfacial properties are those of a pure surfactant, although the liquid films can be stratified (see §2.3.1).

(b) In the case of oppositely charged components, both adsorb at the interface, often through a *synergistic co-adsorption*: a surfactant/polyelectrolyte complex, more surface-active than the components taken separately, first forms in the bulk and then adsorbs to the surface. So even for very low concentrations of each component, the surface tension is lowered significantly. The critical aggregation concentration (*cac*) at which aggregates form in the bulk is much lower than the *cmc* of the

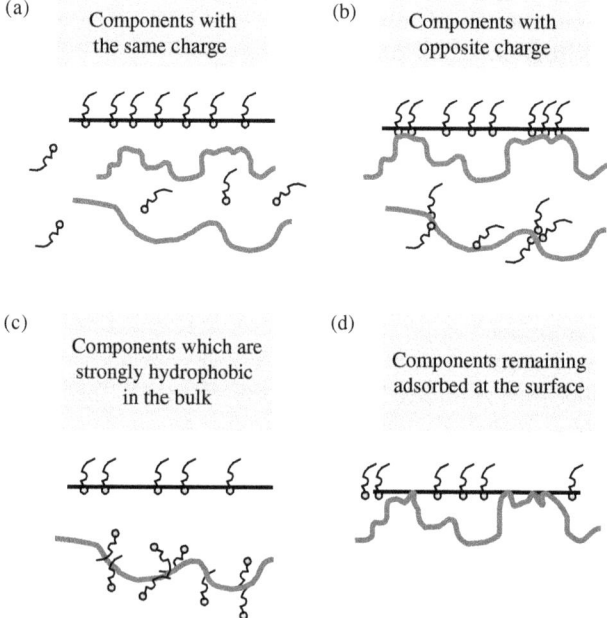

FIG. 3.51: Surfactant/polyelectrolyte mixtures in the bulk and at the surface. The behaviour that arises from the various possible affinities that the molecules exhibit is explained in the text.

PLATE 1: A liquid foam some time after its creation. It is dry at the top, with large polyhedral bubbles, and wet at the bottom, with a smaller average bubble size. The bubbles become spherical when they come into contact with the liquid on which the foam floats. See FIG. 0.1. Photograph courtesy of S. Cohen-Addad, R.M. Guillermic and A. Saint-Jalmes.

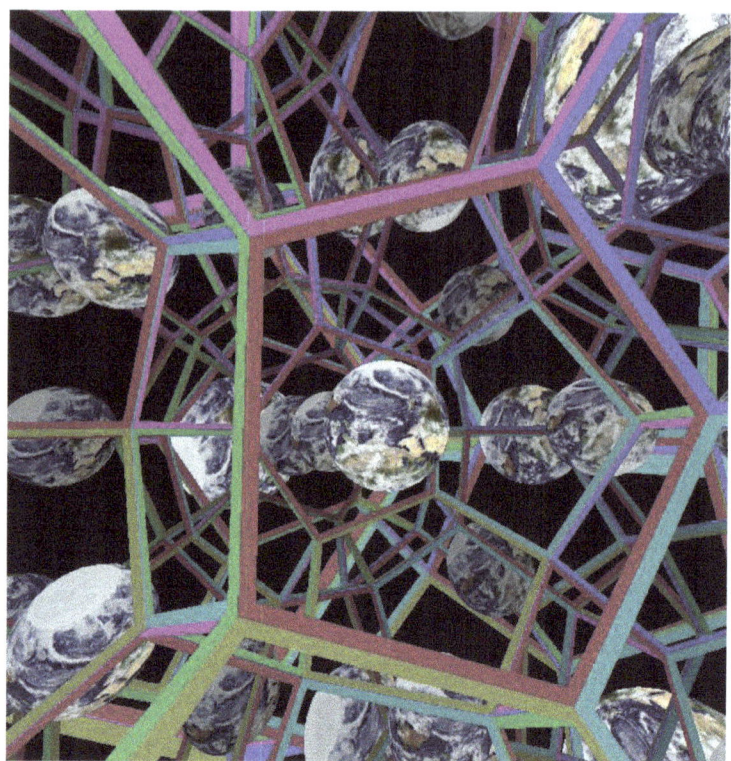

PLATE 2: Some observations at the scale of the universe (10^{23} m) can be interpreted by assuming that the universe is an assembly of dodecahedra, with faces of constant curvature that meet at angles of 120° to form a periodic structure similar to a foam. See FIG. 1.7b, chap. 1. Image obtained by F. Graner and R. Lehoucq by using J. Weeks' *Curved Spaces* software.

PLATE 3: A minimal surface obtained with a soap film. The iridescence produced by the interference between the two interfaces of the film allow its thickness (of the order of microns) to be estimated as it thins under the effect of gravity. See FIG. 2.2b, chap. 2. Photograph courtesy of F. Elias.

PLATE 4: Suspended liquid film over 10 m in height and 2 m in width. The iridescence reveals thickness and velocity fluctuations, enabling the study of hydrodynamic turbulence in two dimensions over large time-scales (the soap film is fed continuously) and length-scales (even up to 20 m in height). See FIG. 2.33, chap. 2. Photograph courtesy of F. Mondot.

PLATE 5: It is not only foams that coarsen. A crystal (composed of monocrystalline domains) evolves in time in a similar way to a dry foam. For example, the grains of a polycrystal of ice, observed here between crossed polarizers, were created as ice accumulated in the Antarctic. Over time, the grains coarsen and deform. Image: 5 × 2.5 cm. See FIG. 3.2, chap. 3. Image courtesy of G. Durand.

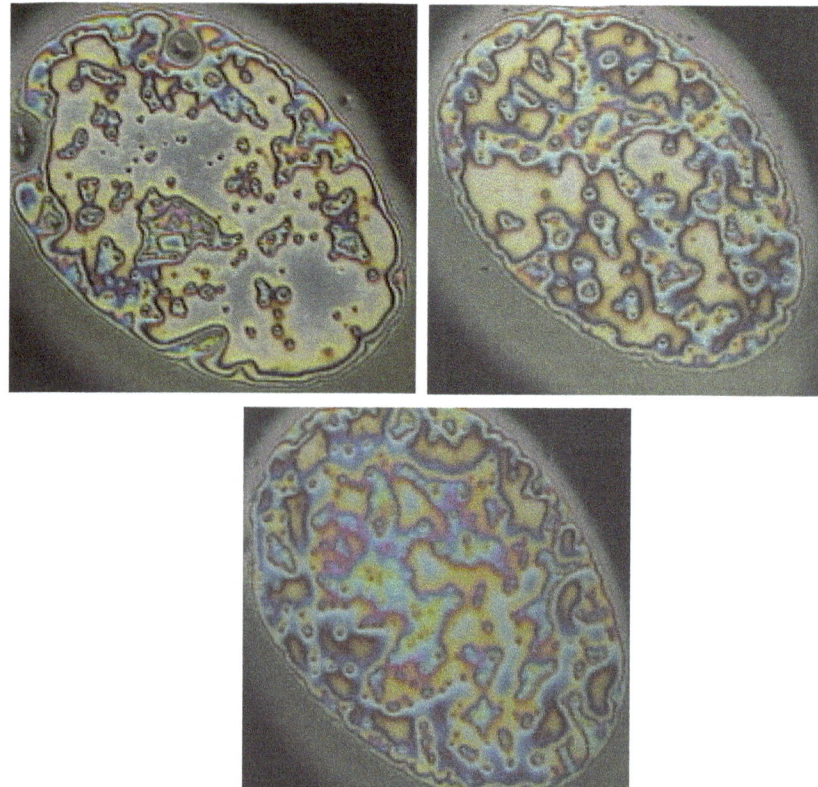

PLATE 6: Inhomogeneous films stabilized by milk proteins, obtained using the film balance technique (see p. 229), showing the texture of the liquid film between two bubbles. The images are for three different protein concentrations. At the lowest concentration (left) the film ruptures quickly and the bubbles coalesce. For the highest concentration (right), the film is much more stable. See FIG. 3.17, chap. 3. Photographs courtesy of A. Saint-Jalmes.

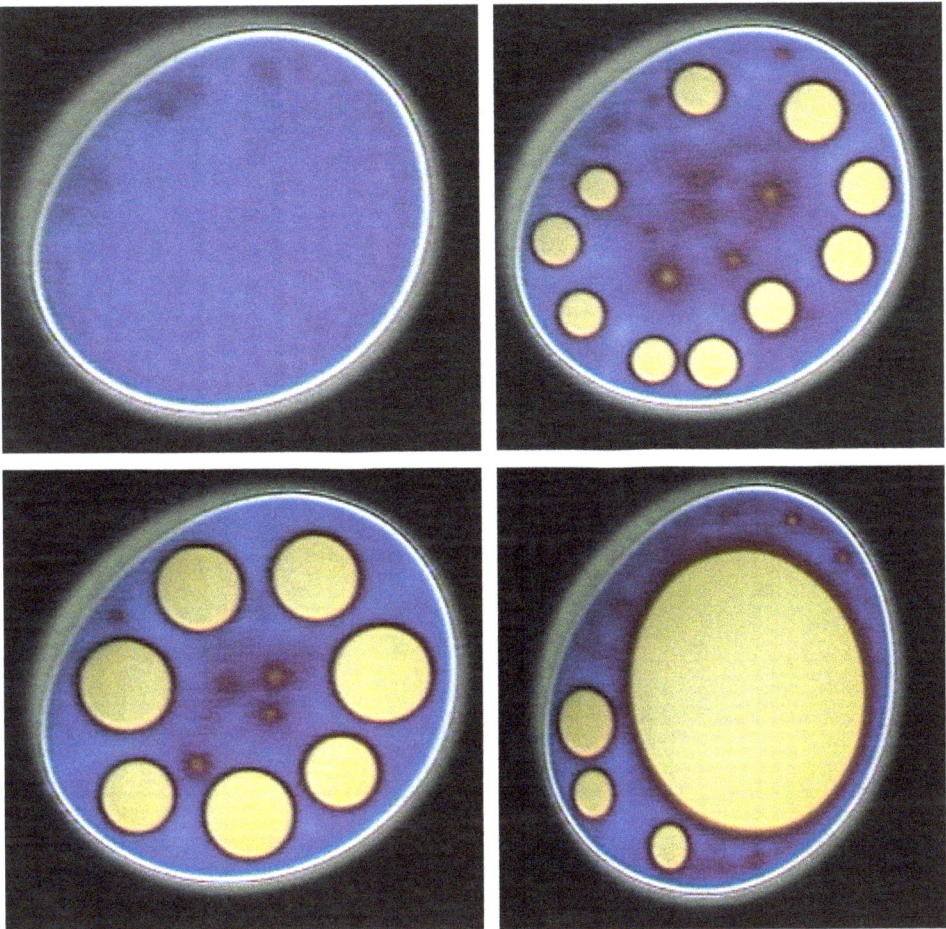

PLATE 7: Images of liquid films stabilized by polymers, obtained using the film balance technique (see p. 229). The images illustrate the dynamic evolution of the film over a period of 30 s (starting at the top left and finishing at the bottom right). The film thins under the effect of capillary suction, induced by the Plateau border which surrounds it. With these polymers the film is stratified in thickness and cannot therefore thin continuously: the thickness jumps from 220 nm (blue) to 130 nm (yellow). See FIG. 3.18, chap. 3. Photographs courtesy of A. Saint-Jalmes.

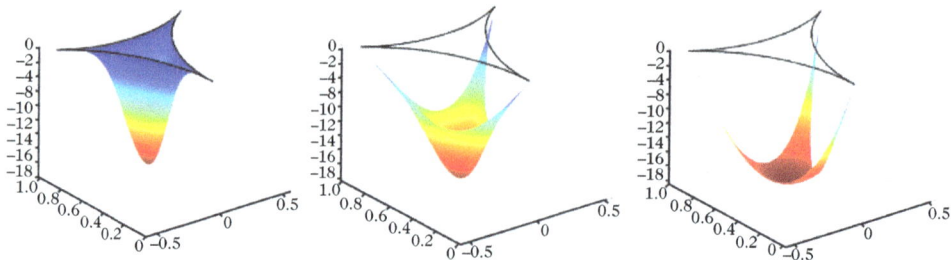

PLATE 8: Numerical simulations of flow through a Plateau border for three interfacial mobilities, characterized by the Boussinesq number Bo. The smaller Bo, and therefore the higher the interfacial mobility, the more the interfaces are entrained by the bulk flow. See FIG. 3.33, chap. 3. Simulations by W. Drenckhan.

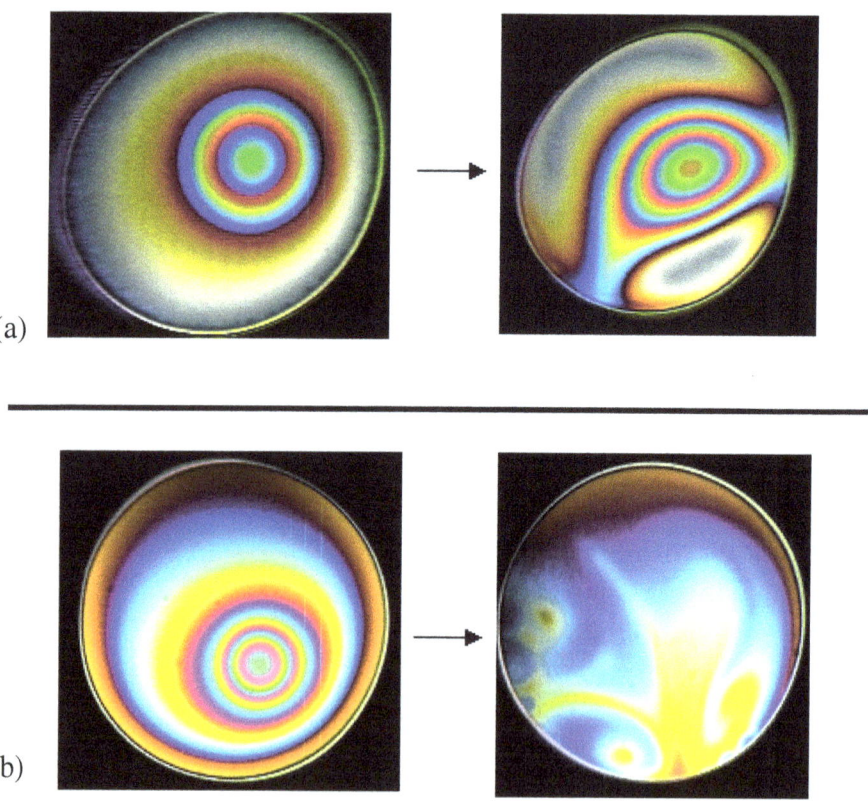

PLATE 9: Two examples of asymmetric drainage of a dimple in a horizontal film with high interfacial mobility, studied with the film balance technique (§1.1.4, chap. 5). (a) A dimple is confined in the film (on the left), then moves and empties into the peripheral Plateau border (on the right). This occurs very rapidly (<1 s) and induces significant flows in the film. (b) Initially we see the dimple in the film, then the flow resulting from the sudden absorption of the dimple into the Plateau border, See FIG. 3.40, chap. 3. Photographs courtesy of A. Saint-Jalmes.

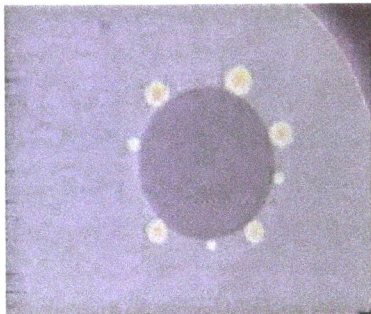

PLATE 10: The thickness of a stratified film (FIG. 3.18) decreases discontinuously. When a thin region develops and grows, its rim may become unstable, forming several large droplets. The film shown consists of a mixture of surfactants and polyelectrolytes (see §6.1, chap. 3), studied with the film balance technique (§1.1.4, chap. 5); its diameter is 0.2 mm and its thickness is 20 nm. See FIG. 3.44, chap. 3. Photograph courtesy of A. Saint-Jalmes.

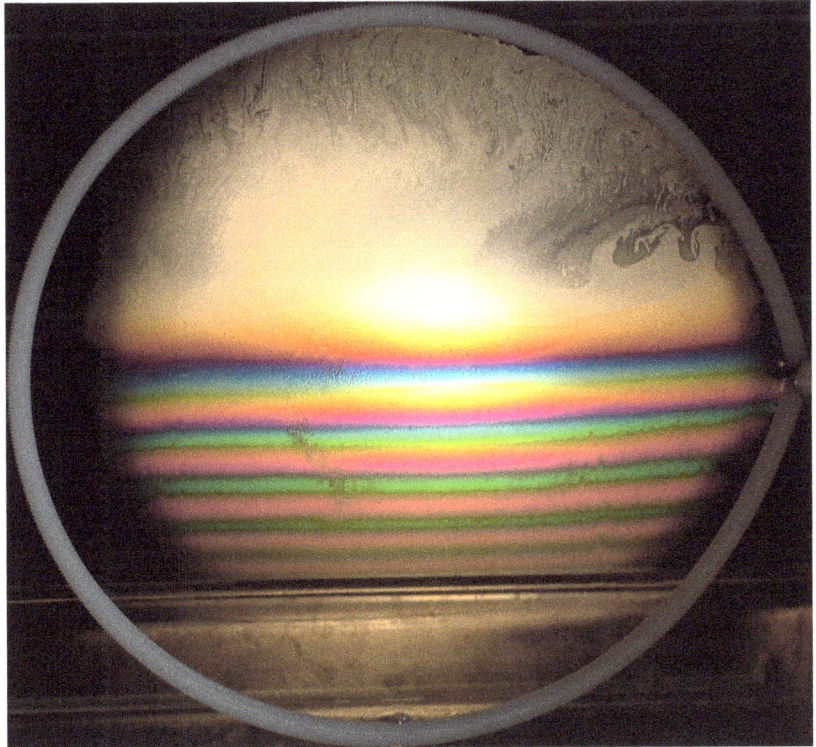

PLATE 11: Soap film suspended on a frame. The iridescence is due to interference of the white light reflected by the two faces of the film. Each colour corresponds to a different film thickness, and the thickness gradient, running from the top to the bottom of the film, is due to gravitational drainage of the liquid in the film. The frame has a diameter of around 15 cm. See FIG. 3.55, chap. 3. Photograph courtesy of F. Elias.

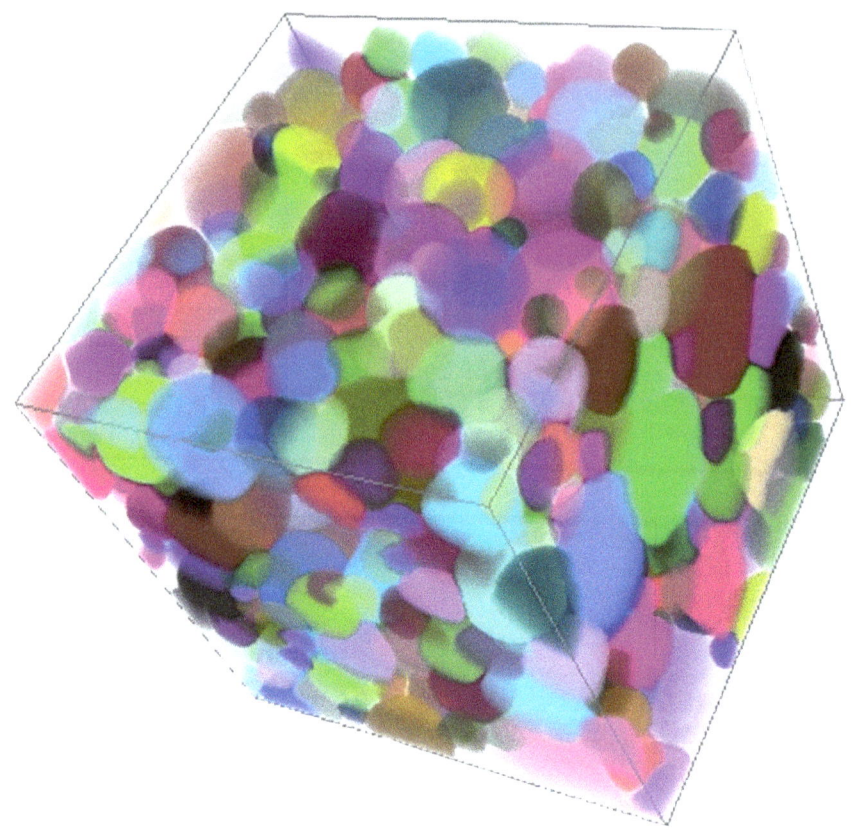

PLATE 12: Reconstruction of the 3D structure of a foam using X-ray tomography. Each bubble is identified and represented in a different colour. The liquid fraction is 15%. See FIG. 5.8, chap. 5. Image courtesy of J. Lambert.

surfactant alone. Due to the mixing at the interfaces, the films obtained are thicker than if they were stabilized by surfactants alone.

(c) In certain cases there are strong attractive interactions between the hydrophobic parts of the polymers and the surfactants, known as *hydrophobic interactions*. The surfactants adsorb onto the polymers, and therefore cover the surface less thoroughly, leading to a reduction in foamability.

(d) Finally, if there is no interaction between the components, the interface is partitioned between them.

6.1.4 Ferrofluids

Overview and physical principles In order to stabilize a foam, as we saw in §2, it is sufficient to add surfactant molecules to the liquid phase. However, another way of stabilizing a foam without adding surfactant is to create repulsive interactions in the liquid phase.

One example is a Langmuir foam (FIG. 2.24a), but it is also possible to produce such foams with a *ferrofluid*, consisting of a colloidal suspension of magnetic particles in a liquid carrier. The particles are 10 nanometres in diameter, and are made of a ferrous or ferromagnetic material (generally maghemite or cobalt ferrite) with a permanent magnetic moment. They consequently act as small nanoscopic magnets in suspension. Under the effect of an external magnetic field of increasing intensity, the magnetic moments gradually orientate themselves in the direction of the applied field, whereas thermal agitation tends to disorientate them. As for a permanent magnet, the interaction between particles depends on their relative position: they attract if the particles are arranged head-to-tail, and repel if they are side-by-side. In order to create a fluid which, when mixed with a second fluid, allows interfaces to form, the interactions must on average be repulsive, so there must be more particles side-by-side than head-to-tail.

Experimental system To create a 2D system, the ferrofluid is confined in a cell consisting of two horizontal plates about 1 mm apart, and the magnetic field is oriented vertically [25], a configuration in which more magnetic particles are side-by-side than head-to-tail. In practice, two immiscible liquids are used: an aqueous ferrofluid and a non-magnetic oil.[4] The oil wets the plates better than the ferrofluid, so that a microscopic film of oil separates the ferrofluid from the plates.

To create a foam, we start with the ferrofluid in a large drop in the middle of the cell and zero applied field. The system is then subjected to an alternating magnetic field, which creates holes in the drop. These holes fill with the oil to make a 2D magnetic cellular structure, as shown in FIG. 3.52. A foam created in this way exhibits many of the characteristics associated with a 2D foam: it respects Plateau's laws (§2.2, chap. 2), the Aboav–Weaire correlation (eq. (5.13)), and Euler's formula (eq. (2.9), so on average its bubbles have six sides) [25]. Moreover, since the oil can move between the

---

[4]Such a mixture of two liquids is in fact an emulsion; here it is called a "ferrofluid foam" because the pattern looks so foam-like.

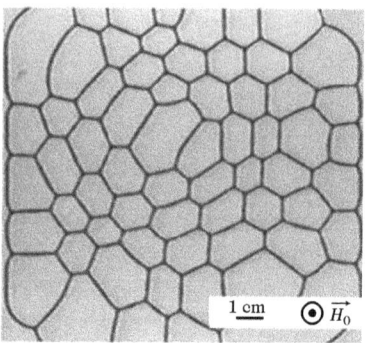

FIG. 3.52: In this 2D ferrofluid foam the bubble walls are made of a ferrofluid (in black) and the bubbles are filled with an immiscible, transparent, and non-magnetic oil which wets the horizontal (upper and lower) plates. The foam is subjected to an external vertical magnetic field $\vec{H}_0$. Photograph courtesy of F. Elias.

bubbles, by flowing through the wetting film, the bubble area can evolve in time, for example due to pressure differences.

Equilibrium situation Unlike soap foams, which are metastable, a ferrofluid foam can exist in equilibrium and therefore be truly stable. In fact, the van der Waals interactions, which are attractive, tend to minimize the surface area of the oil/ferrofluid interface, while the magnetic dipolar interactions, which are repulsive, tend to increase it. So at equilibrium the total surface area of the interface has a value fixed by the competition between these opposing forces. In contrast to a drop of a simple fluid, the surface area is not minimal; instead, the creation of a pattern allows the interfacial area to increase and reach equilibrium [3]. Depending on the way in which the system is prepared, we may obtain labyrinthine patterns, a network of ferrofluid drops, or a cellular structure.

In the latter case, which concerns us here, if the energy of the magnetic dipolar interaction increases, so does the interfacial area and the number of bubbles at equilibrium. Since the ferrofluid is paramagnetic, the magnetic interaction energy is proportional to the square of the amplitude H_0 of the magnetic field, so the number of bubbles at equilibrium increases with H_0. Conversely, the number of bubbles decreases if H_0 is decreased, and the foam is completely destroyed when the field is removed. The increase in the size of the bubbles when H_0 is decreased recalls the coarsening of a soap foam. The amplitude of the applied magnetic field is consequently the *control parameter* for a ferrofluid foam, since it fixes all of its properties at equilibrium.

Out-of-equilibrium situation When H_0 is increased in a given foam configuration, a certain volume of ferrofluid is required at a vertex to allow a new bubble to appear and a new equilibrium state to be found. However this volume of fluid is not always available, especially when the foam is dry. In this case, equilibrium will not be reached and the foam will evolve in time, with some bubbles growing and others shrinking under the effect of the pressure differences between them.

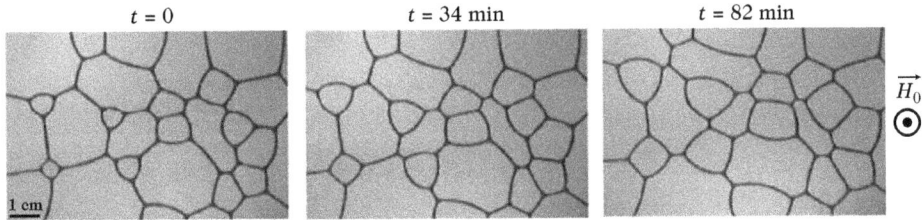

FIG. 3.53: 2D ferrofluid foam in an out-of-equilibrium situation. Observe that bubbles with less than 6 sides grow in time while bubbles with more than 6 sides shrink, in contrast to a coarsening 2D soap foam governed by von Neumann's law. Photograph courtesy of F. Elias [24].

However, unlike in a soap foam, a ferrofluid foam needs *more* interface to reach its equilibrium state, and so the temporal evolution ("coarsening") of cell areas follows an inverted von Neumann's law (§3.1.1): bubbles with less than 6 sides grow at the expense of bubbles with more than 6 sides (and on average the area of a 6 sided bubble doesn't change) [24]. This process, shown in FIG. 3.53, occurs because magnetic dipolar repulsion leads to the existence of a macroscopic *magnetic curvature energy*, which manifests itself as an additional pressure difference between the bubble and its environment. This has the opposite sign to the radius of curvature, making a negative contribution to the Laplace pressure difference.

Note that the interfaces of a soap foam also possess a curvature energy, due to the electrical dipolar moments carried by the surfactant molecules, but the induced pressure difference is negligible compared to the Laplace pressure. For ferrofluid foams this curvature term dominates, and consequently reverses, the evolution of cell area as a function of the number of sides. If the magnetic field increases further, the walls between bubbles sometimes buckle, analogous to the response of a solid foam subjected to compression [23].

One advantage of ferrofluid foams, besides the fact that they possess a control parameter, is the possibility to *locally manipulate* them. It is possible to use a local magnetic field gradient (for example, a magnetic needle pushed into the field H_0) to attract individual vertices, and so to artificially induce T1s [26] (see shaded section on p. 80), or to induce *anti-T2s*, i.e. to create a new bubble (as in FIG. 3.7 viewed from right to left) by gathering sufficient ferrofluid at a vertex.

Finally, note that it is possible to use a magnetic field or force to manipulate a magnetic soap foam created from an aqueous ferrofluid containing surfactant molecules. It is necessary to carefully determine the amount of surfactant to be introduced into the ferrofluid to avoid destabilizing the colloidal suspension (i.e. to prevent the magnetic particles aggregating and precipitating at the bottom of the solution) [15, 22].

6.2 Dissipation due to surfactant motion during the steady expansion of a film

When a soap film of area S is stretched at a constant expansion rate $\dot{\varepsilon} = d(\ln S)/dt$, the surface concentration of surfactant decreases and concentration gradients appear.

Surfactants then move and repopulate the interfaces. In a thick film the bulk solution can act as a reservoir of surfactant and this replenishment can occur in a direction perpendicular to the interface (see FIG. 3.12). However, if a film is very thin, surfactant must migrate from the edges of the film (Plateau borders), which also induces dissipation, the nature of which depends on the properties of the surfactant, and several different modes of transport are possible, as illustrated in FIG. 3.54.

Such surfactant-specific dissipation is in addition to the hydrodynamic dissipation due to the velocity gradients in the bulk of the fluid, associated with the bulk viscosity η (§3.3, chap. 4). Buzza, Lu, and Cates [6] predict the contribution of each mode to the dissipated power $\mathcal{D}$ per unit volume of foam (in W.m^{-3}). In every case, $\mathcal{D}$ is proportional to $\dot{\varepsilon}^2$ (at least in a linear approximation), and the prefactor has the dimensions of a viscosity. Calculating this effective viscosity, or equivalently $\mathcal{D}$, thus provides a convenient way to compare the contributions of different modes, even if we estimate only orders of magnitude.

In the following we estimate for each mode the dissipation in the film per unit surface area. Dividing it by the characteristic size of the bubbles, d (as in chap. 4, § 3.3.1 and 3.3.2), gives the dissipation per unit volume of the foam. Here d is also the characteristic size (not thickness) of the film, and thus sets the scale of gradients in surfactant concentration parallel to the plane of the film. In what follows, η_d, Γ, and j_s are surface quantities, with dimensions of viscosity times length, concentration times length, and flux times length, respectively.

6.2.1 Surface expansion

The first contribution (FIG. 3.54a) is due to the interfacial dilatational viscosity η_d associated with the expansion and compression of the monolayer. This mode dominates when the surface moves together with the bulk solution, and the surface surfactant concentration Γ is not conserved: it is observed over very short time-scales or for surfactants which are insoluble. In this case the dissipated power is

$$\mathcal{D}_{\text{dilat. visc.}} \sim \frac{\eta_d}{d}\dot{\varepsilon}^2. \tag{3.105}$$

6.2.2 Surface diffusion

In the other limit, the surface concentration is kept constant (at a value lower than its equilibrium value) by the flux of surfactants parallel to the plane of the film. Film stretching then requires the transport of $\dot{S}/(Sd_m^2) \sim \dot{\varepsilon}/d_m^2$ surfactants per unit area and time, with d_m^2 the area occupied by one molecule at the surface.

We assume that there is no exchange with the bulk, and that surfactants migrate along the interface with a speed limited by friction with the subsurface layer (FIG. 3.54b). The flux of surfactants in the surface is $\mathbf{j}_s = -D_s \nabla \Gamma$, where D_s is the surface diffusion coefficient of the surfactants. Its 2D divergence is of order $j_s/d \sim \dot{\varepsilon}/d_m^2$. The surfactant chemical potential is $\mu = \mu_0 + k_B T \ln \Gamma$, where k_B is Boltzmann's constant and T the temperature. The associated dissipation per unit area of film is of order $\mathbf{j}_s \cdot \nabla \mu \sim k_B T (\Gamma D_s)^{-1} \mathbf{j}_s^2$ and so the dissipation per unit volume of foam is

$$\mathcal{D}_{\text{surf. diff.}} \sim \frac{k_B T}{\Gamma D_s} \frac{d}{d_m^4} \dot{\varepsilon}^2. \tag{3.106}$$

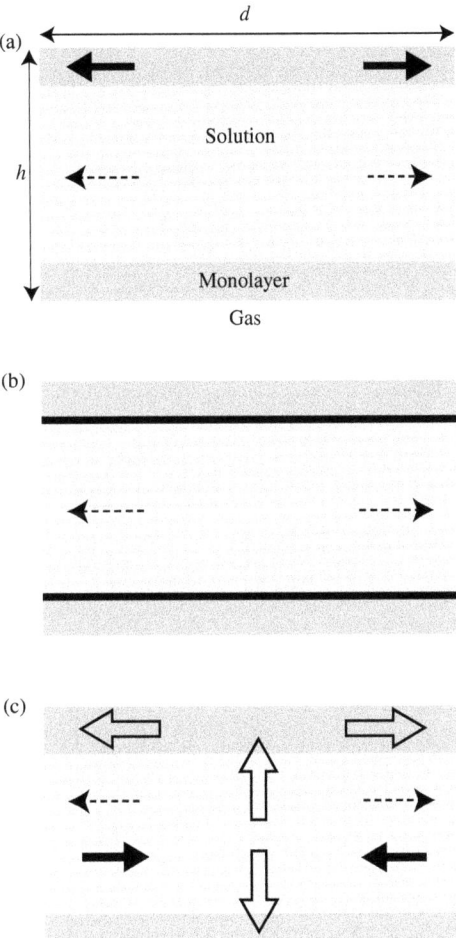

FIG. 3.54: Stretching a film induces both a flow in the liquid (dashed arrows) and motion of surfactants to replenish the newly created interface. Dissipation can arise from three contributions, shown schematically here on a cross-section through a film. (a) Surface expansion: flow of surfactants within the interface (solid arrows). (b) Surface diffusion: diffusion of surfactants within the interface (the most dissipative regions are indicated by the thick solid lines). (c) Bulk diffusion: diffusion of surfactants within the solution (solid arrows), followed by the less-dissipative migration of surfactant towards and along the interface (open arrows).

6.2.3 Bulk diffusion in the plane of the film

When surfactant movement towards and along the interface is rapid enough, the dominant dissipation is due to bulk diffusion in the plane of the film (FIG. 3.54c). The average flux across a section of the film is $\mathbf{j} = D_v \nabla c$, where D_v is the bulk diffusion coefficient and c is the bulk concentration. Typically $j \sim \dot{\varepsilon} d/(h\, d_m^2)$ and the dissipation per unit volume of film is of order $\mathbf{j} \cdot \nabla \mu$, where μ is as above. Multiplying this dissipation by h/d yields the dissipation per unit volume of foam:

$$\mathcal{D}_{\text{bulk diff.}} \sim \frac{k_B T}{c D_v} \frac{d}{h d_m^4} \dot{\varepsilon}^2. \tag{3.107}$$

7 Experiments

7.1 Flow in a soap film

difficulty level: ☻☻	cost of materials: $$
preparation time: 1 hr 30	experimental time: 10 min

<u>Phenomenon demonstrated</u>: *film drainage.*

<u>References in the text</u>: §2.3, §5.1.1 and §1.1.3, chap. 5.

<u>Materials</u>:

- a circular or rectangular frame, 5 to 20 cm in size, attached to a piece of string. Ideally, the frame should be made from a rod with a screw thread, which creates a liquid reservoir and means that the soap film lasts longer than if the frame is smooth
- a bowl deep enough to contain the frame
- a large cardboard box at least twice the size of the bowl
- black paint or black paper
- a retort stand that will fit in the box
- a desk lamp (a source of white light),
- soap solution: 1% washing-up liquid in deionized water
- optional: glycerine.

1. Cut out one of the sides of the cardboard box. Paint the inside black a day in advance, or cover the inside of the box with black paper. Place the box on a table with its opening towards you. The inside of the box will be the stage for the experiment, protected from air currents and external light.
2. Place the bowl containing the soap solution inside the box. Hang the frame from the stand, just above the bowl, as close as possible to the back of the box. The plane of the frame should be facing you, parallel to the opening of the box (FIG. 3.55, left).
3. Place the lamp outside the box, but directed towards the box, with an angle of incidence of around 45 degrees (FIG. 3.55, left).
4. Create a soap film by dipping the frame into the soapy solution, preferably by moving the bowl upwards so as not to move the frame.

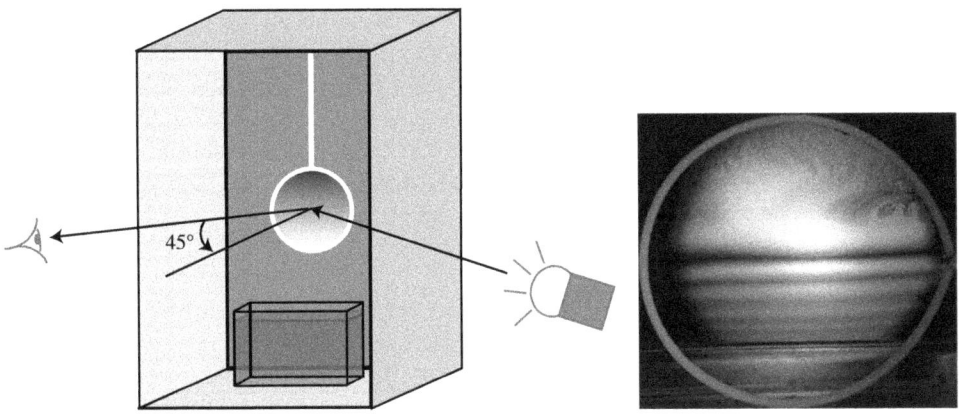

FIG. 3.55: Sketch of the liquid-flow experiment in a vertical soap film (left), and a photograph of a film (right) in a frame of diameter 15 cm. Photograph courtesy of F. Elias. See PLATE 11.

5. Adjust the position of the lamp until you can see the reflection of the bulb on the soap film (the angle of observation should be equal to the angle of incidence). It is possible to place a diffusing screen (for example a sheet of white paper) between the lamp and the frame to increase the illuminated area.

The soap film is at first white because its thickness, which is about ten microns, is greater than the coherence length of white light. After several seconds horizontal bands of colour appear at the top of the film and move downwards (FIG. 3.55, right). These iridescent bands of interference are oriented perpendicular to the thickness gradient. Their colour is determined by the film thickness, between about one hundred nanometres and several microns, and their downwards movement is due to the thinning of the film over time. If the film is sufficiently stable, you will see the appearance of the black film above the coloured bands. It is generally at this point that the soap film bursts.

By carefully observing the soap film close to the edges of the frame during drainage, you should be able to see upward motion. This is *marginal regeneration* (see §5.1.1 and FIG. 3.41).

Comments:

1. Glycerine can be added to the soap solution (around 10%) to increase the life of the soap films and slow down drainage. Instead of using glycerine, the film can be "fed" with soap solution (see experiment 6.3, chap. 2).
2. The larger the frame, the more spectacular the experiment (see FIG. 2.33), but the less stable the film.
3. By illuminating the soap film with a monochromatic light source (sodium vapour or laser) it is possible to precisely measure the thickness and its variations.

156 *Birth, life, and death*

4. If the soap film is created rapidly and protected from air currents, the drainage dynamics will be more easily reproducible.
5. Depending on the washing-up liquid used, and its concentration, it is possible to observe inhomogeneities in film thickness. This is due to micellar phases, or the organization of physico-chemical complexes added to the surfactant molecules by the manufacturer (see FIGS. 3.18 and 3.51). In the first case, you can get rid of them by further diluting the solution. In the second case, you need to change the washing-up liquid!

7.2 Free drainage in a foam and the vertical motion of bubbles

difficulty level: ✹✹	cost of materials: $$
preparation time: 1 hr	experimental time: 1 hr

<u>Phenomenon demonstrated</u>: *Correspondence between downwards motion of the liquid and upward motion of the bubbles during free drainage.*

<u>References in the text</u>: §4.2, §4.7.2

<u>Materials</u>:

- a transparent, plastic cylinder, preferably with a square cross-section, with a height of about 40 cm and a lid
- small pieces of laminated cardboard, about 1 cm in size
- soap solution: 1% washing-up liquid in deionized water
- a graduated ruler of the same height as the column
- adhesive tape
- a stopwatch.

1. Preparation of the cylinder: pierce a slit every 5 cm up one of the sides, in order to be able to insert the cardboard markers in the foam. Fix the ruler to the same side of the column.
2. Preparation of the foam: cover the slits with adhesive tape, pour all the solution into the cylinder, and shake hard to produce a foam.
3. Uncover the slits and insert the pieces of cardboard as positional markers (FIG. 3.56).[5]
4. Measure the height $L(t)$ of drained liquid at the bottom of the column (FIG. 3.56) as a function of time.
5. Measure the vertical movement of bubbles by tracking the position of the markers relative to the slit as a function of time.

Under the effect of gravity, the liquid drains towards the bottom of the foam and the much lighter gas bubbles rise. Using only a simple argument based on volume conservation (the equilibrium between liquid movement downwards and bubble movement

---

[5] As a precaution, the adhesive tape can be put back over the slits, although the foam is unlikely to flow out.

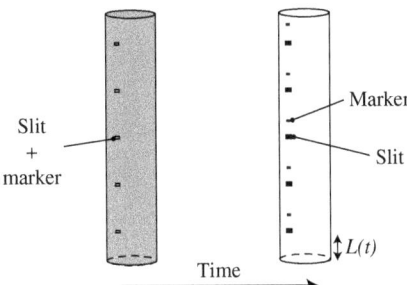

FIG. 3.56: The free drainage experiment.

upwards), the movement of the markers allows the evolution of the liquid fraction at different heights to be determined [52].

7.3 Forced drainage in a foam: observation of the wetting front

difficulty level: ✱✱	cost of material: $$
preparation time: 1 hr	experimental time: 10 min

<u>Phenomenon demonstrated</u>: *forced drainage*

<u>References in the text</u>: §4.3, §4.7.1

<u>Materials</u>:

- a transparent cylinder of 2 to 4 cm in diameter, at least 15 cm high
- a retort stand to keep the tube vertical
- a large bowl
- 1 m of flexible rubber tubing
- a porous frit (an aquarium oxygenator for example) or a hypodermic needle
- an air pump
- a liquid pump
- two clamps to hold the rubber tubing
- soap solution: 1% washing-up liquid in deionized water.

1. Fill the bowl with soapy water. Fix the tube vertically and place its lower end in the liquid (FIG. 3.57).
2. Cut 30 cm of flexible tubing. Connect the air pump to one end of the tubing and the porous medium or needle to the other end. Place the latter in the soapy water, beneath the tube.
3. Fill the cylinder with foam by injecting air through the frit or needle, at constant flow rate (to have bubbles of the same size). If the flow rate is too great it can be reduced by pinching the rubber tubing with a clamp.
4. Once the cylinder is full, wait a few minutes for the foam to drain.
5. Connect some tubing to the water pump to take soapy water from the bowl.

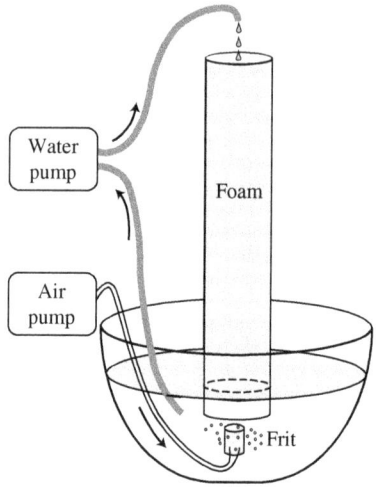

FIG. 3.57: The forced drainage experiment.

6. Set the liquid flow to about 10 mL.min^{-1}, pinching the tube with the second clamp if necessary.
7. Quickly place the free end of the tube at the top of the foam and keep it in this position while the flow of soapy water continues.

A wet foam front should form at the top of the foam, and moves downwards at constant speed. You have successfully carried out a forced drainage experiment.

Comments:

1. Flexible tubing, forceps, air, and liquid pumps and aerators can all be found in shops selling aquarium products.
2. By illuminating the column of foam homogeneously and filming the light transmitted through the foam, you can measure optically the speed of progress of the wetting front u_{front} (see FIGS. 3.30 and 3.36). Next, redo the experiment with a different liquid flow rate Q. Provided that the foam is very dry at the start of the experiment (which might require a long waiting time, depending on the properties of the foam), you can then obtain a measure of the liquid fraction in the foam ($\phi_l = Q/(S u_{\text{front}})$, where S is the cross-sectional area of the cylinder), and you can test the theory of foam drainage (§4.7.1).

7.4 Life and death of a foam measured by electrical conductivity

difficulty level: ◉◉◯	cost of materials: $$$
preparation time: 4 hrs	experimental time: 4 hrs

Phenomena demonstrated: *free drainage, coarsening, coalescence, electrical conductivity of a foam.*

References in the text: §3, §4 and §5, and §1.2.6, chap. 5.

Materials:

- materials necessary to fill a cylinder with foam, as in experiment 7.3
- a lid for the foam column
- 2 brass screws of the same diameter (about 5 mm)
- glue (preferably slow-setting Araldite)
- 2 crocodile clips
- a resistor with resistance close to 1 kΩ
- 5 cables with banana plugs
- 1 coaxial cable (BNC)
- 2 banana/BNC adapters (female/female)
- 1 male BNC T junction
- a low-frequency generator
- an oscilloscope
- optional: a computer to plot and analyse the results.

1. The day before the experiment, pierce two holes opposite each other in the cylinder about two thirds of the way up. Insert a brass screw in each hole, so that the end of the screw is flush with the inside wall of the column. Glue the screws in this position: these will be the two electrodes.
2. Place the column vertically so that it is ready to be filled with foam (see experiment 7.3). The electrodes must be in the upper part of the tube.
3. Set up the electrical wiring (see FIG. 3.58) by connecting the generator in series with the foam and the resistor. Connect the oscilloscope to the terminals of the resistor (channel 2). In channel 1 of the oscilloscope, directly record the signal emitted by the generator.
4. To avoid electrolysis of the foam at the terminals, use an alternating current. Set the generator to pulse at a frequency of 1 kHz and amplitude of around 5 V.
5. Fill the cylinder with foam, adjust the oscilloscope over a long time base, and watch the amplitude of the signal change on channel 2 of the oscilloscope.
6. Stop the bubbling and place the lid on top of the foam column.

Plot the signal over a long time (at least one hour) after the bubbling has stopped. Over time the liquid drains out of the foam, so its conductivity decreases (electrical resistance increases) and so does the amplitude of the signal being measured. In fact, as shown on the electrical diagram in FIG. 3.58, the foam not only has an electrical resistance but also a capacitance, which depends on the amount of liquid in the foam. The complex impedance Z of the foam is thus $Z = |Z|e^{i\varphi}$, where $|Z| = R/\sqrt{1 + (RC\omega^2)}$ and $\varphi = \text{arctanh}\,(RC\omega)$. The capacitive effects can therefore be ignored at low frequency and the resistance R of the foam is

$$R = \frac{(U_0 - U_{\text{meas}})R_{\text{meas}}}{U_{\text{meas}}} \tag{3.108}$$

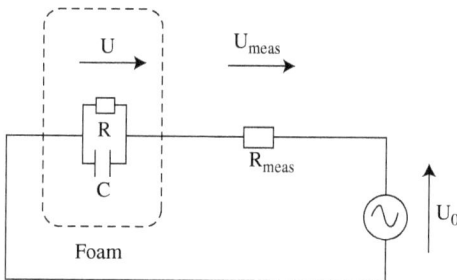

FIG. 3.58: Electrical set-up for measuring the electrical conductivity of a foam. The generator provides a voltage U_0, and the foam has a resistance R and a capacitance C in parallel. In channel 2 of the oscilloscope the voltage U_{meas} is measured at the terminals of the measurement resistor (shunt) with resistance R_{meas}. At low frequency, U_{meas} doesn't depend on C, enabling the measurement of R. The conductivity of the foam is equal to the distance between the electrodes in contact with the foam, normalized by their cross-sectional area, divided by the resistance R. In practice a calibration run is performed with the foaming solution so that it is not required to know the geometrical configuration of the electrodes precisely.

If you are using a computer, plot the amplitude of the signal as a function of time on log-log scales. You should observe several different regimes (i.e. straight line segments with different gradients). At short times (typically one minute) the signal remains constant, because the liquid at the top of the column feeds the foam situated at the level of the electrodes. The next regime corresponds to the thinning of the Plateau borders under the effect of drainage. Finally, if the foam is unstable, the last regime will reflect coalescence.

Comments:

1. If the signal-to-noise ratio is too low, you can add a little salt to the foam to increase its conductivity (but not so much that you screen the interactions between surfactant molecules, which would cause the films to rupture), or electrically isolate the electrodes (screws) with thermo-retractable sheaths.
2. Experimentally, you can confirm that capacitive effects are negligible by varying the frequency and confirming that the amplitude of the signal doesn't change.
3. It is possible to link the conductivity to the liquid fraction (see §1.2.6, chap. 5). It is then necessary to measure the conductivity of the soap solution, for example by filling the cylinder with the solution and measuring the signal at the electrodes.
4. In the laboratory, for a more detailed analysis and a less noisy signal, we use a lock-in amplifier instead of an oscilloscope.
5. This experiment enables you to determine the complete drainage dynamics by installing pairs of electrodes along the cylinder, and scanning the signal at each pair of electrodes. It is then necessary to completely automate the experiment, as the scanning time must be short compared to the evolution time of the foam.

8 Exercises

Solutions to the exercises are available on request through the following webpage: http://www.oup.co.uk/academic/physics/admin/solutions.

8.1 Exponent in the scale-invariant regime

Consider a 2D foam.

(a) Integrate von Neumann's law, eq. (3.26), to give an expression for the characteristic time over which an individual bubble i, with n_i sides, grows or shrinks (i.e. the 2D analogue of eq. (3.34)).

(b) Give an expression for the rate dN/dt at which bubbles with a positive topological charge disappear.

(c) Assume that the foam has coarsened enough to reach the self-similar growth regime, in which the various statistical measures of the foam no longer change in time (see p. 105). Demonstrate that:

$$\text{2D:} \quad \frac{dN}{dt} \propto -N^2 \tag{3.109}$$

where the proportionality factor is a constant.

(d) Show that eq. (3.109) is equivalent to $N \propto t^{-1}$. Show that the average bubble area is proportional to t and so the average radius is proportional to $t^{1/2}$.

NB: the 3D case (eq. (3.40)), not described here, is similar but the formulae are more involved. We refer instead to the approximate equations (3.37) and (3.39).

8.2 Frumkin equation of state

The Langmuir model describes the dynamics of adsorption of surfactant molecules at the surface of a solution in the ideal case where molecules don't interact with each other (see eq. (3.3)). Frumkin corrected this model to account for non-negligible interactions between molecules and when adsorption (or desorption) of a molecule at the surface is associated with an adsorption (or desorption) energy E_a (or E_b, respectively). The coefficients a and b in eq. (3.7) are

$$a = a_0 \exp\left(-\frac{E_a^0 + E_a \Gamma}{k_B T}\right), \quad b = b_0 \exp\left(-\frac{E_b^0 + E_b \Gamma}{k_B T}\right). \tag{3.110}$$

(a) Establish the relationship between bulk and surface concentrations by introducing the concentration $c_a = (b_0/a_0) \exp[(E_a^0 - E_b^0)/k_B T]$ and the measure of non-ideality $H_F = (E_b - E_a)\Gamma_\infty/2$.

(b) Use the Gibbs equation (3.2) to establish the relationship between the surface pressure and the surface concentration.

8.3 Foam drainage and equilibrium height

Consider a foam with initially uniform liquid fraction ϕ_{l0}, height Z, constant bubble size d, and constant cross-section S.

(a) Calculate the initial volume of liquid.
(b) Calculate the volume at equilibrium.
(c) For what initial foam height are the initial and equilibrium liquid volumes equal?
(d) Calculate this foam height with $\phi_{l0} = 0.01$ or 0.1, $d = 0.1$ mm or 10 mm, and a surface tension of 35 mN.m^{-1}.

8.4 Drainage in the bulk and at the wall

Using FIG. 3.34, estimate the proportion of liquid flowing through bulk Plateau borders and through surface Plateau borders in a cylinder of foam with radius a. Use the assumption that in any given cross-section of the cylinder there is, on average, one Plateau border per bubble of diameter d and consider the cases $Bo = 0.1$ and $Bo = 10$.

8.5 Free drainage: characteristic times and liquid fraction profiles

Eq. (3.95) gives the characteristic time τ for free drainage in the Plateau border-dominated case.

(a) Based on the calculation presented in §4.7.2 for a foam of height Z, in particular eq. (3.96), show that in the vertex-dominated case the characteristic time (assuming that the initial liquid fraction is uniform, $\phi_l = \phi_{l0}$) corresponds to the time at which $2/3$ of the total volume of liquid has drained out of the foam.
(b) Show that $\tau = 2\eta Z/(3C_v \rho_l g R_V^2 \phi_{l0}^{1/2})$. It may help to recall that the liquid fraction profile is $\phi_l/\phi_{l0} = A_v(t)(Z-z)^2$, where $A_v(t)$ is a coefficient that depends on time and z is the vertical coordinate, oriented upwards.

8.6 The true 3D pressure and 2D surface pressure

For a monolayer with (2D) surface pressure Π_l, estimate the (3D) pressure within the monolayer (pay attention to units!), and compare it with atmospheric pressure.

References

[1] A.W. ADAMSON and A.P. GAST, *Physical Chemistry of Surfaces*, Wiley, New York, 6th ed. 1997.
[2] A.A. ARADIAN, *Quelques problèmes de dynamique d'interfaces molles*, Thèse de l'Université Paris 6, 2001. http://tel.archives-ouvertes.fr/tel-00001386.
[3] J.-C. BACRI and F. ELIAS, in *Morphogenesis, origins of patterns and shapes*, P. BOURGINE and A. LESNE (eds.), Springer-Verlag, Berlin, 2011.
[4] V. BERGERON, *J. Phys.: Condens. Matter*, **11**, R215, 1999.
[5] A. BHAKTA and E. RUCKENSTEIN, *Adv. Colloid Interface Sci.*, **70**, 1, 1997.
[6] D. BUZZA, C.-Y. LU, and M.E. CATES, *J. Phys. II*, **5**, 37, 1995.
[7] V. CARRIER, S. DESTOUESSE, and A. COLIN, *Phys. Rev. E*, **65**, 061404, 2002.
[8] A. CERVANTES-MARTINEZ, E. RIO, G. DELON, A. SAINT-JALMES, D. LANGEVIN, and B.P. BINKS, *Soft Matter*, **4**, 1531, 2008.

[9] S. COHEN-ADDAD and R. HÖHLER, *Phys. Rev. Lett.*, **86**, 4700, 2001.
[10] S.J. COX, G. BRADLEY, S. HUTZLER, and D. WEAIRE, *J. Phys.: Condens. Matter*, **13**, 4863, 2001.
[11] F.E.C. CULICK, *J. Appl. Phys.*, **31**, 1128, 1960.
[12] K.D. DANOV, P.A KRALCHEVSKI, N.D. DENKOV, K.P. ANANTHAPADMANABHAN, and A. LIPS, *Adv. Colloid Interface Sci.*, **119**, 17, 2006.
[13] P.-G. DE GENNES, F. BROCHARD-WYART, and D. QUÉRÉ, *Capillarity and Wetting Phenomena: Drops, Bubbles, Pearls, Waves*, Springer, New York, 2004.
[14] G. DEBRÉGEAS, P. MARTIN, and F. BROCHARD-WYART, *Phys. Rev. Lett.*, **75**, 3886, 1995.
[15] W. DRENCKHAN, F. ELIAS, S. HUTZLER, D. WEAIRE, E. JANIAUD, and J.-C. BACRI, *J. Appl. Phys.*, **93**, 10078, 2003.
[16] W. DRENCKHAN, H. RITACCO, A. SAINT-JALMES, D. LANGEVIN, P. MCGUINNESS, A. SAUGEY, and D. WEAIRE, *Phys. Fluids*, **19**, 102101, 2007.
[17] Z. DU, M. BILBAO-MONTOYA, B.P. BINKS, E. DICKINSON, R. ETTELAIE, and B.S. MURRAY, *Langmuir*, **19**, 3106, 2003.
[18] J. DUPLAT, B. BOSSA, and E. VILLERMAUX, *J. Fluid Mech.*, **673**, 147, 2011.
[19] G. DURAND, F. GRANER, and J. WEISS, *Europhys. Lett.*, **67**, 1038, 2004.
[20] M. DURAND and D. LANGEVIN, *Eur. Phys. J. E*, **7**, 35, 2002.
[21] D.A. EDWARDS, H. BRENNER, and D.T. WASAN, *Interfacial Transport Processes and Rheology*, Butterworth Heinemann, Stoneham, 1991.
[22] F. ELIAS, J.-C. BACRI, C. FLAMENT, E. JANIAUD, D. TALBOT, W. DRENCKHAN, S. HUTZLER, and D. WEAIRE, *Colloids Surf. A*, **263**, 65, 2005.
[23] F. ELIAS, I. DRIKIS, A. CEBERS, C. FLAMENT, and J.-C. BACRI, *Eur. Phys. J. B*, **3**, 203, 1998.
[24] F. ELIAS, C. FLAMENT, and J.-C. BACRI, *Magnetohydrodynamics*, **35**, 303, 1999.
[25] F. ELIAS, C. FLAMENT, J.-C. BACRI, F. GRANER, and O. CARDOSO, *Phys. Rev. E*, **56**, 3310, 1997.
[26] F. ELIAS, C. FLAMENT, J.A. GLAZIER, F. GRANER, and Y. JIANG, *Phil. Mag. B*, **79**, 729, 1999.
[27] W.E. EWERS and K.L. SUTHERLAND, *Aust. J. Sci. Res.*, **5**, 697, 1952.
[28] D. EXEROWA and P.M. KRUGLYAKOV, *Foam and Foam Films—Theory, Experiment, Application*, Elsevier, Amsterdam, 1998.
[29] I. FORTUNA, G. L. THOMAS, R.M.C. DE ALMEIDA, and F. GRANER, *Phys. Rev. Lett.*, **108**, 248301, 2012.
[30] F. G. GANDOLFO and H. L. ROSANO, *J. Colloid Interface Sci.*, **194**, 31, 1997.
[31] P.R. GARRETT, in *Defoaming: Theory and Industrial Applications*, P.R. Garrett (ed.), Surfactant Science Series, Marcel Dekker, New York, **45**, 1, 1993.
[32] J. GLAZIER, *Phys. Rev. Lett.*, **70**, 2170, 1993.
[33] J.A. GLAZIER, S.P. GROSS, and J. STAVANS, *Phys. Rev. A*, **36**, 306, 1987.
[34] I.I. GOL'DFARB, K.B. KHAN, and I.R. SHREIBER, *Fluid Dyn.*, **23**, 244, 1988.
[35] F. GRANER, B. DOLLET, C. RAUFASTE, and P. MARMOTTANT, *Eur. Phys. J. E*, **25**, 349, 2008.
[36] S. HILGENFELDT, S.A. KOEHLER, and H.A. STONE, *Phys. Rev. Lett.*, **86**, 4704, 2001.

[37] S. HUTZLER, S.J. COX, E. JANIAUD, and D. WEAIRE, *Colloids Surf. A.*, **309**, 33, 2007.
[38] S. HUTZLER and D. WEAIRE, *Phil. Mag. Lett.*, **80**, 419, 2000.
[39] S. HUTZLER, D. WEAIRE, and R. CRAWFORD, *Europhys. Lett.*, **41**, 461, 1998.
[40] I.B. IVANOV, *Thin Liquid Films: Fundamentals and Applications*, Surfactant Sciences Series Vol. 29, CRC Press, Boca Raton, 1988.
[41] Y. JAYALAKSHMI, L. OZANNE, and D. LANGEVIN, *J. Colloid Interface Sci.*, **170**, 358, 1995.
[42] S.A. KOEHLER, S. HILGENFELDT, and H.A. STONE, *J. Colloid Interface Sci.*, **276**, 420, 2004.
[43] S.A. KOEHLER, S. HILGENFELDT, and H.A. STONE, *Langmuir*, **16**, 6327, 2000.
[44] J. LAMBERT, I. CANTAT, R. DELANNAY, R. MOKSO, P. CLOETENS, J.A. GLAZIER, and F. GRANER, *Phys. Rev. Lett.*, **99**, 058304, 2007.
[45] J. LAMBERT, R. MOKSO, I. CANTAT, P. CLOETENS, J.A. GLAZIER, F. GRANER, and R. DELANNAY, *Phys. Rev. Lett.*, **104**, 248304, 2010.
[46] D. LANGEVIN, *Curr. Opin. Colloid Interface Sci.*, **3**, 600, 1998.
[47] R.A. LEONARD and R. LEMLICH, *AIChE J.*, **11**, 18, 1965.
[48] E. LORENCEAU, N. LOUVET, F. ROUYER, and O. PITOIS, *Eur. Phys. J. E*, **28**, 293, 2009.
[49] N. LOUVET, *Etude multiéchelle du transport de particules dans les mousses liquides*, Thèse de l'Université Paris-Est Marne-la-Vallée, 2009. http://tel.archives-ouvertes.fr/tel-00541198
[50] J. LUCASSEN and M. VAN DER TEMPEL, *Chem. Eng. Sci.*, **27**, 1283, 1972.
[51] A.H. MARTIN, K. GROLLE, M.A. BOS, M.A. COHEN-STUART, and T. VAN VLIET, *J. Colloid Interface Sci.*, **254**, 175, 2002.
[52] G. MAURDEV, A. SAINT-JALMES, and D. LANGEVIN, *J. Colloid Interface Sci.*, **300**, 735, 2006.
[53] W.R. MCENTEE and K.J. MYSELS, *J. Phys. Chem.*, **73**, 3018, 1969.
[54] D. MONIN, A. ESPERT, and A. COLIN, *Langmuir*, **16**, 3873, 2000.
[55] W. MÜLLER and J.M. DI MEGLIO, *J. Phys.: Condens. Matter*, **11**, L209, 1999.
[56] W.W. MULLINS, *Acta Metall.*, **37**, 2979, 1989.
[57] W.W. MULLINS, *J. Appl. Phys.*, **59**, 1341, 1986.
[58] K.J. MYSELS, K. SHINODA, and S. FRENKEL, *Soap Films, Studies of their Thinning*, Pergamon Press, London, 1959.
[59] S.J. NEETHLING, H.T. LEE, and J.J. CILLIERS, *J. Phys.: Condens. Matter*, **14**, 331, 2002.
[60] O. PITOIS, C. FRITZ, and M. VIGNES-ADLER, *J. Colloid Interface Sci.*, **282**, 458, 2005.
[61] O. PITOIS, E. LORENCEAU, N. LOUVET, and F. ROUYER, *Langmuir*, **25**, 97, 2009.
[62] O. PITOIS, N. LOUVET, and F. ROUYER, *Eur. Phys. J. E*, **30**, 27, 2009.
[63] H.M. PRINCEN and S.G. MASON, *J. Colloid Interface Sci.*, **20**, 353, 1965.
[64] A. PRINS in *Food Emulsions and Foams*, E. Dickinson (ed.), Royal Society of Chemistry Special Publication, **58**, 30, 1986.
[65] D. REINELT and A. KRAYNIK, *J. Rheol.*, **44**, 453, 2000.
[66] A. SAINT-JALMES, *Soft Matter*, **2**, 836, 2006.

[67] A. SAINT-JALMES and D. LANGEVIN, *J. Phys.: Condens. Matter*, **14**, 9397, 2002.
[68] A. SAINT-JALMES, S. MARZE, H. RITACCO, D. LANGEVIN, S. BAIL, J. DUBAIL, L. GUINGOT, G. ROUX, P. SUNG, and L. TOSINI, *Phys. Rev. Lett.*, **98**, 058303, 2007.
[69] A. SAINT-JALMES, Y. ZHANG, and D. LANGEVIN, *Eur. Phys. J. E*, **15**, 53, 2004.
[70] A. SALONEN, M. IN, J. EMILE, and A. SAINT-JALMES, *Soft Matter*, **6**, 2271, 2010.
[71] G. SINGH, G.J. HIRASAKI, and C.A. MILLER, *J. Colloid Interface Sci.*, **184**, 92, 1996.
[72] J. STAVANS, *Rep. Progr. Phys.*, **56**, 733, 1993.
[73] H.A. STONE, S.A. KOEHLER, S. HILGENFELDT, and M. DURAND, *J. Phys.: Condens. Matter*, **15**, S283, 2003.
[74] C. STUBENRAUCH and R. VON KLITZING, *J. Phys.: Condens. Matter*, **15**, R1197, 2003.
[75] G.L. THOMAS, R.M.C. DE ALMEIDA, and F. GRANER, *Phys. Rev. E*, **74**, 021407, 2006.
[76] N. VANDEWALLE, J.F. LENTZ, S. DORBOLO, and F. BRISBOIS, *Phys. Rev. Lett.*, **86**, 179, 2001; N. VANDEWALLE and J.F. LENTZ, *Phys. Rev. E*, **64**, 021507, 2001; N. VANDEWALLE, H. CAPS, and S. DORBOLO, *Physica A*, **314**, 320, 2002.
[77] M.U. VERA and D.J. DURIAN, *Phys. Rev. Lett.*, **88**, 088304, 2002.
[78] G. VERBIST, D. WEAIRE, and A. KRAYNIK, *J. Phys.: Condens. Matter*, **8**, 3715, 1996.
[79] J. VON NEUMANN, in *Metal Interfaces*, American Society for Metals, Cleveland, 1952, p. 108
[80] A. VRIJ, *Discuss. Faraday Soc.*, **42**, 23, 1966; A. VRIJ and J.TH. OVERBEEK, *J. Am. Chem. Soc.*, **90**, 3074, 1968.
[81] A.F.H. WARD and L.J. TORDAI, *J. Chem. Phys.*, **485**, 63, 1946.
[82] D. WEAIRE, S. HUTZLER, G. VERBIST, and E. PETERS, *Adv. Chem. Phys.*, **102**, 315, 1997.
[83] D. WEAIRE and V. PAGERON, *Phil. Mag. Lett.*, **62**, 417, 1990.
[84] D. WEAIRE and N. RIVIER, *Contemp. Phys.*, **25**, 59, 1984.
[85] A.J. WEBSTER and M.E. CATES, *Langmuir*, **17**, 595, 2001.

4
Rheology

Rheology is the study of the deformation and flow of "complex" fluids which exhibit both liquid and solid behaviour. Such materials are common in our everyday lives (cf. §1, chap. 1). Foams, although mostly air and water, are certainly complex fluids. A small amount of shaving foam or whipped cream doesn't spread out under its own weight, but keeps the shape it is given, yet it can flow if sufficient force is applied. This complex behaviour forms the basis of this chapter.

1 Introduction

The mechanical properties of foams are important in numerous industrial applications (cf. §3, chap. 1). In oil recovery, for example, aqueous foams are used as a lubricant and to keep solid debris in suspension when drilling stops. Moreover, a foam's viscosity decreases as the strain rate increases, which decreases the dissipative losses. In addition, the low density of foam reduces the hydrostatic pressure in the wellbore, a significant advantage for wells that are several kilometres deep.

Aqueous foams typically exhibit a combination of the following three types of behaviour:

- *elastic:* an aqueous foam can deform reversibly; it can store up mechanical energy that it will release on returning to its initial shape.
- *plastic:* beyond a certain point the deformation is irreversible and a foam will acquire and retain a new shape. The amount of energy dissipated depends on how much of the deformation is irreversible, but not on the rate of deformation.
- *viscous:* an aqueous foam can flow like a liquid, dissipating an amount of energy that depends on the deformation rate.

The special rheological properties of a foam are a consequence of their high interfacial area. Due to the many different length- and time-scales, foam rheology is particularly difficult to model. At the macroscopic scale (i.e. large compared to the bubble size), the relationships between stress, deformation, and flow are described with constitutive laws, which take into account the characteristic time-scales of the external forcing (for example the strain rate of a steady flow, or the frequency of an oscillatory shear). However, it is also essential to understand how the behaviour at the (local) scale of the films and bubbles is linked to the macroscopic response. This enables a better description and control of the rheological behaviour of a foam, and the possibility to give to it the characteristics required by a given industrial process [13].

This chapter is divided into three parts. The first (§2) consists of a brief overview of topics in the rheology of complex fluids that are relevant to foams. We will introduce constitutive laws and common rheological tests, and show how to define the strain (especially at large deformations) in a complex fluid and the stress tensor in a mixture of two immiscible fluids (especially water and air).

In the second part (§3) we consider the foam at the film-scale, in order to determine the origin of the local processes that give rise to the macroscopic rheological response. The elasticity of a foam is due to the fact that a small applied stress increases the film area, and thus the surface energy, per unit volume. Beyond the elastic limit, an applied stress induces irreversible bubble rearrangements (T1s) and hence plasticity. We will show how to calculate the change in film area in a dry foam and then we will calculate the elastic modulus and the yield stress, both of which depend on the foam structure. Finally, we will describe the various processes by which a foam loses mechanical energy, whether through hydrodynamic dissipation in the liquid (films, Plateau borders) or dissipation linked to the movement of surfactants at the interfaces.

The third and final part of the chapter (§4) is concerned with the link, as it is currently understood, between mechanisms at the local scale and the macroscopic rheological properties. We will see why a foam is much softer and flows more easily when it is wet. For small slow deformations, the response is dominated by elasticity and the shear stress is linear in the strain. Yet there are also non-linear elastic effects, for example the normal stress difference varies quadratically in the strain, as for an elastic solid. We will discuss the effects of plasticity at large strains and, noting that a foam is elastic over short times and liquid-like over long times, the dependence of the rheological response of a foam on the characteristic time-scale of the perturbation. We will identify different mechanical relaxation processes and show how they are coupled to coarsening. Finally, we will explain why the steady flow of a foam may be inhomogeneous.

2 Overview of the rheological behaviour of complex fluids

We will give a general overview of the rheology of complex fluids, restricting ourselves to those parts which are useful for describing the mechanical behaviour of liquid foams. The reader can find further details in specialist books [41, 43].

2.1 Constitutive laws

In order to model the macroscopic rheological behaviour of a complex fluid consisting of a large number of bubbles, it is necessary to treat the fluid at a length-scale sufficiently large that the details of the microstructure can be ignored. That is, we must consider the foam sample as a continuous medium with a constitutive law that relates stress, strain, and strain rate. Other parameters may appear in this relationship, for example temperature, liquid fraction, etc. We start by considering shear stresses before giving general expressions for constitutive laws in §2.3 and §2.4.

Simple shear can be described with respect to Cartesian axes x, y, z, as shown in FIG. 4.1. Each point of the cube is displaced in the x direction while its y and z coordinates remain the same. The displacement is linear in y, and we define the shear strain as $\varepsilon = u/d$. The shear stress σ_{xy} is the force per unit surface area which acts

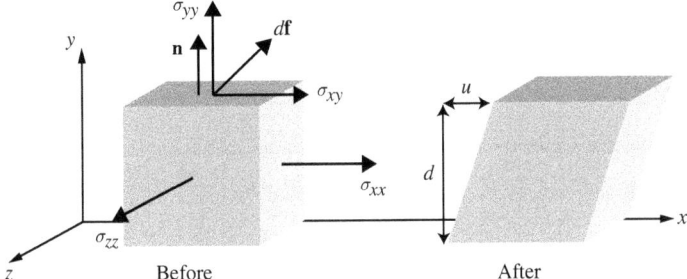

FIG. 4.1: Simple shear of a cube of material in the x direction. The strain is given by $\varepsilon = u/d$. The normal stresses σ_{xx}, σ_{yy}, and σ_{zz} correspond to the forces per unit surface area which act perpendicular to the faces of the cube in each of the three directions. **n** is a unit vector normal to the upper face of the cube, and **df** indicates the force which acts on this face.

tangentially on the top surface of the sample (i.e. the surface with normal in the y direction) in the x direction. The relationship between σ and ε can be linear or non-linear, as we shall see.

Linear response The linear static response of an elastic solid is described by Hooke's law,

$$\sigma = G\,\varepsilon, \tag{4.1}$$

where the constant G is the static shear modulus. G is related to the elastic energy density $\mathcal{E}$ by

$$G = \left.\frac{\partial^2 \mathcal{E}}{\partial \varepsilon^2}\right|_{\varepsilon=0}. \tag{4.2}$$

Elastic behaviour is represented symbolically with an elastic spring (FIG. 4.2).

On the other hand, a liquid is called Newtonian if the shear stress is proportional to the shear rate $\dot{\varepsilon}$, that is

$$\sigma = \eta\,\dot{\varepsilon}, \tag{4.3}$$

where the constant η is the viscosity. A Newtonian liquid is represented as a dashpot (FIG. 4.2). In an elastic solid, an oscillatory stress and the oscillatory strain that it stimulates are in phase, whereas for a liquid we observe a phase shift of $\pi/2$ between them. We can describe the relationship between stress and strain in terms of phase and amplitude using the complex shear modulus G^* (§2.2) whose real and imaginary parts, written G' and G'', describe, respectively, the elastic part and the viscous part of the response of a material.

In general, whether a complex fluid behaves like a solid or a liquid depends on the characteristic time-scale of the external forcing. For example, a Maxwell liquid

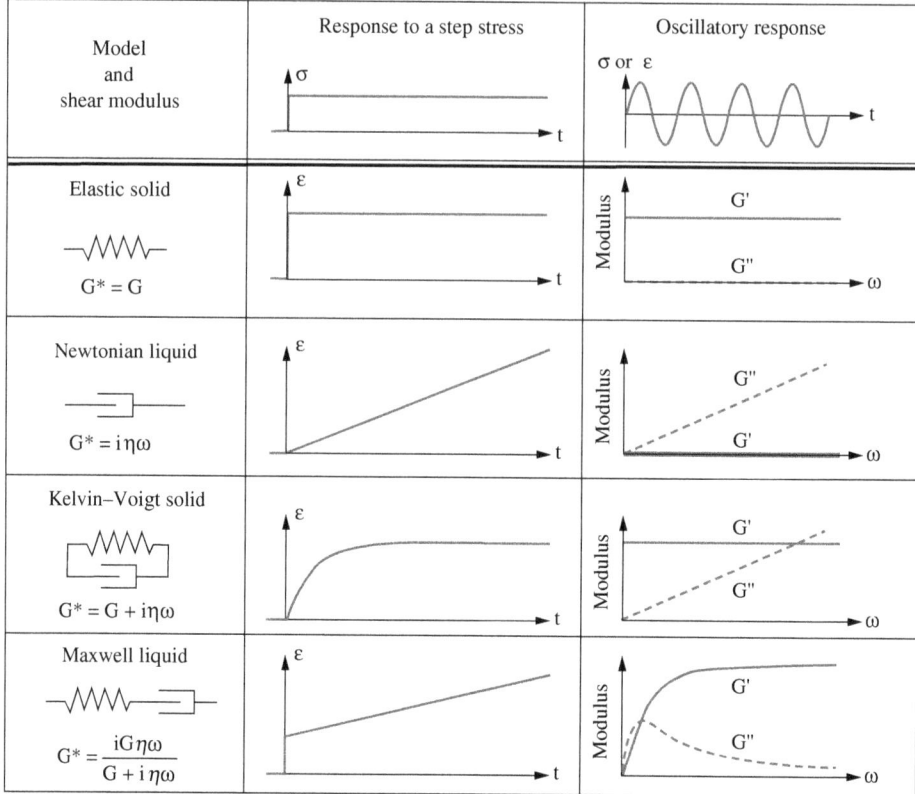

FIG. 4.2: Examples of linear rheological behaviour, showing the time and frequency response of elementary models (see eqs. (4.8) and (4.9)). For the Maxwell liquid, the maximum of G'' corresponds to the frequency $\omega = G/\eta$, with the characteristic relaxation time $\tau = \eta/G$.

is elastic at short times (high frequency or large applied strain rate) and viscous at long times (low frequency or small applied strain rate). On the other hand, a Kelvin–Voigt solid is viscous at short times and elastic at long times. These two material responses are modelled by placing a spring and a dashpot either in series or in parallel (FIG. 4.2).

Non-linear response If we gradually increase the strain applied to a material, the elastic stress becomes a non-linear function of the strain and in the end the material yields. If it is fragile, then it fractures, whereas if it is ductile, it undergoes a permanent *irreversible* deformation ε_p, known as plastic deformation. If the applied strain increases rapidly, there is an additional stress due to viscous friction. Elastic, viscous, and plastic behaviour can be modelled with springs, dashpots, and sliding friction elements.

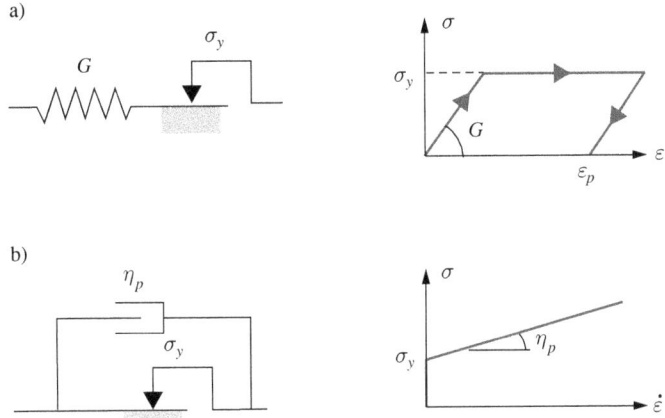

FIG. 4.3: Examples of non-linear rheological behaviour. On the left, the spring represents an elastic response, the sliding element represents solid friction, and the dashpot represents a viscous response. The mechanical response is shown on the right. (a) Response of an elasto-plastic solid to an applied stress that increases to the yield stress and is then reduced back to zero. (b) Representation and flow curve of a Bingham fluid.

For *quasi-static* applied strains, i.e. strains that are applied sufficiently slowly that the *viscous* stresses in the complex fluid are negligible, the response is "elasto-plastic", a combination of elastic and plastic responses. This is represented by a spring and a sliding friction element in series (FIG. 4.3a). The yield stress σ_y, above which the friction element slides, is the maximum static stress that the material can withstand before it flows. The corresponding value of strain is the yield strain ε_y. Above the yield strain, the strain can be separated into two contributions,

$$\varepsilon = \varepsilon_e + \varepsilon_p, \qquad (4.4)$$

the sum of a plastic part ε_p and an elastic part ε_e; the latter is reversible and is thus $\varepsilon_e = \sigma_y/G$, at least in the linear elastic regime.

In order to model the *steady* flow of a "yield stress fluid" (i.e. a material that behaves as a solid or a liquid depending on the applied stress) over a large range of strain rates, we must account for viscous stresses. The Bingham model is often used. It is represented by a sliding friction element in parallel with a viscous dashpot that has a "plastic" viscosity η_p (FIG. 4.3b). The constitutive law is

$$\begin{aligned}\dot{\varepsilon} &= 0 \quad &\text{if} \quad \sigma < \sigma_y \\ \sigma &= \sigma_y + \eta_p \dot{\varepsilon} \quad &\text{if} \quad \sigma > \sigma_y.\end{aligned} \qquad (4.5)$$

When $\sigma > \sigma_y$, the effective viscosity $\sigma/\dot{\varepsilon}$ of a Bingham fluid decreases as the shear rate increases. This is known as *shear-thinning* behaviour.

The empirical Herschel–Bulkley constitutive law adds to the Bingham model a further free parameter β, which allows a more accurate fit to the shear-thinning response found experimentally. Above the yield stress it is given by

$$\sigma = \sigma_y + \eta_p \dot{\varepsilon}^\beta \quad \text{if} \quad \sigma > \sigma_y. \tag{4.6}$$

Since for $\beta \neq 1$ the units of η_p are no longer those of a viscosity, η_p is sometimes known as the *consistency* of the fluid.

More complex configurations of springs, viscous dashpots, and frictional elements can be constructed to describe the relationships between stress, strain, and strain rate, not only of foams but of many other complex materials [3, 9, 55]. In §3 we will identify the microstructural origin of these elements.

2.2 Shear tests

We next describe a few shear tests that are used to probe the rheological response of complex fluids, particularly foams. The equipment used for these experiments is described in §1.2.7, chap. 5.

Shear start-up This test consists of applying to the material an increasing strain, $\varepsilon(t) = \dot{\varepsilon} t$, at constant shear rate $\dot{\varepsilon}$, and measuring the resulting stress to give both the yield stress and the effective viscosity, $\sigma/\dot{\varepsilon}$.

Steady flow A steady flow experiment is determined by both the stress and the strain rate. We set one, wait until a steady state is reached, then measure the other. This results in a flow curve $\sigma(\dot{\varepsilon})$. For a Newtonian liquid the flow curve is a straight line through the origin with a slope that corresponds to the viscosity η (eq. (4.3)). The flow curve for a Bingham fluid (FIG. 4.3b) is also a straight line, but it does not pass through the origin: instead it intersects the stress axis at the yield stress σ_y (eq. (4.5)). For a Herschel–Bulkley fluid the flow curve also intersects the stress axis at the yield stress, but it is not straight (eq. (4.6)).

Relaxation and creep In a relaxation test a strain ε is applied and the change in the stress $\sigma(t)$ is measured as a function of time. The response is characterized by the *relaxation modulus*, $G(t) = \sigma(t)/\varepsilon$. For a perfectly elastic solid, $G(t)$ is constant and equal to the static shear modulus G (eq. (4.1)).

Conversely, during a creep experiment the stress σ is prescribed and the change in the strain $\varepsilon(t)$ is measured as a function of time (FIG. 4.2). The mechanical response is characterized by the *compliance* $J(t) = \varepsilon(t)/\sigma$. For a perfectly elastic solid, J is constant and equal to $1/G$. For a viscoelastic material, $J(t)$ differs from $1/G(t)$ (eq. (4.10)), and a creep experiment is an appropriate method for studying *slow* mechanical relaxations, as we shall see in §4.1.5.

Oscillatory response The rheological properties of a viscoelastic material may be probed over a given time-scale by applying *oscillatory shear*, that is, a sinusoidal strain of amplitude ε_0 and frequency ω, and measuring the resulting stress. If the amplitude of the strain is sufficiently small, the stress will also be sinusoidal, with an amplitude

σ_0 proportional to ε_0, and a phase shift of δ with respect to the strain.[1] In complex notation this is written

$$\varepsilon(t) = \varepsilon_0 \mathcal{R}[e^{i\omega t}] \quad \sigma(t) = \sigma_0 \mathcal{R}[e^{i(\omega t+\delta)}], \tag{4.7}$$

where $\mathcal{R}[\ ]$ denotes the real part. The constitutive law,

$$\sigma(t) = \varepsilon_0\, \mathcal{R}[G^*(\omega)e^{i\omega t}], \tag{4.8}$$

is expressed in terms of the complex shear modulus

$$G^*(\omega) = G'(\omega) + iG''(\omega). \tag{4.9}$$

The storage modulus G' describes the capacity of a material to store elastic energy, and the loss modulus G'' describes its dissipative character. For an elastic solid, G^* is real and equal to the static shear modulus $G' = G$, $G'' = 0$ (eq. (4.1)), while the complex modulus of a Newtonian liquid is purely imaginary: $G' = 0$ and $G'' = \eta\,\omega$. FIG. 4.2 illustrates these viscoelastic moduli for several common models.

In the linear regime the information obtained from a relaxation (or creep) experiment is equivalent to that obtained by applying an oscillatory shear. In fact, in this case the relaxation modulus and the compliance are linked to G^* by an integral transform [23], for example

$$\frac{1}{G^*(\omega)} = i\omega \int_0^\infty J(t)e^{-i\omega t}\, dt. \tag{4.10}$$

If we apply a sinusoidal stress of frequency ω and an amplitude σ_0 or ε_0 that increases from zero, we go beyond the linear regime, and even exceed the yield strain. Consequently the response includes not only a Fourier component at the fundamental frequency ω, but also a whole spectrum of harmonics at multiples of ω. Under these conditions, the complex shear modulus is defined in terms of the fundamental component and depends on the strain amplitude ε_0:

$$G^*(\omega, \varepsilon_0) = \frac{\omega}{\pi\,\varepsilon_0} \int_0^{2\pi/\omega} \sigma(t)e^{-i\omega t} dt. \tag{4.11}$$

Then $G^*(\omega, \varepsilon_0)$ does not completely characterize the non-linear rheological response because the harmonics are not taken into account.

Finally, note that if $G'(\omega, \varepsilon_0) > G''(\omega, \varepsilon_0)$ the predominant behaviour is that of an elastic solid, whereas if $G'(\omega, \varepsilon_0) < G''(\omega, \varepsilon_0)$ it is that of a viscous liquid.

2.3 Small and large strains

We have seen how to describe simple shear by its amplitude ε (FIG. 4.1). In order to characterize any general strain, we must specify, for a point with initial position

---

[1] An equivalent experiment (in the linear regime) consists of applying an oscillatory stress and measuring the strain.

vector **x**, the new position **x**'(**x**) that results from applying the strain. We can then characterize the displacement field **X**(**x**) = **x**' − **x**, as a function of the initial positions. This is simplified by considering the displacement field at a much larger scale than that of the microstructure. The material can then be regarded as a homogeneous, continuous medium.

An *infinitesimal* deformation is then described by the infinitesimal strain tensor

$$\varepsilon_{ij} = \frac{1}{2}\left(\frac{\partial X_i}{\partial x_j} + \frac{\partial X_j}{\partial x_i}\right) \quad \text{with} \quad i,j = 1,2,3, \tag{4.12}$$

where we identify the 1 direction with x, the 2 direction with y, and the 3 direction with z. In this notation, a simple shear strain ε (cf. FIG. 4.1) is 2 ε_{xy}.

This is appropriate in the linear regime. At greater strains, in the non-linear regime, linear functions of ε_{ij} are no longer sufficient to describe the kinematics. To characterize the strain here we introduce the deformation gradient tensor F_{ij} and the Finger tensor B_{ij}:

$$F_{ij} = \frac{\partial x'_i}{\partial x_j} \quad \text{and} \quad B_{ij} = \sum_k F_{ik} F_{jk}. \tag{4.13}$$

Non-linear elastic constitutive laws are expressed in terms of B_{ij} [43]. Beyond the yield strain, this total strain differs from the elastic strain (eq. (4.4)) due to plastic deformation.

2.4 Stress tensor in a complex fluid

2.4.1 Definition To define the stress tensor in 3D we consider one face of a small cube of material (FIG. 4.1) of area dS and unit normal vector **n** pointing outwards. The total force $d\mathbf{f}$ acting on the cube across the surface dS has components

$$df_i = \sum_j \sigma_{ij}\, n_j\, dS, \tag{4.14}$$

where σ_{ij} is the stress tensor (and a shear stress corresponds to the element σ_{xy}, cf. p. 169). The pressure p exerts a stress $-p\,\delta_{ij}$, but for an incompressible medium, the pressure affects neither the shape nor the behaviour of the material. We therefore subtract from the stress an *isotropic part* $-p\,\delta_{ij}$ and the rest, $\sigma_{ij} + p\,\delta_{ij}$, is known as the *deviatoric stress tensor*. Its off-diagonal elements represent shear, and its three diagonal elements are the differences of the normal stresses from their average value $p = (\sigma_{xx} + \sigma_{yy} + \sigma_{zz})/3$. The latter are the forces per unit surface area acting perpendicularly to the faces of the cube, in each of the three coordinate directions (FIG. 4.1). Hooke's law (of which eq. (4.1) is a simplified version) in tensor form is

$$\sigma_{ij} = 2\,G\,\varepsilon_{ij} - p\,\delta_{ij}, \tag{4.15}$$

and Newton's law (eq. (4.3)) can be generalized to

$$\sigma_{ij} = 2\,\eta\,\dot{\varepsilon}_{ij} - p\,\delta_{ij}. \tag{4.16}$$

To characterize the normal stresses in a way which does not depend on changes in ambient pressure, we construct two independent *normal stress differences*:

$$N_1 = \sigma_{xx} - \sigma_{yy} \quad \text{and} \quad N_2 = \sigma_{yy} - \sigma_{zz}. \tag{4.17}$$

For example, consider the compression of a cube in the y direction (FIG. 4.1), leaving its lateral faces free. Then N_1 and N_2 are independent of the pressure and vary linearly with the (uniaxial) strain according to Hooke's law. But, remarkably, normal stress differences can also appear during shear. This *non-linear* effect will be discussed in §4.1.2.

2.4.2 Relationship between stress and structure Understanding the microstructure of a material is key to understanding the origin of its stresses. For example, in a crystal, macroscopic stresses are the consequence of particular interatomic forces. Similarly, the rheological properties of a foam are essentially governed by the interfaces.

Batchelor [2] established the general expression for the stress induced in a mixture of two immiscible phases by a quasi-static strain. The stress tensor averaged over a volume V, denoted σ_{ij}, is linked to the fluid microstructure, the pressures in the dispersed phase, and the interfacial tension γ according to

$$\sigma_{ij} = -\frac{1}{V} \sum_{\text{bubbles } k} V_k P_k \delta_{ij} - \phi_l p \delta_{ij} + \frac{\gamma}{V} \int_{S_{\text{int}}} (\delta_{ij} - n_i n_j) \, dS. \tag{4.18}$$

The first two terms represent the average pressure in the dispersed and continuous phases, with V_k the volume of bubble k, P_k its pressure, p the pressure in the liquid, and $V\phi_l$ its volume. The last term represents the contribution of the interfaces of tension γ (recall that a liquid film has two interfaces). The integration domain S_{int} consists of the total interface in a volume V, and $\delta_{ij} - n_i n_j$ represents projection on to the surface with local normal $\mathbf{n}$; the orientation of the normal can be chosen arbitrarily since the product $n_i n_j$ does not depend on it. When we apply a macroscopic strain, the areas and the orientations of the interfaces change, and eq. (4.18) predicts the resulting stress. The shaded section below derives the stress tensor for the case of a dry 2D foam.

The Batchelor expression shows that if we evaluate the average stress in a volume V smaller than a bubble, the result will fluctuate strongly depending on where this volume is located. For this reason, the rheology of a foam is often modelled at an intermediate scale between that of the bubble and that of the whole foam. This scale must be sufficiently large that the stress is independent of the microstructure, and thus continuous and differentiable; however, it must remain small enough to demonstrate the spatial variations of stress within a sample. Finally, we note that the total stress in a flowing foam can include, in addition to this static stress, dynamic contributions due to viscous forces.

The stress tensor for a dry 2D foam

For a dry foam, the second term in eq. (4.18) vanishes. The first term gives the average gas pressure in the bubbles and the last term is the stress due to the surface tension of the soap films. We demonstrate here its derivation in 2D, where a stress is a force per unit length, and the line tension λ is a force (cf. §5.2, chap. 2).

Due to the microstructure of a foam, the stress is not defined locally and eq. (4.14) is only meaningful with a finite value of dS, much larger than the typical bubble size. To calculate the average stress in a rectangular foam of size Δx by Δy and area A, we consider the force acting across an arbitrary line $\mathcal{L}$ in the foam (FIG. 4.4). We denote by $\mathbf{f}(x_\mathcal{L})$ the total force exerted by the foam to the right of $\mathcal{L}$ on the foam to the left. If the foam were continuous, $f_j(x_\mathcal{L})$ would be given by eq. (4.14),

$$f_j(x_\mathcal{L}) = \sigma_{1j}\, \Delta y. \tag{4.19}$$

The average stress tensor depends on the value of $\mathbf{f}(x_\mathcal{L})/\Delta y$, averaged over position $x_\mathcal{L}$. To obtain the components of the stress tensor we must therefore integrate over $x_\mathcal{L}$:

$$\sigma_{1j} = \frac{1}{\Delta x\, \Delta y} \int_0^{\Delta x_1} f_j(x_\mathcal{L})\, dx_\mathcal{L}. \tag{4.20}$$

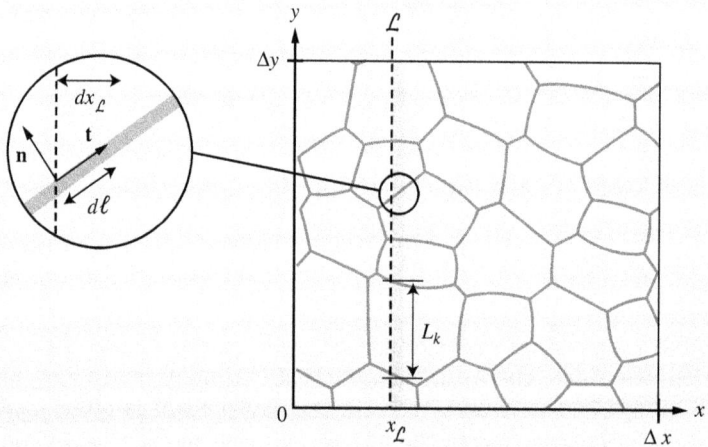

FIG. 4.4: A 2D foam with area $A = \Delta x \Delta y$. The zoom shows a film that crosses the straight line $\mathcal{L}$, with unit tangent and normal vectors $\mathbf{t}$ and $\mathbf{n}$, respectively.

We first estimate the contribution due to interfacial forces by writing eq. (4.20) as an integral along the films. The length $d\ell$ of a film element which crosses a strip of width $dx_\mathcal{L}$ (FIG. 4.4) satisfies

$$dx_\mathcal{L} = t_x \, d\ell, \qquad (4.21)$$

where $\mathbf{t} = (t_x, t_y)$ is a unit vector tangent to the film where it intersects $\mathcal{L}$. The interfacial force exerted by this film on the foam to the left of $\mathcal{L}$ is $2\lambda \mathbf{t}$. Using eq. (4.20), we find the surface tension contributions to the stress:

$$\sigma_{xx} = \frac{1}{A} \int_A 2\lambda \, t_x \, t_x \, d\ell, \qquad (4.22)$$

$$\sigma_{xy} = \frac{1}{A} \int_A 2\lambda \, t_x \, t_y \, d\ell. \qquad (4.23)$$

The integrals are taken over the total length of films in the area A. A similar reasoning gives $\sigma_{yx} = \sigma_{xy}$ (the stress tensor is clearly symmetrical, as expected [43]), and

$$\sigma_{yy} = \frac{1}{A} \int_A 2\lambda \, t_y \, t_y \, d\ell. \qquad (4.24)$$

Now consider the pressure P_k in a bubble k, which intersects $\mathcal{L}$ over a length L_k (which may be zero). The gas to the right of $\mathcal{L}$ exerts on the line segment L_k a force $-P_k L_k$ in the x direction. Since the area of bubble k is $A_k = \int L_k dx_1$, we obtain

$$\sigma_{xx} = \sigma_{yy} = -\frac{1}{A} \sum_{\text{bubbles } k} P_k A_k. \qquad (4.25)$$

(The reasoning required to obtain this expression for σ_{yy} is the same, but with $\mathcal{L}$ parallel to the y axis.)

We recover Batchelor's expression in 2D (eq. (4.18)) by adding up the contributions due to interfacial tension (eqs. (4.22), (4.23), (4.24)) and to pressures (eq. (4.25)), and by noting that in 2D we have $t_i t_j = \delta_{ij} - n_i n_j$.

2.4.3 Trapped stresses

Even if a complex material is at mechanical equilibrium and the average deviatoric stress is zero, it can nevertheless experience variations in stress at the scale of structural heterogeneities.[2] These trapped stresses result from the history of the strains applied to the sample, for example during its manufacture.

---

[2] Likewise, in atomic solids, trapped stresses are often found in the vicinity of defects such as dislocations.

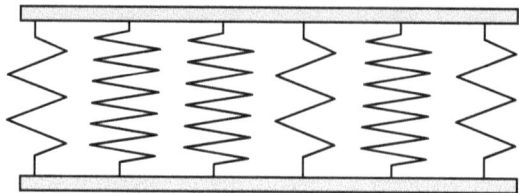

FIG. 4.5: Trapped elastic stresses in a network of springs, each with the same stiffness and a different relaxation length.

To illustrate this phenomenon, we consider a group of springs whose ends are fixed to two parallel bars (FIG. 4.5). If each spring has a different rest length, they can never all be relaxed simultaneously, no matter what the distance between the bars. So there are always trapped elastic forces that can't be relaxed without modifying the structure of the system. This idea can be generalized to assemblies of springs with different lengths in a regular network, or to identical springs in a disordered network. At a macroscopic scale, the mechanical state of such a network is characterized by trapped stresses. By analogy, it is the bubbles in a foam which play the role of springs and are subject to trapped stresses.

3 Local origin of rheological properties

We seek to understand the origin of the mechanical properties of a foam at a local scale (of bubbles or soap films). Why is it elastic? What is the value of strain or stress above which it flows? How is mechanical energy dissipated? In §4 we will describe the link between the mechanisms at the local scale and the macroscopic rheological response.

3.1 Elastic shear modulus of a dry monodisperse foam

In the absence of an applied stress, the structure of a dry foam at equilibrium corresponds to a minimum of surface energy, since the volume of gas in each bubble is fixed. In order to deform a foam, it is necessary to apply a stress. As long as this stress is not too great, the structure returns to its initial state as soon as the stress is removed: this is a foam's *elastic response*.[3] An elastic response is easily observed experimentally, as long as care is taken to avoid T1s (cf. §1.2.2, chap. 3). In this section, we consider simple shear of small amplitude (linear regime) in the quasi-static limit, i.e. at each instant, the structure, although deformed, is at equilibrium (a more precise definition is given in the shaded section on p. 186). This includes elongational strains at constant volume, since at small amplitude simple shear is in effect equivalent to a uniaxial strain followed by a rotation.

---

[3]This is the usual elasticity, which should not be confused with Gibbs' surface elasticity (cf. eq. (3.11)). Here, surface tension is assumed constant.

3.1.1 Dimensional analysis and orders of magnitude The elastic response of a foam is influenced by pressure, capillary forces, and the characteristic length-scale of the structure, although pressure does not affect the shear stress (cf. eq. (4.18)). In the dry limit, Plateau border radii and film thicknesses are negligible compared to the bubble size and it can be shown that they do not influence the static response (although in wet foams the liquid fraction does play an important role, see §4.1.1). The shear modulus G (eq. (4.1)) thus depends primarily on surface tension γ and the characteristic size of the bubbles d. As a modulus is an energy density or a pressure, we thus expect $G \sim \gamma/d$. We will show that this scaling is indeed appropriate for a range of dry foams (2D, 3D, ordered, and disordered), with a different numerical coefficient for each structure.

Note that for 2D foams, the line tension (with the dimensions of a force of the order of γe, cf. §5.2, chap. 2) is denoted λ, $\hat{G}$ is the 2D shear modulus (also in N.m^{-1}) and $\hat{\sigma}$ is the 2D stress (in N.m^{-1}). For simplicity, the $\hat{\ }$ will usually be omitted.

The shear modulus γ/d is of the order of 300 Pa for $d \approx 100\,\mu$m and $\gamma \approx 30$ mN.m^{-1}. Therefore foams are certainly "soft matter" compared to ordinary solids (metal, glass, etc.) which have a modulus of the order of 10^{10} to 10^{11} Pa! When a sample of aqueous foam is squashed, it flattens like a soft incompressible material. In fact, the bulk (compression) modulus of a foam is dominated by the response of the gas, being of the order of the external pressure ($\approx 10^5$ Pa) and certainly greater than G (see exercise 7.6). In practice compressive deformations arise for example during the propagation of sound[4] or shock waves (which we do not discuss here).

3.1.2 Ordered 2D foam In his pioneering work on foam rheology, Henry Princen considered a model 2D foam consisting of an ordered network of regular, identical hexagons. This simple model gives a good idea of the effect of shear at the bubble scale [49]. FIG. 4.6 shows how the shape of the bubbles changes as an increasing

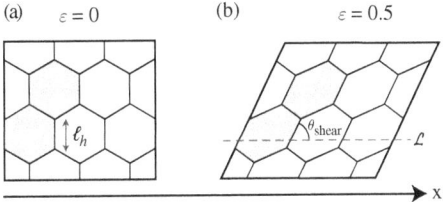

FIG. 4.6: Quasi-static shear of a 2D hexagonal foam with edge length ℓ_h at rest. The structure remains ordered as the strain ε increases from (a) 0 to (b) 0.5. All bubbles therefore have the same pressure, and the edges are straight and meet at 120°. The stress exerted across the line $\mathcal{L}$, calculated in the text, is non-zero if θ_{shear} is different from 90°, i.e. for $\varepsilon \neq 0$.

---

[4] Measuring the speed of sound in a dry foam is one way of determining its liquid fraction, since the two are strongly correlated. At a frequency of a few kHz, it is about 50 m.s^{-1} for $\phi_l = 0.1$ and reaches a minimum of the order of 20 m.s^{-1} for a bubbly liquid which is half water and half air [19, 25, 47]. It is much lower than the speed of sound in water (1500 m.s^{-1}) or in air (340 m.s^{-1}).

shear strain ε is applied. The disordered case is dealt with in more detail in [37] and exercise 7.2.

Stress equilibrium The two states in FIG. 4.6 both satisfy Plateau's laws (§2.2, chap. 2). In fact, the foam is free to relax locally to a minimum energy structure. However, the applied shear reduces the number of degrees of freedom of the foam by fixing the points where the edges meet the top and bottom walls of the "box". A shear stress σ is exerted across lines parallel to the x direction. In this way equilibrium states with non-zero stress ($\sigma \neq 0$), distinct from the stress-free state ($\sigma = 0$ and $\varepsilon = 0$), are found.

Stress and shear modulus The foam located on one side of a line $\mathcal{L}$ parallel to the x axis (FIG. 4.6b) exerts a shear stress σ on the foam on the other side, as a result of the edges which are "cut" by $\mathcal{L}$ and pull in the θ_{shear} direction. As the strain ε increases, the angle θ_{shear} decreases, and so σ grows. Writing ℓ_h for the length of an edge when $\varepsilon = 0$ and λ for the line tension, the 2D stress $\hat\sigma$ (the force per unit length) is [40]

$$\hat\sigma = \frac{2\lambda}{\sqrt{3}\ell_h}\frac{\varepsilon}{\sqrt{\varepsilon^2+4}}. \tag{4.26}$$

Linearizing this expression gives the shear modulus $G = \sigma/\varepsilon$:

$$\hat G_{\text{hex}} = \frac{\lambda}{\sqrt{3}\,\ell_h} = 0.52\,\frac{\lambda}{R_A} = \frac{\lambda\,L_{\text{int}}}{4\,A_{\text{foam}}}. \tag{4.27}$$

R_A is the radius of a disc of the same area as a hexagon with edge length ℓ_h ($R_A \approx 0.909\,\ell_h$) and $L_{\text{int}}/A_{\text{foam}}$ is the length of interface per unit area when $\varepsilon = 0$, $L_{\text{int}}/A_{\text{foam}} = 4/(\sqrt{3}\ell_h)$. $\hat G$ is clearly of the order of the line tension divided by the bubble size. Note that in the linear regime, the energy, and thus the total quantity of interface, varies as ε^2 (p. 169). In this example, the modulus $\hat G$ is independent of the shear direction [60]. As we will see, this is not the case for a Kelvin foam.

3.1.3 Ordered 3D foam

The Kelvin cell forms the unit cell of an ordered dry foam (see shaded section on p. 50) in 3D, as the hexagon does in 2D. Under the effect of shear, the area and orientation of the faces change (FIG. 4.7). Due to the cubic symmetry, the magnitude of these variations depends on the direction in which the shear is oriented. Consequently, the elastic response of a Kelvin foam is anisotropic. Using the *Surface Evolver* (§2.1.2, chap. 5), Kraynik and Reinelt showed that the shear modulus averaged over all orientations is [33]

$$G_{\text{Kelvin}} = 0.15\frac{\gamma\,S_{\text{int}}}{V_{\text{foam}}} = 0.50\frac{\gamma}{R_V}, \tag{4.28}$$

where S_{int} is the total interfacial area and V_{foam} the total volume of foam. Here again, the prefactor is of order one, as in eq. (4.27) for a hexagonal foam. Depending on the shear direction, the prefactor varies between 0.35 and 0.60.

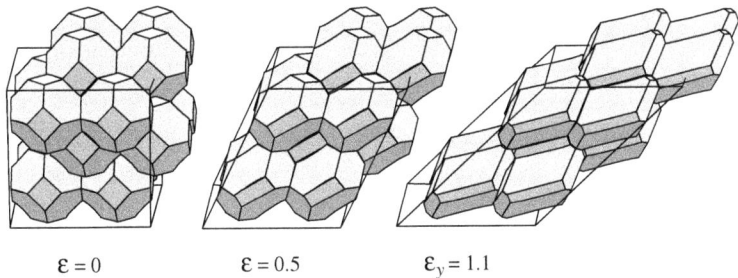

$\varepsilon = 0$ $\varepsilon = 0.5$ $\varepsilon_y = 1.1$

FIG. 4.7: A numerical simulation using the *Surface Evolver* gives the shape of a Kelvin cell for different strains ε. The shear direction corresponds here to the (100) direction of a Kelvin foam's bcc lattice.

3.1.4 Disordered 3D foam: affine deformation When a 3D foam is disordered, it is difficult to predict the precise deformation of each film under the effect of a shear. The simplest hypothesis assumes that the deformation is affine: a point initially situated at position $A(x, y, z)$ is found at position $A'(x + \varepsilon y, y, z)$ after a shear strain ε. In an isotropic foam the films are oriented randomly, which leads to Derjaguin's expression [18]:

$$G_{\text{affine,3D}} = \frac{4}{15} \frac{\gamma S_{\text{int}}}{V_{\text{foam}}} = 0.26 \frac{\gamma S_{\text{int}}}{V_{\text{foam}}}. \tag{4.29}$$

Numerical simulations with the *Surface Evolver* show that in a disordered monodisperse dry 3D foam [35] $S_{\text{int}}/V_{\text{foam}} \approx 3.3/R_V$ (cf. eq. (2.19)).[5] From this we deduce

$$G_{\text{affine,3D}} = 0.88 \frac{\gamma}{R_V}. \tag{4.30}$$

An affine deformation does not satisfy Plateau's laws—the angles between films are no longer 120° and the angle at which edges meet at a vertex is no longer 109.5°—so that the resulting configuration does not consist of minimum energy surfaces; we will see that the modulus of a disordered 3D foam (eq. (4.35)) is actually lower than $G_{\text{affine,3D}}$.

Affine deformation of a 2D foam
The 2D version of Derjaguin's expression for the shear modulus (eq. (4.29)) is easily derived for an isotropic foam. We assume that the foam consists of a collection of straight edges whose initial orientations are random and uncorrelated with their length. The deformation modifies the orientation and the length of each edge affinely

---

[5]Contrast this with the case of spherical bubbles, for which $S_{\text{int}}/V_{\text{foam}} = 3/R_V$.

(FIG. 4.8), and the shear stress $\hat{\sigma}$ can be calculated using Batchelor's expression (cf. shaded section on p. 176):

$$\hat{\sigma} = \frac{2\lambda}{A_{\text{foam}}} \sum_{\text{edges } i} \ell'_i \cos\theta'_i \sin\theta'_i, \qquad (4.31)$$

where the sum is over all the edges in the sample. Using the geometrical calculation illustrated in FIG. 4.8 we find

$$\hat{\sigma} = \frac{2\lambda}{A_{\text{foam}}} \sum_{\text{edges } i} \ell_i \sin\theta_i \cos\theta_i \left(1 + \varepsilon \frac{\sin^3\theta_i}{\cos\theta_i}\right) = \frac{3\lambda\varepsilon}{4 A_{\text{foam}}} \sum_{\text{edges } i} \ell_i. \qquad (4.32)$$

The second equality in eq. (4.32) is obtained by averaging over θ_i, assuming that the angles are uniformly distributed in $[0, \pi]$ and uncorrelated with the edge lengths.[6] With $\sum_{\text{edges } i} \ell_i = L_{\text{int}}/2$, we deduce that the shear modulus $G = \sigma/\varepsilon$ is

$$\hat{G}_{\text{affine, 2D}} = 0.375 \frac{\lambda L_{\text{int}}}{A_{\text{foam}}}. \qquad (4.33)$$

As in 3D (eq. (4.30)), the Derjaguin approximation overestimates the shear modulus.

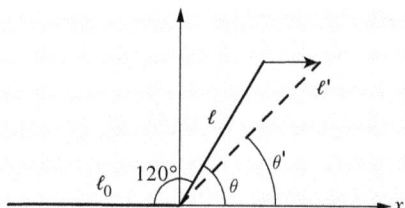

FIG. 4.8: Affine shear (with strain ε) of two edges in a 2D foam, of lengths ℓ_0 and ℓ, that initially intersect at 120°. The edge ℓ_0, parallel to the direction of shear, remains unchanged. The length and orientation of ℓ become, to leading order in ε: $\ell' = \ell\,(1 + \varepsilon \cos\theta \sin\theta)$, $\sin\theta' = \sin\theta\,(1 - \varepsilon \cos\theta \sin\theta)$, and $\cos\theta' = \cos\theta\,(1 + \varepsilon \sin^3\theta/\cos\theta)$.

3.1.5 Disordered 3D foam: non-affine deformation

To estimate the shear modulus more precisely it is necessary to take into account Plateau's laws: during a quasi-static shear, the angles between films and the angle at which edges meet must remain constant and equal to 120° and 109.5°, respectively (cf. p. 30). In the Stamenovic

---

[6] For the same reason, the stress is initially zero (eq. (4.31) with $\theta'_i = \theta_i$).

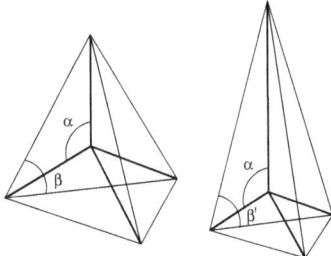

FIG. 4.9: Non-affine deformation of a tetrahedron. The bottom triangle is reduced by a factor $1 - \varepsilon/2$ in both directions of the plane, and the total height of the tetrahedron increases by $1 + \varepsilon$. The central vertex is placed in the position that preserves the tetrahedral angles (denoted α). The angle β, which is not the angle between edges, is not preserved.

model the foam is described as a collection of tetrahedra which are deformed according to Plateau's laws (FIG. 4.9). A geometric calculation predicts the variation in film surface area, and therefore the surface energy density, giving [57]

$$G_{\text{tet}} = \frac{1}{6} \frac{\gamma S_{\text{int}}}{V_{\text{foam}}}. \tag{4.34}$$

Expressing S_{int} in terms of the bubble size (p. 181) gives

$$G_{\text{tet}} = 0.55 \frac{\gamma}{R_V}. \tag{4.35}$$

This is currently the best prediction for disordered monodisperse dry foams, and is in close agreement with numerical calculations using the *Surface Evolver* [34] which give $G = 0.51\gamma/R_V$.

3.2 The elastic limit of a dry foam

At larger deformations, the response of a foam is no longer elastic and it begins to flow. We consider here the plastic regime under quasi-static deformation, in which the bubbles slide past each other in a succession of small jumps. Each of these plastic events is a T1 topological process (§1.2.2, chap. 3), the mechanical consequences of which will be discussed. Experiment 6.1 illustrates the T1s which accompany 2D foam flow.

3.2.1 The elastic limit In general, bubbles are not greatly deformed, that is, their aspect ratio remains close to 1. Consequently, the *yield strain* ε_y (defined in §2.1), also known as the *elastic limit*, is of order one for a dry foam. In consequence the yield stress σ_y, which is approximately $G\varepsilon_y$ if the linear regime extends up to ε_y, scales in the same way as the shear modulus:

$$\sigma_y \sim \frac{\gamma}{d}. \tag{4.36}$$

184 *Rheology*

Ordered 3D foam   A Kelvin foam (p. 50) leaves the elastic regime when the length of an edge first goes to zero (FIG. 4.7). A vertex is then formed at which six edges meet; this is unstable (by Plateau's second law, §2.2.2, chap. 2) and the structure relaxes towards a new equilibrium configuration through a cascade of T1 topological changes [51, 52]. The strain at this point is the *yield strain* ε_y, which depends strongly on the shear direction because of the anisotropy of the structure [51]. Beyond ε_y, in the plastic regime, the stress varies discontinuously with the strain due to the jumps at each T1.

Ordered 2D foam   A more detailed calculation is possible for a hexagonal 2D foam (FIG. 4.10). Under the effect of shear, some edges stretch and others shorten. When the length of an edge goes to zero, a vertex with four edges forms (FIG. 4.10C). This is unstable (by Plateau's first law, §2.2.2, chap. 2): there is a T1, and then the foam relaxes irreversibly (§3.2.3) to a new equilibrium state (FIG. 4.10D). Point C corresponds to the yield strain. In this case ε_y can be calculated geometrically (cf. exercise 7.2), and then we may deduce $\hat{\sigma}_y$ using eq. (4.26). For the shear direction in FIG. 4.10, we find

$$\varepsilon_y = \frac{2}{\sqrt{3}} \quad \text{and} \quad \hat{\sigma}_y = \frac{\lambda}{\sqrt{3}\ell_h}. \tag{4.37}$$

3.2.2 Energy landscapes, basins of attraction, and hysteresis   The example of a hexagonal 2D foam enables us to illustrate some important concepts in plasticity. FIG. 4.11a shows how the shear stress $\hat{\sigma}$ changes as a function of strain ε (eq. (4.26)); the letters correspond to the different states of the foam shown in FIG. 4.10. FIG. 4.11b shows the energy,[7] or *energy landscape*, of the foam, again as a function of strain ε. The landscape consists of two (quasi-parabolic) potential wells which partially overlap. A and D are the stress-free equilibrium states, B is an equilibrium state (p. 180), and C is an out-of-equilibrium state. If we release the stress at B, the foam relaxes spontaneously towards the energy minimum at A. Generally, the points situated on the parabola centred at A return towards A and constitute the *basin of attraction*

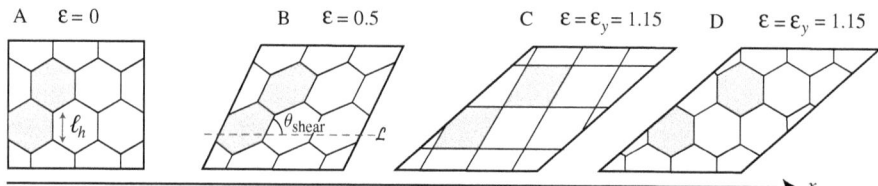

FIG. 4.10: Quasi-static shear of a hexagonal 2D foam in the x direction. The edges meet at 120°, except in C where unstable four-fold vertices appear. In D, T1s occur at each vertex and neighbouring layers are shifted by one bubble in relation to each other, as the two grey bubbles illustrate.

[7] which can be calculated as $E = 6\lambda \ell_h + A \int \hat{\sigma}\, d\varepsilon$, where A is the bubble area.

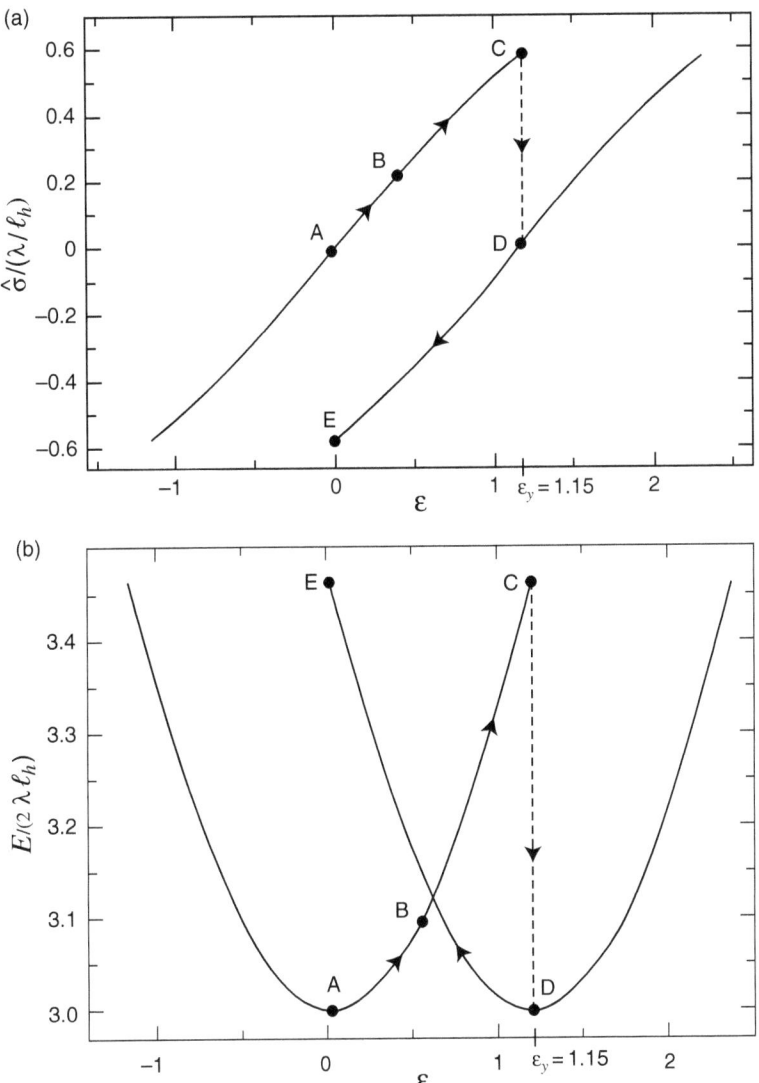

FIG. 4.11: Stress and energy in a hexagonal 2D foam. (a) 2D shear stress (eq. (4.26)), normalized by λ/ℓ_h, as a function of the strain ε. Points A, B, C, and D (with strains $\varepsilon_A = 0, \varepsilon_B = 0.5, \varepsilon_C = \varepsilon_D = 1.15$) correspond to FIG. 4.10. Point E is a state with the same magnitude of elastic strain as C. (b) Energy per bubble, normalized by the energy $2\lambda\ell_h$ of an undeformed edge.

of A. State B is in this basin of attraction and would relax towards A, even if ε_B were closer to ε_D than ε_A. Only a T1 transformation can lead to a change of the basin of attraction (as from C to D). The abscissa of the graphs (FIG. 4.11) is the total strain of the foam. It coincides with the elastic strain for the points situated on the parabola centred at A, but not for the parabola centred at D, where the elastic strain is $\varepsilon_e = \varepsilon - \varepsilon_y$, the plastic part of the strain is $\varepsilon_p = \varepsilon_y$, and we again have $\sigma = G\varepsilon_e$.

There are several possible structures for a given strain, which leads to the phenomenon of hysteresis. If, having arrived at D, we reduce the applied strain, the foam does not return via the previous structures. Instead, the structure in D is deformed without changing the topology until E, at which point T1s occur and the foam returns to A.

3.2.3 Energy dissipated by a T1

A T1 is always accompanied by dissipation of mechanical energy. In order to understand this, we return to the transition from state C to D in a hexagonal foam (FIG. 4.10). This transition consists of an instantaneous change in topology followed by a relaxation. The T1 dissociates the unstable four-fold vertex into two stable three-fold vertices then, during the ensuing relaxation, the structure deforms to reach the equilibrium state D. The relaxation happens without the boundaries of the system moving, and thus without exchange of work with the outside. The difference in energy between C (just at the yield strain) and D is thus entirely dissipated as heat during the relaxation. In this hexagonal case we have $\Delta E = E_C - E_D \approx 0.46 \times 2\lambda \ell_h \approx E_{\text{bub}}/6$, where $E_{\text{bub}} = 6\lambda \ell_h$ is the energy per bubble; for a bubble of edge-length ℓ_h and a foam of thickness e, both of the order of a millimetre, and a surface tension 25 mN.m^{-1} we thus find $\Delta E \sim 3 \times 10^{-8}$ J.

The amount of energy dissipated per bubble is fixed by the energy accumulated during the elastic loading, which itself depends only on the equilibrium properties of the foam (see exercise 7.5). Thus the dissipation, per unit volume and for a strain of fixed amplitude ε, doesn't tend towards zero in the limit of small strain rates. It is proportional to the number of T1s (and thus, on average, to ε) and independent of $\dot\varepsilon$. We can therefore calculate the dissipation per unit volume *and per unit time*, i.e. the dissipation rate. In the quasi-static regime, summarized in the shaded section below, this rate is proportional to $\dot\varepsilon$ as in solid friction, and not to $\dot\varepsilon^2$ as in viscous friction or, more generally, as in any of the dissipative modes at the scale of the films and interfaces, which will be described later (§3.3.2). This is because we control the macroscopic deformation rate but not the relaxation time (and therefore the local strain rate) of the foam between C and D. This time, which is intrinsic to the foam, depends on local modes of dissipation.

Quasi-static flow
The following is a summary of the conditions necessary for *quasi-static* flow:
- the foam must be at mechanical equilibrium at each instant, except during the brief relaxation after a T1;

- the relaxation time must be negligible compared to the characteristic time of the macroscopic deformation ($1/\dot{\varepsilon}$);
- the new equilibrium structure reached after a T1 must not depend on dissipation at the local scale (films and interfaces).

Under these conditions, the succession of structures adopted by the foam is independent of the applied strain rate and, in this sense, the notion of time is no longer relevant. Note that in a macroscopic sample, consisting of a large number of bubbles, there is almost always a relaxation taking place: the first condition is thus rarely strictly respected. In order to resolve this difficulty, it is necessary to introduce the notion (as yet poorly defined) of the *range* of a T1, cf. FIG. 3.5 [14], and to determine the extent to which T1s "overlap".

During rapid flow, many T1s occur and a foam never reaches an equilibrium state. The imposed strain, which tends to elongate the bubbles, dominates the relaxation after a T1, which tends to reduce the bubble deformation. The yield strain thus increases with the strain rate [53] (§4.2.2).

3.3 Dissipative processes

As we have just seen, liquid foams are only capable of accumulating a finite amount of energy: the elastic energy density (eq. (4.2)) is at most of the order of $\sigma_y^2/G \sim \gamma/d$, i.e. several J.m^{-3} for bubbles of diameter $d \sim 1$ mm. The kinetic energy per unit volume of foam is of the order of $\rho_l \phi_l v^2$—i.e. around 1 J.m^{-3} for a liquid fraction of $\phi_l = 10\%$ and a speed v of about 10 cm.s^{-1}. The additional energy supplied to the system, in order to impose a steady flow for example, is continually dissipated as heat. In the quasi-static limit, it was possible to predict the *amount* of energy dissipated (§3.2.3); here, we discuss *the way* in which it is dissipated.

The processes that dissipate energy in the rapid flow regime or during the fast relaxation following a T1 in a quasi-static flow are the same. The dissipation modes can be grouped into two categories: the hydrodynamic dissipation in the liquid phase due to the liquid shear viscosity; and the dissipation associated with surfactant motion due to the interfacial viscosity of the monolayer or the diffusivity of the surfactants [6]. The coupling between these different modes is complex and there is currently no simple and general criterion to identify the dominant modes in any given situation. A weakly dissipative mode contributes little to the production of heat, while a strongly dissipative mode is probably not stimulated and does not contribute either, in which case we speak of a *frozen* mode. In the following we extend the discussion of §6.2, chap. 3 to give for each process the order of magnitude of the dissipative power per unit volume of foam, denoted $\mathcal{D}$, as if they were independent of each other.

3.3.1 Dissipation in the liquid The viscous dissipation $\mathcal{D}$ in a Newtonian liquid subject to a local shear rate $\dot{\varepsilon}_l$ is $\eta \dot{\varepsilon}_l^2$, where η is the viscosity, of the order of 10^{-3} Pa.s for foaming solutions. The dissipation in a foam is difficult to evaluate, as the macroscopic shear rate $\dot{\varepsilon}$, determined by the movement of the boundaries at the scale of several bubbles, is not directly related to the velocity gradients in the liquid phase. The solution is sheared locally, over the thickness of the films and over the cross-section of Plateau borders, and in general $\dot{\varepsilon}_l$ is different from $\dot{\varepsilon}$. Moreover, even if the exact shape of the bubbles (itself difficult to measure *in situ* in the flow) were known precisely at each moment, the problem still would not be resolved since for a given evolution of the structure, there may be several possible velocity fields in the liquid.

For a small strain of amplitude ε, in the absence of topological changes, the relative change in surface area $\Delta S/S$ of an individual film (which may be positive or negative) is proportional to ε. For the whole foam, since those films which shrink and those which expand almost compensate each other, we have $\Delta S_{\text{int}}/S_{\text{int}} \sim \varepsilon^2$ (p. 180). We disregard this global expansion.

We first distinguish two modes of deformation at the local scale, assuming affine flow of the bubbles:

Mode A If an adsorbed surfactant monolayer is incompressible, for example if the dilatational surface modulus E_s^* is high (cf. eq. (3.13)), strong surface stresses make it impossible to create or destroy interface. This is known as a *rigid* or *immobile* interface. These interfaces behave like flexible inextensible membranes, moving at the speed of the bubble that they enclose and sliding past one another (FIG. 4.12A). The relative speed of two interfaces situated on either side of a film is thus of the order of the difference in velocity between two bubbles in contact, $\dot{\varepsilon} d$. The velocity gradient in a film of thickness h is therefore of the order of $\dot{\varepsilon} d/h$. In a volume d^3 (the bubble size), the volume of sheared liquid is hd^2, which implies that the dissipation rate per unit volume of foam is

$$\mathcal{D}_{\text{hydro,A}} \sim \eta \left(\frac{\dot{\varepsilon} d}{h}\right)^2 \frac{hd^2}{d^3} \sim \eta \frac{d}{h}\dot{\varepsilon}^2. \tag{4.38}$$

Mode B In contrast, for a perfectly compressible interface the surface stress is due only to the equilibrium surface tension γ. This is known as a *fluid* or *mobile* interface,[8] which can deform as represented in FIG. 4.12B. The film length increases (or decreases) at a rate $d\dot{\varepsilon}$ inducing velocity gradients of the order of $\dot{\varepsilon}$ in a film of volume d^2h. The dissipation rate per unit volume of foam is then

$$\mathcal{D}_{\text{hydro,B}} \sim \eta\dot{\varepsilon}^2 \frac{d^2 h}{d^3} \sim \eta\dot{\varepsilon}^2 \frac{h}{d}. \tag{4.39}$$

Mode B induces less hydrodynamic resistance to the applied strain than mode A, but it also requires that a significant amount of interface is created on one side of a Plateau border and destroyed on the other. It is therefore associated with a transport of surfactants from within the film, which also induces dissipation. Mode A, on the

---

[8]In the case of drainage (p. 119), interfaces are classified as mobile or immobile depending on the interfacial shear viscosity (or, more precisely, the Boussinesq number Bo). Here, the mobility or rigidity of an interface refers to its response to compression and expansion; we thus prefer to refer to perfectly compressible or incompressible interfaces.

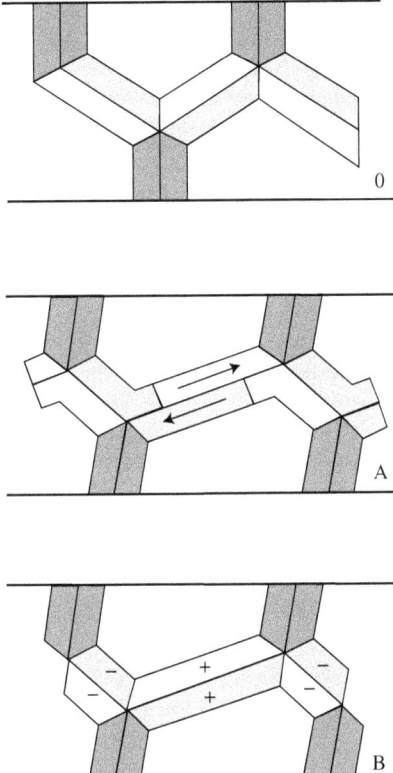

FIG. 4.12: Movement of the gas/liquid interfaces in a sheared 2D foam. The interfaces are notionally separated into different parts represented by the shading, to enable a schematic visualization of local deformations and displacements during shearing. State 0 is the reference state: it can be deformed according to either mode A or B. In mode A, the length of each part of the interfaces is fixed (to first order in ε) and they slide over one another, which induces velocity gradients in the films. In mode B, the interfaces change length (according to the sign + or –), the interfaces on each side of a film undergo the same displacements, and the velocity gradient in each film is zero.

other hand, does not induce surfactant-related dissipation. In practice, it is therefore the competition between hydrodynamic flows and surfactant transport which dictates the flow.

The expressions for the dissipation (eqs. (4.38) and (4.39)) correspond to rather unrealistic velocity fields, but illustrate the coupling between the liquid and the surfactant. In particular, they don't take into account the fact that film thickness h itself depends on the flow, and can vary over several orders of magnitude. Moreover, thickness variations are coupled to strong heterogeneities in velocity gradients within the same film. Schwartz and Princen [56], taking this effect into account, have suggested that the dissipation comes mainly from the movement of liquid from the Plateau border into the film. They predict that the liquid is entrained over a distance of the order of $x^* \sim rCa^{1/3}$, with a film of thickness $h^* \sim rCa^{2/3}$, where r is the radius of curvature of the Plateau border, and $Ca = \eta\dot{\varepsilon}d/\gamma$ is the capillary number, which compares viscous to capillary effects. For a bubble of volume d^3, the sheared volume is thus x^*h^*d, with a shear rate $\dot{\varepsilon}d/h^*$. The bulk dissipation rate is then

$$\mathcal{D}_{\text{S.P.}} \sim \eta \left(\frac{\dot{\varepsilon}d}{h^*}\right)^2 \frac{x^*h^*d}{d^3} \sim \eta\dot{\varepsilon}^2 Ca^{-1/3} \sim \eta\dot{\varepsilon}^2 \left(\frac{\eta\dot{\varepsilon}d}{\gamma}\right)^{-1/3}. \tag{4.40}$$

This process predicts that the shear stress is a non-linear function of the strain rate, $\sigma \sim \dot{\varepsilon}^{2/3}$, which contradicts the linear response observed macroscopically in the limit of small strains.

3.3.2 Dissipation associated with surfactant motion We next determine the dissipation associated with the transport of surfactants and the viscosity of a monolayer.

Shear The upper and lower films in FIG. 4.12 (mode A or B) don't change in area, but are subject to the global shear rate $\dot{\varepsilon}$. The surface area per unit volume is of the order of $1/d$, so the dissipation associated with shear can be estimated from a scaling argument:

$$\mathcal{D}_{\text{shear}} \sim \frac{\eta_s}{d}\dot{\varepsilon}^2, \tag{4.41}$$

where η_s is the surface shear viscosity defined in §2.2.3, chap. 3.

Expansion and compression Expansion and compression are important in mode B of FIG. 4.12, where the interfaces undergo relative changes in area at a rate of the order of $\dot{\varepsilon}$. The surface area per unit volume is again of the order of $1/d$, so the dissipation is

$$\mathcal{D}_{\text{exp/comp}} \sim \frac{\eta_{d,\text{film}}}{d}\dot{\varepsilon}^2, \tag{4.42}$$

where $\eta_{d,\text{film}}$ is the dilatational viscosity of the film. As discussed in §2.2.3, chap. 3, a film may expand in different ways, associated with the exchange of surfactants between the interface and film. These different modes and the corresponding dissipation are explained in FIG. 3.54, §6.2, chap. 3. Comparing those expressions (eqs. (3.105), (3.106), and (3.107)) with eq. (4.42) allows the effective viscosity $\eta_{d,\text{film}}$ corresponding to each mode to be estimated.

Other models, based on surfactant-linked dissipation, are also proposed in the literature [4, 5, 20].

3.3.3 Dissipation at the wall Lastly, we investigate the dissipation due to the wetting film which is present between a foam and a wall. Two limiting cases arise, depending on whether the gas/liquid interface is perfectly compressible or incompressible, as defined in the footnote in §3.3.1 [8, 16, 58]:

Perfectly compressible interface In this case, a surface Plateau border slides along the wall without causing the wetting film to move; the bubble "rolls without sliding" (FIG. 4.13). The velocity gradient is localized in the region where the surface Plateau border and the wetting film meet (i.e. the region where the liquid is expelled or absorbed by the Plateau border). Although the characteristic size of this region increases with the velocity, the following argument, analogous to that of Schwartz and Princen (p. 190), shows that the dissipation is not a quadratic function of the imposed shear rate.

Consider a surface Plateau border (with radius of curvature r), which slides at speed $U\mathbf{e}_x$ along a wall.[9] The flow is quasi-parallel, and a lubrication analysis is appropriate [1]. The velocity is zero at the wall and the tangential stress is zero at the interface (because we are assuming that the surfactants are mobile). The pressure depends only on x, and is directly related to the curvature by the Young–Laplace law: $p(x) = P - \gamma(\partial^2 h/\partial x^2)$. The pressure gradient induces a parabolic velocity field in which the flux per unit length is $Q \sim \nabla p \, h^3/\eta \sim (\gamma/\eta)(\partial^3 h/\partial x^3)h^3$. Conservation of mass requires that $\partial Q/\partial x = \partial h/\partial t$. Finally, at steady state the space and time derivatives are linked by $\partial h/\partial t = -U \partial h/\partial x$ and so

$$U\frac{\partial h}{\partial x} \sim -\frac{\gamma}{\eta}\frac{\partial}{\partial x}\left(h^3 \frac{\partial^3 h}{\partial x^3}\right). \qquad (4.43)$$

We can determine the properties of $h(x)$ without solving this differential equation, but simply by examining the orders of magnitude of its terms. The thickness h is negligible and constant in the wetting film, and has constant curvature $1/r$ in the surface Plateau border. The distance over which we go from one to the other is x^*, the length of the transition zone. In this zone, $h(x)$ is of the order of h^*, and its nth derivatives are of order h^*/x^{*n}. We must determine h^* and x^*. Eq. (4.43) gives

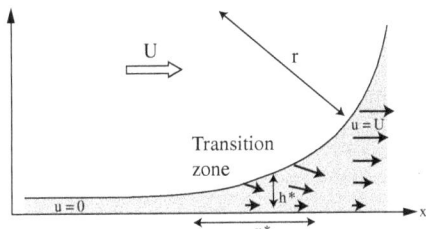

FIG. 4.13: A surface Plateau border sliding along a wall in the case of a perfectly compressible interface. The arrows represent the local velocity field u in the liquid. The surface Plateau border—of which only half is shown—moves with velocity U; the wetting film is at rest on the wall ($y \leq 0$). There is a strongly dissipative region where the wetting film meets the Plateau border, known as the transition zone, of thickness h^* and length x^*. The flow is assumed invariant along the axis of the Plateau border (in the z direction, perpendicular to the plane of the figure).

---

[9]The problem is analogous to that of a plate withdrawn from a reservoir of liquid [15].

the relationship $h^*/x^* \sim (\eta U/\gamma)^{1/3}$. Expressed as a function of the capillary number $Ca = \eta U/\gamma$, this is $h^*/x^* \sim Ca^{1/3}$.

On the other hand, the continuity of curvature between the transition zone and the surface Plateau border dictates that $h^*/x^{*2} \sim 1/r$. Thus $h^* \sim rCa^{2/3}$ and $x^* \sim rCa^{1/3}$. From this we may deduce the order of magnitude of the dissipation associated with the motion of a surface Plateau border, which is localized in the transition zone. The velocity gradients are $\partial u/\partial y \sim U/h^*$ and the volume of the zone is $\Omega \sim h^* x^* d$ for a surface Plateau border of length d. Then the dissipation per unit surface area of foam in contact with the wall is

$$\mathcal{D}_{\text{wall, m}} \sim \frac{1}{d^2} \int_\Omega \eta \left(\frac{\partial u}{\partial y}\right)^2 d\Omega \sim \frac{\eta U^2 Ca^{-1/3}}{d}. \tag{4.44}$$

So in the case of a perfectly compressible interface the effective viscosity of a foam sliding over a solid wall decreases with velocity as $Ca^{-1/3}$. This is because the faster the foam goes, the more the films inflate, facilitating the slippage. Nevertheless, the dissipation increases with the speed. A similar reasoning leads to a (dimensional) expression for the shear stress on the wall:

$$\sigma_{\text{wall, m}} \sim \frac{1}{d^2} \int_{x^* d} \eta \frac{\partial u}{\partial y} dS \sim \frac{\gamma Ca^{2/3}}{d}. \tag{4.45}$$

Incompressible interface If the monolayer is incompressible, then interface is neither created nor destroyed. The bubbles undergo uniform translation (the bubble "slides without rolling") and the gas/liquid interface slides at the speed of the bubble. This increases the velocity gradient in the wetting film (which has characteristic size d^2): the side of the wetting film in contact with the wall is stationary, while the other side moves at the speed of the bubble, so that the dominant contribution to the hydrodynamic dissipation comes from the shear of the wetting film. The flow in the film induces an overpressure which leads to an increase in film thickness[10] and leads once again to a non-quadratic dependence of the dissipation on velocity [17]. The experimental data are well fitted by the expression

$$\mathcal{D}_{\text{wall, i}} \sim \frac{\eta U^2 Ca^{-1/2}}{d}. \tag{4.46}$$

In this case the stress at the wall is:

$$\sigma_{\text{wall, i}} \sim \frac{\gamma Ca^{1/2}}{d}. \tag{4.47}$$

FIG. 4.14 shows experimental results that illustrate both behaviours [16]. The missing prefactors in eqs. (4.45) and (4.47) depend strongly on the liquid fraction and on the nature of the surfactants, but not on the viscosity of the solution, which is accounted for in the capillary number.

[10] This is the principle of aquaplaning.

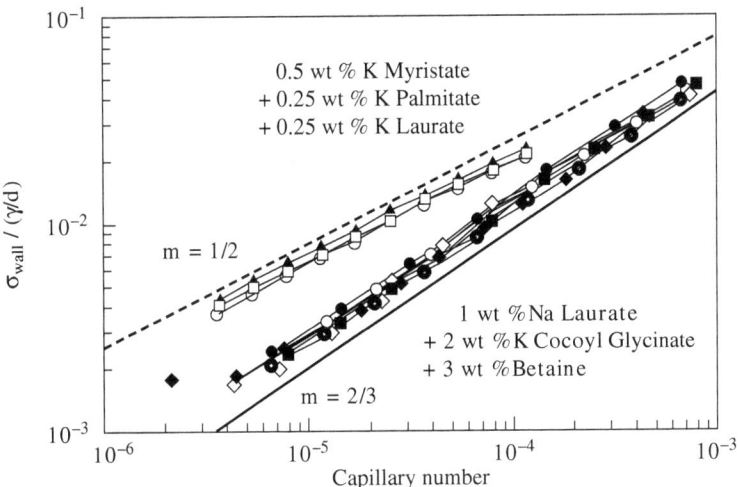

FIG. 4.14: Wall stress as a function of capillary number for two different foaming solutions. The different symbols for each surfactant mixture correspond to different glycerol concentrations. Solutions with perfectly compressible interfaces exhibit a power law with exponent $m = 2/3$ (lower curves, eq. (4.45)) and solutions with incompressible interfaces exhibit a power law with exponent $m = 1/2$ (upper curves, eq. (4.47)). The wall stress for a given surfactant mixture barely depends on the glycerol concentration, indicating that the role of solution viscosity is correctly represented in the predictions [16].

4 The multiscale character of foam rheology

A foam behaves as a solid or as a liquid depending on the liquid fraction, the applied stress, or the typical time-scale of a perturbation (FIG. 4.15). As we saw in §3, elasticity is due to the surface tension of the gas/liquid interfaces, and plasticity is associated

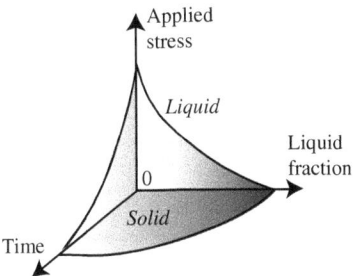

FIG. 4.15: Schematic diagram of the mechanical behaviour of a foam as a function of the applied stress, the liquid fraction, and the characteristic time-scale of a mechanical perturbation.

194 *Rheology*

with T1s. During flow, the dissipation of mechanical energy is dominated by creep or plasticity at low shear rates (quasi-static regime), and by (surface or liquid) viscosity at high shear rates. We now look at how the mechanisms acting at the scale of films and bubbles influence the elastic, plastic or viscous behaviour of "real" foams, which have a finite liquid fraction and are often polydisperse.

4.1 Solid behaviour

4.1.1 Static linear elasticity We have seen (§3.1) that dry foams, whether ordered or not, become softer (lower shear modulus) as the bubbles become larger, in both 2D and 3D. We now turn to the shear modulus of *wet* polydisperse foams, and the non-linear elastic response at large strains, for example due to normal stresses.

Effect of polydispersity The shear modulus G of a monodisperse foam depends on the inverse of the bubble radius (§3.1.1). Which characteristic length determines G in a polydisperse foam? The average bubble radius $\langle R_V \rangle$ (with $R_V = (3V/4\pi)^{1/3}$) is not always the best choice. This is clear in the case of a bidisperse foam constructed by replacing each vertex of a monodisperse foam with a small three-sided bubble (in 2D) or a tetrahedron (in 3D): this construction significantly changes the average bubble size but does not modify the structure of the surrounding foam.[11] Shearing this foam barely deforms the small bubbles (see exercise 7.2), and so the shear modulus of this decorated foam is close to that of a foam consisting only of large bubbles. For this special case the right length-scale needed to estimate the shear modulus is hence the size of the large bubbles.

For real disordered foams, there is no clear rule. When a foam is sheared, the interfacial area increases in proportion to S_{int}, the area of the interfaces at rest. In dimensional terms, the surface energy density $\mathcal{E}$ varies in proportion to the surface to volume ratio $\langle R_V^2 \rangle / \langle R_V^3 \rangle$, and we know that $\mathcal{E} = G\varepsilon^2$ (eq. (4.2)). We therefore introduce the *Sauter mean radius*, $R_{32} = \langle R_V^3 \rangle / \langle R_V^2 \rangle$ (see §3.2.2, chap. 5), which is

FIG. 4.16: Small tetrahedral bubble decorating a vertex of a large bubble. Photograph courtesy of V. Labiausse.

---

[11] by analogy with the 2D decoration theorem (§5.3, chap. 2).

often about 10% to 20% greater than R_V for real foams (FIG. 5.19 and eq. (2.19)). Then for a disordered 3D foam $G \approx \gamma/R_{32}$. The 2D case is dealt with in more detail in reference [37] and exercise 7.2.

Effect of liquid fraction A dry foam is stiffer than a wet foam for a given bubble size. When the liquid fraction increases, G decreases, and finally goes to zero at the volume fraction of close packing, denoted ϕ_l^*. In fact, at this point, the contacts between bubbles disappear since the bubbles are spherical (FIG. 4.17 gives the value of ϕ_l^* for different structures). Moreover, at ϕ_l^*, under the effect of shear, the bubbles remain spherical, the interfacial energy is constant, and no restoring force opposes the deformation: the foam loses its elasticity.

By reasoning in 2D, we can understand how G decreases with ϕ_l. We consider a hexagonal foam "decorated" with Plateau borders (FIG. 4.18). Batchelor's expression

Structure	φ_l^*
Ordered 2D (hexagonal)	$1 - \dfrac{\pi}{2\sqrt{3}} \approx 0.09$
Disordered 2D	0.16
Ordered 3D (face-centred cubic)	$1 - \dfrac{\pi}{3\sqrt{2}} \approx 0.26$
Disordered 3D	0.36

FIG. 4.17: Liquid fractions at which monodisperse foams lose their elasticity. For disordered structures, ϕ_l^* corresponds to the density of a random close packing of solid discs (in 2D) or spheres (in 3D).

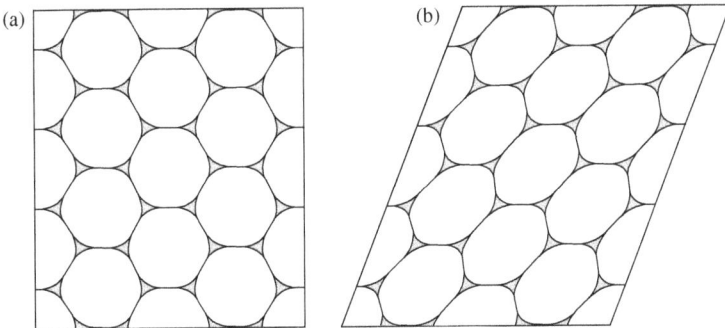

FIG. 4.18: A hexagonal 2D foam "decorated" at a liquid fraction of a few percent. (a) At rest. (b) After shear: unlike the films, the Plateau borders are not affected by the shear.

for the shear stress due to the interfaces (eq. (4.18)) consists of two contributions: one associated with the films (the contact zones between bubbles) and one with the Plateau borders. However, in this case the latter contribution is zero because of their three-fold symmetry (which is preserved during shear, see exercise 7.2). Consequently, only the films contribute to the shear modulus. As ϕ_l increases, the bubble shape changes from a hexagon to a disc, and the film length decreases whilst the Plateau border areas increase (§4.1, chap. 2). From this we deduce that G decreases with ϕ_l.

For a *hexagonal* wet foam (FIG. 4.18), Princen [49] gave a geometric argument to show that the variation in the modulus with liquid fraction is

$$\hat{G}_{\text{hex}} = 0.52 \frac{\lambda}{R_A} (1 - \phi_l)^{\frac{1}{2}} \quad \text{for} \quad \phi_l < \phi_l^*, \quad (4.48)$$

where R_A is the equivalent circular radius ($R_A = (A/\pi)^{1/2}$ for a bubble of area A), and λ is the line tension. As expected, the modulus decreases as ϕ_l increases. At $\phi_l = \phi_l^*$ all the contacts between bubbles disappear and $\hat{G}$ drops discontinuously to zero (FIG. 4.19).

For a *disordered* 2D foam, the loss of contacts is a statistical phenomenon: the foam continuously loses its elasticity with increasing ϕ_l and G eventually vanishes at ϕ_l^*, when the network of contacts between bubbles is no longer capable of transmitting a static stress [21, 61]. Likewise, for disordered 3D foams, G decreases continuously until ϕ_l^*. Experimental observations [46] of the variation of G with ϕ_l and with the Sauter mean radius R_{32} are well described by the phenomenological expression (FIG. 4.19)

$$G = 1.4 \, (1 - \phi_l)(\phi_l^* - \phi_l) \frac{\gamma}{R_{32}}. \quad (4.49)$$

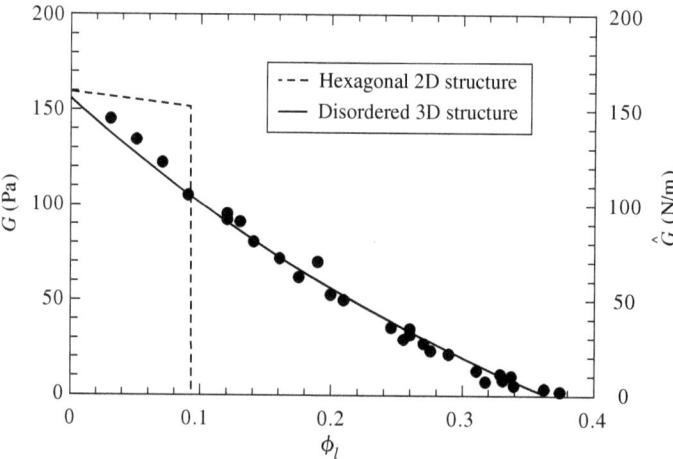

FIG. 4.19: Shear modulus G as a function of liquid fraction ϕ_l for a disordered 3D foam ($R_{32} = 66$ μm and $\gamma = 20$ mN.m^{-1}, data from [54]). The continuous line is eq. (4.49), and the dotted line is for a 2D hexagonal foam (eq. (4.48)) calculated with $R_A = 66$ μm and $\lambda = 20$ mN (right-hand axis).

This expression is in good agreement with numerical simulations for dry 3D foams [34] ($\phi_l = 0$ and $\phi_l^* = 0.36$) which give $G = 0.51\gamma/R_{32}$.

Physical origin of the rigidity loss at ϕ_l^* In the vicinity of ϕ_l^*, the coupling between strain and surface energy depends on the average number Z_c of contacts between bubbles, and on the force required to bring the bubbles together (the "contact force"), opposing the repulsion between bubbles as they seek to reduce their surface area and be as round as possible. To make this simpler, we describe this repulsion schematically using a model of a compressed spring in the shaded section below. The spring constant k and Z_c decrease when the liquid fraction (and hence the distance between bubble centres) increases. In an ordered 2D foam, the springs are harmonic and of constant stiffness $k \sim \gamma/R_A$. As all the contacts disappear simultaneously at ϕ_l^* (Z_c jumps from 6 to 0), the modulus is discontinuous at this liquid fraction (FIG. 4.19). In an ordered 3D foam, the contacts also disappear at ϕ_l^*, and G goes to zero, but we expect a continuous variation in G because of the anharmonic character of the capillary forces; indeed, k depends on the distance between bubbles and goes to zero as ϕ_l tends towards ϕ_l^*.

Interaction between neighbouring bubbles in a foam
When $\phi_l < \phi_l^*$, neighbouring bubbles repel each other because of the surface energy required to deform them. This is really an interaction between many bubbles, but to understand it at a given liquid fraction,[12] we take an initially circular 2D bubble of radius R_A and squeeze it between two parallel planes (to imitate the contact between bubbles) whilst conserving its area (FIG. 4.20). When equilibrium is reached, the perimeter of the bubble is again minimal, implying that the two ends are circular arcs with radius L; if L' is the length of contact with each plane (exaggerated in the figure), we have to leading order ($\delta/R_A \ll 1$) $L'L \sim \delta R_A$. Consequently, the force required is $F \sim (\lambda/L)L' \sim \lambda\delta/R_A$, with λ the line tension. Close to ϕ_l^* therefore, the repulsive force between neighbouring bubbles can be represented as a harmonic spring with stiffness $k \sim \lambda/R_A$. Here again, we observe that a foam is stiffer when the bubbles are smaller.

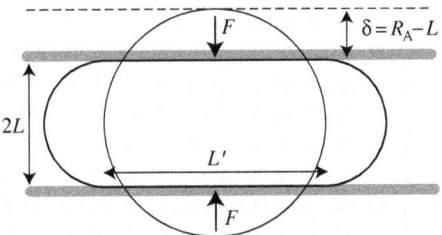

FIG. 4.20: A 2D circular bubble of initial radius R_A squeezed between two planes, keeping the area constant, with a force per unit length F.

---

[12]The changes in surface energy associated with a change in liquid fraction are described by the osmotic pressure of the foam in §4.2, chap. 2.

In 3D, an initially spherical bubble squeezed between two planes adjusts its two radii of curvature in order to minimize its surface area at constant volume. A numerical calculation shows that for small deformations ($\delta/R_V \ll 1$), the energy (per unit surface area) required to trap the bubble is [39] $\Delta E \sim \gamma (\delta/R_V)^\alpha$ where $\alpha > 2$ ($\alpha = 2.5$ for a monodisperse wet foam with a face-centred cubic structure). Thus in 3D, the contact between two bubbles is described by anharmonic springs with stiffness $k \sim \gamma (\delta/R_V)^{\alpha-2}$ that goes to zero at ϕ_l^*, the liquid fraction at which $\delta = 0$.

4.1.2 Static non-linear elasticity A shear deformation close to the yield strain but insufficient to trigger T1s induces stretches and rotations of films which creates normal stress differences (p. 175). In general, the first normal stress difference (eq. (4.17)) induced by a quasi-static shear in an isotropic elastic material obeys Poynting's law (exercise 7.3):

$$N_1 = \sigma \, \varepsilon = G \, \varepsilon^2. \tag{4.50}$$

In a sheared foam, the soap films are stretched and oriented in a preferential direction (FIG. 4.21a). When the stress $\sigma = \sigma_{xy}$ is large, the film normals tend to be aligned in the y direction. Consequently, the normal stresses σ_{xx} and σ_{zz}, due to the surface tension of the films, are large compared to σ_{yy}. As the x and z directions are equivalent in this respect, we have $\sigma_{xx} \approx \sigma_{zz}$, and we expect that qualitatively [40] $N_1 \approx \sigma_{xx}$ and $N_2 \approx -\sigma_{zz} \approx -N_1$. This effect is different from that observed for polymers, in which the chains tend to be aligned only in the direction of the applied shear (FIG. 4.21b), so that σ_{xx} is large compared to σ_{yy} and σ_{zz}, and $N_1 \approx \sigma_{xx}$ and $N_2 \approx 0$.

A model based on individual films, generalizing the approach of Derjaguin (p. 181), leads to the following tensorial constitutive law, valid as long as the yield strain is not exceeded [30]:

$$\sigma_{ij} = -p \, \delta_{ij} + \frac{G}{7}(B_{ij} - 6B_{ij}^{-1}). \tag{4.51}$$

In this (Mooney–Rivlin) law, finite deformations are described by the Finger tensor B_{ij} (eq. (4.13)). It predicts a linear relationship between shear stress and strain, accompanied by a non-linear variation of normal stresses. For a finite applied shear, the expression for B_{ij} is given in exercise 7.3, and eq. (4.51) predicts

$$\sigma = G\varepsilon \qquad N_1 = G\varepsilon^2 \qquad N_2 = -\frac{6}{7}G\varepsilon^2. \tag{4.52}$$

As expected, N_2 is of the same order of magnitude as N_1 and of opposite sign. The quadratic dependence of N_1 on ε is confirmed in numerical simulations (FIG. 4.22a) and experimentally up to the yield strain (FIG. 4.23). Moreover, eq. (4.52) predicts that the normal stress difference varies in proportion to the elastic shear modulus G. Consequently, in foams which are soft and have a low yield strain, N_1 is always

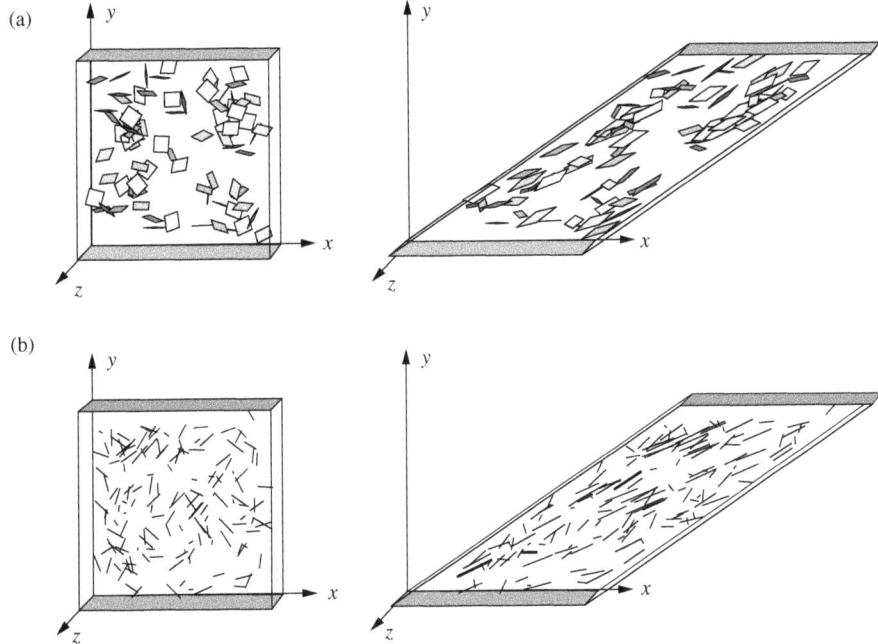

FIG. 4.21: Shear applied to a cube containing elastic elements that are randomly oriented in space and represented by: (**a**) facets which represent the films in a foam; (**b**) strands which represent polymer chains. Under the effect of shear, the facets tend to be aligned parallel to the (x, z) plane and the strands tend to align in the x direction.

small compared to the shear stress. Typically for $G \approx 100$ Pa and $\varepsilon \approx 0.1$, we have $N_1 \approx 1$ Pa! Polymers, which are more elastic (with G of the order of MPa) and can undergo strains ε of order one (think of rubber), show much greater normal stress differences, since we can have $N_1 \approx \sigma$. Finally, we note that these normal stresses also induce an effect in steady shear flow: the Weissenberg effect, observed in polymeric liquids and, to a lesser extent, in foams, in which the fluid climbs the rotating rod driving the flow.

4.1.3 Linear viscoelasticity

Frequency response In §3 we identified the different modes of dissipation associated with bubble motion, either with or without topological changes (T1s): the viscosity of the solution, interfacial viscosity, and surfactant transport. In the limit of small strains, much lower than the yield strain, they cause a foam to behave as a linear viscoelastic medium whose response depends on the time-scale of the perturbation, of the order of $1/\omega$ for an oscillatory experiment with frequency ω. The response is characterized by the complex shear modulus $G^*(\omega)$ (eq. (4.9)), and the variations of $G^*(\omega)$ are similar to those observed in other disordered soft materials such as concentrated emulsions and pastes. They show a large spectrum of relaxation times, associated with

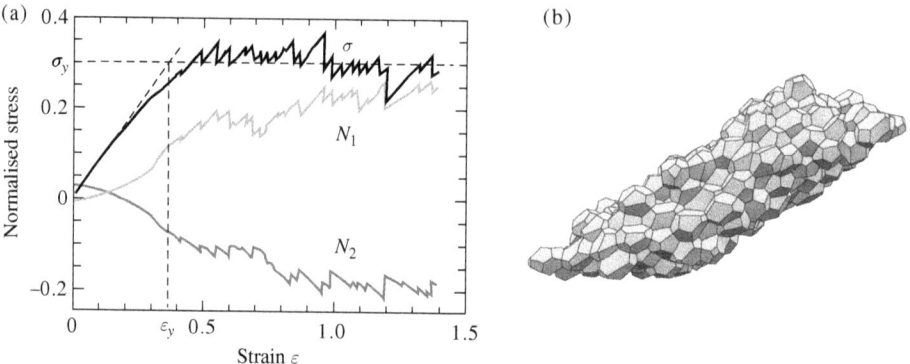

FIG. 4.22: **(a)** Shear stress and first and second normal stress differences as a function of the (quasi-static) strain for a monodisperse, disordered, dry 3D foam (*Surface Evolver* simulation [34]). The stresses are normalized by $\gamma/V^{1/3}$, where V is the bubble volume. Below the yield strain ε_y, we observe that σ is linear in ε, while N_1 and N_2 vary as ε^2 (eq. (4.52)). Beyond ε_y, the stress fluctuates around the yield stress σ_y due to T1s (see §5). **(b)** The structure of a foam of 216 bubbles at a strain of $\varepsilon = 1.4$. Image courtesy of A. Kraynik.

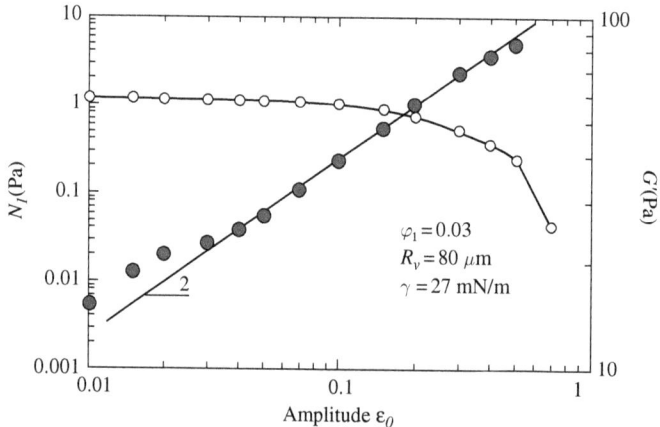

FIG. 4.23: First normal stress difference N_1 (●) and shear modulus G' (○) as a function of the amplitude of the applied strain ε_0 (the yield strain is $\varepsilon_y \approx 0.4$). N_1 and G' are measured by an oscillatory method at a frequency of 2 Hz (cf. FIG. 5.11). G' is constant if the amplitude is much lower than the yield strain ε_y, then decreases beyond that. N_1 is quadratic in ε (eq. (4.52)), except at very low amplitudes where the deviation is due to residual trapped stresses (cf. p. 177). Data from [38].

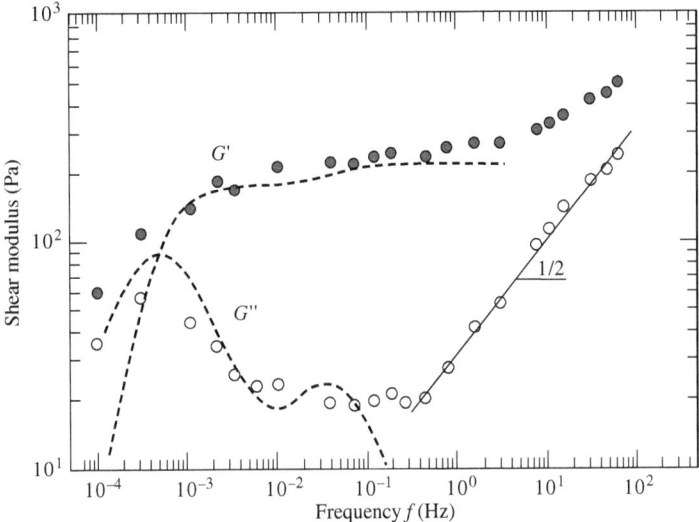

FIG. 4.24: Complex shear modulus of a foam (eq. (4.9)) as a function of frequency, in the linear regime. The sample is a shaving foam that has been left to coarsen for 100 min, with $\phi_l = 0.08$. The dashed curves are the Laplace transform of the creep compliance (eq. (4.54)) whilst the straight line of slope 1/2 is eq. (4.53). Data from [11, 12, 27].

different dissipative mechanisms, which have been measured over approximately six decades in frequency. FIG. 4.24 shows that G' grows monotonically (with a plateau at intermediate frequencies) while G'' goes through a maximum, then a minimum, then a power law increase.[13] We next discuss these different responses one-by-one.

Very rapid relaxations In order to estimate the shortest relaxation time, we consider an individual soap bubble in air. A bubble of size d behaves roughly like a spring with stiffness $G \sim \gamma/d$ (cf. the shaded section on p. 197). If the movement is restricted by the liquid flow in the film, the effective viscosity is $\eta_{\text{eff}} \sim \eta d/h$ (eq. (4.38)), where h is the thickness of the soap film and η the viscosity of the solution (cf. §3.3.1). The relaxation time is $\tau \sim \eta_{\text{eff}}/G$ (cf. FIG. 4.2), i.e. $\tau \sim \eta d^2/\gamma h$; it is of the order of 10^{-6} to 10^{-3} s for $\gamma = 30$ mN.m^{-1}, $\eta = 1$ mPa.s, $d = 30$ µm and h between 20 nm and 1 µm, which corresponds to a characteristic frequency $f = 1/(2\pi\tau)$ of 10^2 to 10^5 Hz. Such a relaxation cannot be observed in the measurement window of FIG. 4.24.

Rapid relaxations At high frequency (between 1 and 10^2 Hz on FIG. 4.24), we can fit the data by

$$G^*(\omega) = G(1 + \alpha \sqrt{i\omega}), \qquad (4.53)$$

---

[13] Note that G'' does not give information about the viscosity of a foam in steady flow, which we will discuss in §4.3.1.

where G is the static elastic (shear) modulus (of the order of 250 Pa in FIG. 4.24). The dependence of G^* on $\omega^{1/2}$ has been suggested to be a consequence of collective processes at the scale of a few bubbles; that is, the structural disorder creates some weak, anisotropic regions, somewhat like cracks in a continuous medium. During shear, the bubbles slide past the weak regions and energy is dissipated by viscous friction. The random orientation of these regions causes a distribution of relaxation times, which leads to square-root dependence [42].

A very different process, due to the interfacial dilatational elasticity E_s^* (§2.2.3, chap. 3), could also contribute to the dissipation. In fact, dimensionally, G^* varies as E_s^*/d. If we assume that the interfacial elasticity is controlled by the bulk diffusion of surfactants, E_s^* is given by eqs. (3.17) and (3.18), and, if the period of the deformation is sufficiently long that the surfactants have time to diffuse and repopulate the interface, we find $E_s^* \sim \sqrt{i\omega}$. We then presume that this contribution is added to the static modulus G to give eq. (4.53) [4].

Elastic plateau At intermediate frequencies (10^{-3} to 1 Hz), we observe in FIG. 4.24 a plateau in the storage modulus G' and a minimum in G''. In this regime, elasticity dominates the response and G' corresponds to the static shear modulus G (p. 178). In describing static elasticity (§4.1.1), we have implicitly assumed that this regime fully describes the low-frequency response.

Slow relaxations The quasi-static regime (seen in §4.1.1) does not apply at very low frequencies ($f < 10^{-3}$ Hz). Here G' tends towards zero and G'' reaches a maximum before tending towards zero. This maximum of G'' is the signature of a process in which the characteristic time $\tau = 1/(2\pi f)$ (which is around 300 seconds in FIG. 4.24) is linked to coarsening. At these long times, the foam coarsens during each oscillation, and it is therefore difficult to decouple the ageing from the relaxation.[14] A better way to probe the long-time response is to perform a creep test (cf. p. 172) as we will see in the next paragraph.

Time response In the linear regime, the frequency and time responses are equivalent (§2.2). Consequently, a creep test is an appropriate way to examine the slow relaxations of a foam as it coarsens: it can be short enough to give information about the viscoelasticity at a specific foam age. Recall that this test probes the linear response by applying to the sample a step stress σ lower than the yield stress, so as not to trigger T1s (§2.2). The strain $\varepsilon(t)$ is then measured over a sufficiently short time that the properties of the foam (bubble size, elastic modulus, rate of coarsening-induced T1s) remain constant, but long enough that coarsening induces a significant number of T1s.

FIG. 4.25a shows the compliance $J(t) = \varepsilon(t)/\sigma$ measured for three foams with different coarsening rates in response to a step stress applied for $\Delta t = 100$ s. In all three cases, $J(t)$ exhibits an instantaneous compliance $J(0) = 1/G$. Then we observe

---

[14] We presume that the effects of drainage are negligible, which a comparison of the orders of magnitude (eq. (2.48)) suggests is possible in practice for a sample of this height over these timescales, cf. FIG. 2.23. Coarsening can be considerably slowed down by adding insoluble gases, see FIG. 3.27 and §3.3.4, chap. 3.

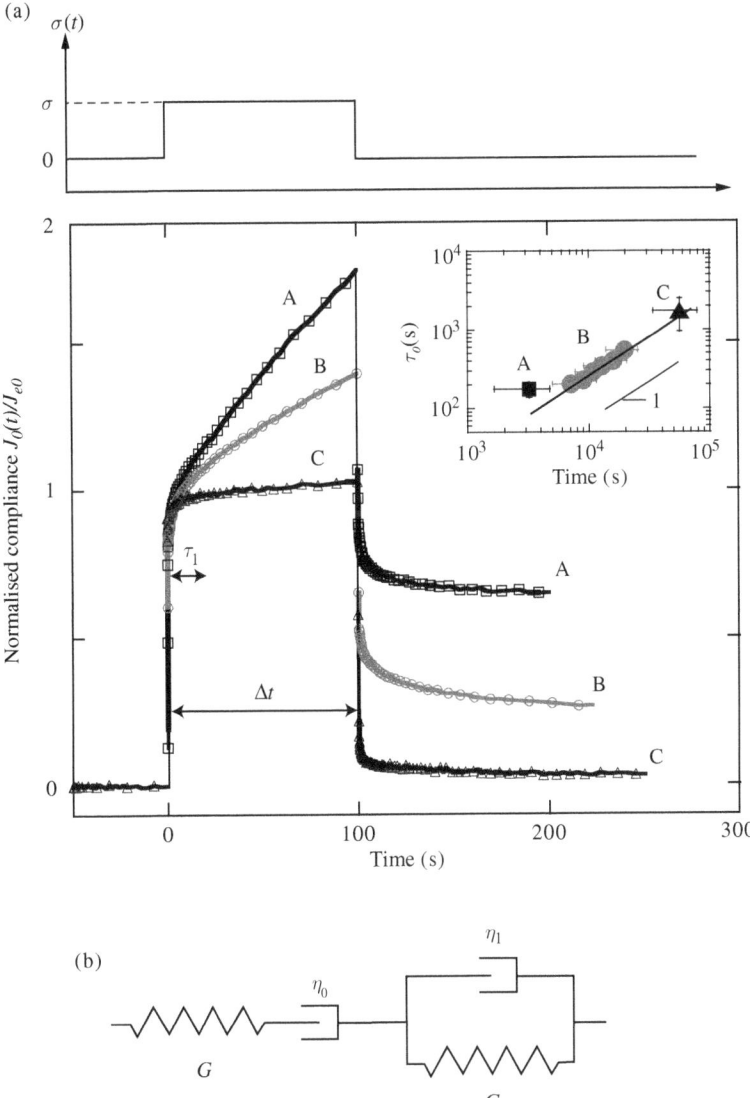

FIG. 4.25: (a) Linear creep response of a foam subjected to a constant stress $\sigma \ll \sigma_y$ for $\Delta t = 100$ s. The three foams A, B, and C (with $\phi_l = 0.07$) are made with gases of different solubilities, in such a way that the coarsening rate, and therefore the rate of topological changes induced by coarsening, decreases from A to C. The compliance is normalized by $J_{e0} = (1/G + 1/G_1)$ (eq. (4.54)). The inset shows the time $\tau_0 = \eta_0 J_{e0}$ deduced from the experiment as a function of the average time T between T1s in a volume $(2R_V)^3$ measured independently by multiple light scattering (see §1.2.5, chap. 5). Data from [12]. (b) Diagrammatical representation of the behaviour observed in (a).

a transient relaxation lasting for a time τ_1, of the order of a few seconds. The flow then reaches a steady state in which $J(t)$ increases linearly with time, more quickly if the foam coarsens quickly. Finally, at $t = \Delta t$, the stress is brought to zero. The foam recovers the elastic part of the strain and retains a residual strain due to the flow. In the time interval $0 < t < \Delta t$, the compliance behaves as

$$J(t) = \frac{1}{G} + \frac{t}{\eta_0} + \frac{1 - \exp(-tG_1/\eta_1)}{G_1}. \tag{4.54}$$

A Laplace transform of $J(t)$ gives the complex modulus $G^*(\omega)$ at low frequency (p. 173). This is in good agreement with oscillatory experiments, as in FIG. 4.24, which is a signature of the linearity of the response. More specifically, $J(t)$ may be represented by a Maxwell liquid (G, η_0) and a Kelvin–Voigt solid (G_1, η_1) in series (FIG. 4.25b). In the stationary regime (over a long time), the foam flows slowly with a viscosity η_0 such that

$$J(t) = J_{e0} + \frac{t}{\eta_0} \tag{4.55}$$

with $J_{e0} = (1/G + 1/G_1)$. In this way the creep test generates two relaxation processes, with characteristic times $\tau_1 = \eta_1/G_1$ for the transient process (τ_1 is of the order of a few seconds in FIG. 4.25a), and $\tau_0 = \eta_0 J_{e0} \approx \eta_0/G$ for steady creep (τ_0 is typically a few minutes in FIG. 4.25a). The parameters G, G_1, η_0, and η_1 depend on the physico-chemical characteristics of the foam. Experimentally, we find that: (1) $G \propto \gamma/R_V$, in agreement with eq. (4.49); (2) $G_1 \propto \gamma/R_V$, but with a larger prefactor ($G_1 \gg G$); (3) steady flow slows down (η_0 increases) when coarsening slows; (4) $\eta_1 \propto 1/R_V$. Although η_0 and η_1 have the dimensions of a viscosity, neither depends significantly on the viscosity of the solution [12, 45]. Likewise, τ_0 increases during coarsening but τ_1 is constant. These two relaxations therefore have very different origins at the scale of bubbles and films, as we will now describe.

4.1.4 Coupling between viscoelasticity of a foam and interfacial rheology

Under stress, a foam is deformed rapidly and acquires a non-equilibrium structure which is governed by viscous forces and in which surface area is not minimized. The transient regime with characteristic time τ_1 observed in FIG. 4.25a corresponds to a relaxation towards an equilibrium structure, during which the area and orientation of the films evolves in order to minimize the interfacial energy. The forces exerted by the films on the Plateau borders drive the dynamics: a film of length L pulls on a border with a force of the order of the surface tension γ, and resists the deformation principally by its interfacial dilatational viscosity η_d (eq. (3.10)). A scaling argument for the dynamics of the film suggests that $\gamma \sim (\eta_d/L)\, dL/dt$ [22] with a characteristic time $\tau_1 \sim \eta_d/\gamma$ independent of bubble size. Typically, $\gamma \approx 30$ mN.m^{-1} and $\eta_d \approx 10^{-3}$ to 10^{-2} kg.s^{-1} for interfaces with high dilatational modulus, which gives a time τ_1 of the order of a fraction of a second, close to the value measured for shaving foam (around 1 s). Note that this is similar to the dynamics of the relaxation of a film after a T1 [5, 20]. Depending on the foaming solution, other dissipative processes associated with the surfactants can play a role in the coupling between the viscoelasticity of the foam and that of its interfaces (§3.3.2).

4.1.5 Coupling between slow relaxation and coarsening-induced T1s

Gas diffusion between neighbouring bubbles intermittently leads to unstable mechanical configurations (cf. FIG. 3.3) and the bubbles undergo T1s (to minimize the surface energy locally) with an average frequency characteristic of coarsening (FIG. 3.25). A T1 causes a jump in the macroscopic strain: in a random direction in the absence of stress, but biased in the shear direction if a stress is applied. Thus, at steady state a foam slowly flows under a low stress like a Maxwell liquid (cf. p. 204). A scaling argument suggests that the creep rate $\dot{J}(t) = \dot{\varepsilon}(t)/\sigma$ is proportional to the compliance J_{e0}, the average volume V_r of the foam affected by a T1, and to the number ω_r of T1s induced by coarsening per unit time and volume [12, 59]:

$$\dot{J}(t) \approx J_{e0}\,\omega_r\,V_r, \tag{4.56}$$

where we have omitted a prefactor of order one.

The phenomenon of creep was described at the macroscopic scale on p. 202. Comparing the creep rates given by eqs. (4.55) and (4.56) allows us to link the macroscopic viscosity to the bubble scale process, giving

$$\frac{1}{\eta_0} = J_{e0}\,\omega_r\,V_r. \tag{4.57}$$

According to this relationship, the characteristic creep time, $\tau_0 = \eta_0 J_{e0}$, varies in proportion to the average time interval $1/\omega_r$ between T1s in a unit volume of foam. By simultaneous *in situ* measurement of $J(t)$ by rheometry and ω_r by diffusing-wave spectroscopy (DWS, see §1.2.5, chap. 5), this linear variation can be observed, illustrated in the inset to FIG. 4.25. We deduce that, for a foam of 7% liquid fraction, V_r is of the order of $(6R_V)^3$ which is a volume of around 50 bubbles, but it is not yet understood why [14].

Scaling law for the complex shear modulus During coarsening, the average bubble size R_V slowly increases (eq. (3.38)) and the frequency ω_r of T1s decreases. As the age of the foam t_a increases, we observe in FIGS. 3.27 and 3.25 that $R_V \sim t_a^{0.5}$ and $\tau^{-1} \sim \omega_r V_r \sim t_a^{-0.66}$. The (Maxwell-type) relaxation induced by T1s is coupled to coarsening and the foam softens, with $G \sim \gamma/R_V \sim t_a^{-0.5}$. From eq. (4.57), with $G_1 \gg G$ (p. 204), the characteristic relaxation time $\tau_0 \approx \eta_0/G = 1/(\omega_r V_r)$ increases as $t_a^{0.66}$, which shifts the maximum of G'' (located at $f = 1/(2\pi\tau_0)$) towards lower frequencies (FIG. 4.24). We therefore find a simple scaling law for the evolution of the complex shear modulus as a function of frequency and age, illustrated in FIG. 4.26:

$$G^*(\omega, t_a) = b(t_a, t_0)\, G^*(\omega\, a(t_a, t_0), t_0). \tag{4.58}$$

The factors $a(t_a, t_0)$ (for the frequency) and $b(t_a, t_0)$ (for the moduli), determined in comparison to a reference age t_0, allow the moduli at a later time t_a to be determined. See the shaded section overleaf. The linear viscoelasticity of foams at low frequency thus shows a "time–age" equivalence, reminiscent of the equivalence between time and temperature.[15]

---

[15]It is known as the Williams–Landel–Ferry law in polymer rheology [41].

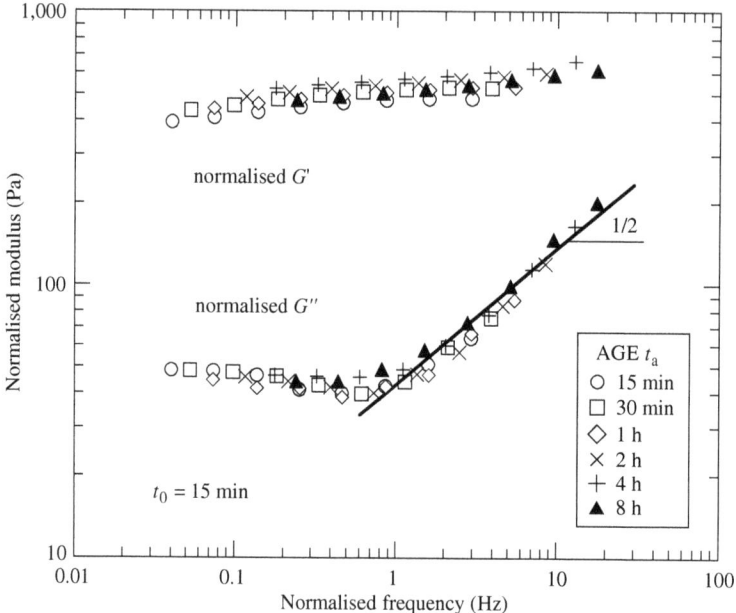

FIG. 4.26: Master curves for the complex shear modulus (eq. (4.58)), representing the normalized moduli $G'(\omega, t_a)/b(t_a, t_0)$ and $G''(\omega, t_a)/b(t_a, t_0)$ as a function of normalized frequency $f\, a(t_a, t_0)$. The sample is a shaving foam with $\phi_l = 0.08$, and the moduli are measured at ages from 15 min (reference age) to 8 hours. Data from [30].

"Time–Age" equivalence

At low frequency, the complex shear modulus $G^*(\omega, t_a)$ depends on the frequency ω and the age t_a (eq. (4.58)). By definition, the factors $a(t_a, t_0)$ and $b(t_a, t_0)$ are equal to one at the reference age $t_a = t_0$ (FIG. 4.27). In this regime, if we ignore the relaxation due to the interfaces (§4.1.4), the complex modulus is that of a Maxwell element, with elastic modulus G and viscosity η_0:

$$G^*(\omega, t_a) \approx G \left(\frac{i\omega \eta_0 / G}{1 + i\omega \eta_0 / G} \right). \tag{4.59}$$

Comparing eqs. (4.58) and (4.59) gives

$$a(t_a, t_0) \propto \frac{\eta_0(t_a)}{G(t_a)} \quad \text{and} \quad b(t_a, t_0) \propto G(t_a), \tag{4.60}$$

suggesting that, according to eq. (4.57), the factor a increases with the average time between T1s induced by coarsening, $1/(\omega_r V_r)$, and that b decreases with $1/R_V$. It is an effect that can be observed experimentally (FIG. 4.27).

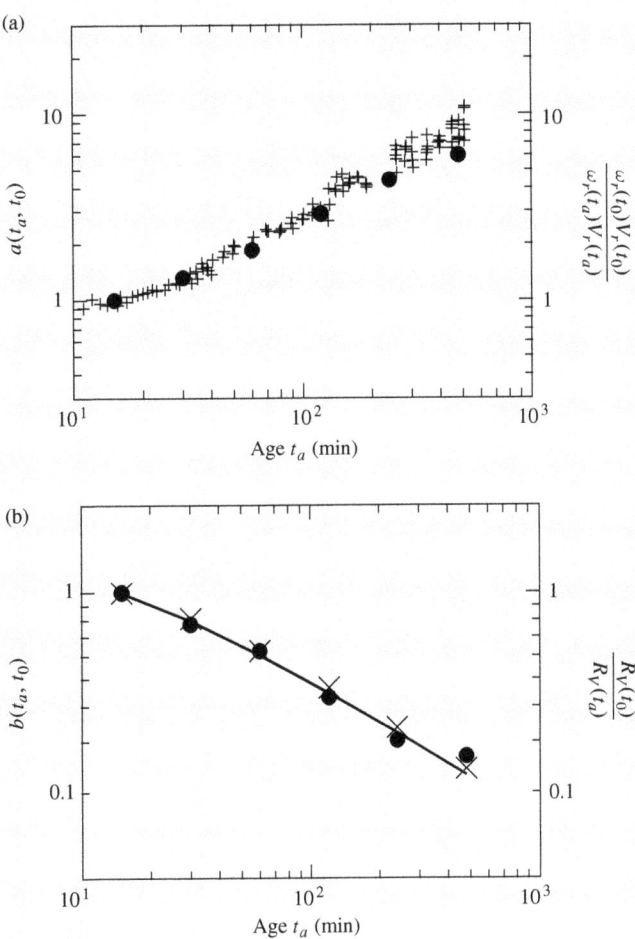

FIG. 4.27: Scaling for frequency $a(t_a, t_0)$ and for modulus $b(t_a, t_0)$ corresponding to the master curve in FIG. 4.26. **(a)** $a(t_a, t_0)$ (•) is compared to $\omega_r(t_0) V_r(t_0)/(\omega_r(t_a) V_r(t_a))$ (+), measured by DWS (§1.2.5, chap. 5). **(b)** (•) $b(t_a, t_0)$ is compared to $R_V(t_0)/R_V(t_a)$ (×), measured by optical microscopy. Data from [30].

4.1.6 Memory and structure Foams, like many other complex fluids, exhibit a rheological "memory" in their microstructure. So the rheological properties of a foam and the T1 dynamics depend strongly on the strain history. This memory effect fades away progressively as the foam coarsens (another example of memory loss with increasing age!).

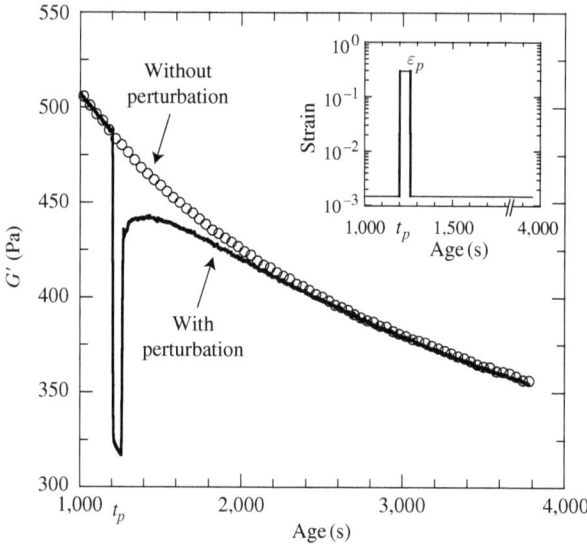

FIG. 4.28: Effect of rheological memory in shaving foam with $\phi_l = 0.08$. The evolution of the shear modulus in time, either during coarsening (○) or following a perturbation according to the protocol described in the inset (continuous line). Data from [31].

In order to demonstrate this property, we allow a foam to coarsen over a time t_p, perturb it briefly with a shear of large amplitude ε_p (of the order of ε_y), then leave it to coarsen again (FIG. 4.28). The shear alters the microstructure without changing the liquid fraction or bubble size. The elastic modulus decreases in response to the perturbation, then increases to reach the unperturbed value. This transient softening is also observed in the glassy dynamics of soft paste, which is interpreted as a rejuvenation of the system by a temporary increase in the rate of T1s, reducing the capacity of the system to store elastic energy, and therefore reducing G'. In foams, on the other hand, the T1 rate ω_r decreases after the perturbation, then relaxes towards the unperturbed value. The changes in G' and ω_r which accompany this "solidification" of the foam occur more slowly as the waiting time t_p becomes longer. The duration of the recovery, or memory, is determined by the time required to restore the characteristic structure and dynamics of a coarsening foam through diffusive gas exchange between bubbles.

4.2 Transition from solid to liquid behaviour

4.2.1 Definition of yield stress and yield strain
At the macroscopic scale, the yield stress σ_y is defined as the maximum shear stress that can be applied to the sample without causing it to flow; the corresponding strain is the yield strain ε_y (p. 171). Shearing a disordered dry 3D foam quasi-statically suggests that at low strain a foam behaves as a linear elastic solid ($\sigma = G\varepsilon$), which deforms plastically when ε is large enough to induce T1 topological changes (cf. §3.2), and finally, when $\varepsilon > \varepsilon_y$, σ fluctuates around σ_y and the foam flows. The characteristics of the stress fluctuations

caused by T1s in the flow regime are described in §5. These are visible in the numerical simulations shown in FIG. 4.22a; in that example the yield stress is

$$\sigma_y \approx 0.3\frac{\gamma}{V^{1/3}} \approx 0.2\frac{\gamma}{R_V} \tag{4.61}$$

and $\varepsilon_y \approx 0.37$. For bubbles of radius $R_V = 30$ μm and surface tension $\gamma = 30$ mN.m^{-1}, this gives $\sigma_y = 200$ Pa. Compared to polycrystalline metals, for which $\varepsilon_y \approx 10^{-3}$ to 10^{-2} and $\sigma_y \approx 10^7$ to 10^9 Pa, foams are much more elastic and flow at lower applied stress.

4.2.2 Experimental methods to measure yielding Various experimental methods (see §1.2.7, chap. 5) are used in order to characterize the yield stress σ_y and yield strain ε_y of a foam:

- Observe a layer of foam slide down an inclined plane under its own weight (see experiment 6.2). The layer thins and the flow stops when the thickness h is such that $\sigma_y = \phi_l \rho_l g h \sin\alpha$, where ρ_l is the density of the solution and α is the angle at which the plane is inclined from the horizontal.
- Measure stress as a function of strain during start-up flow at constant shear rate $\dot{\varepsilon}$ (FIG. 4.29a). The maximum stress is σ_y and the corresponding strain is ε_y. The values of σ_y and ε_y obtained do not depend on $\dot{\varepsilon}$ if it is small enough.
- Measure the limiting steady-state stress when a decreasing shear rate is applied (FIG. 4.29b). Extrapolating the flow curve $\sigma(\dot{\varepsilon})$ to $\dot{\varepsilon} = 0$ gives an estimate of σ_y, known as the dynamic yield stress.
- Measure the complex modulus as a function of the amplitude of an applied oscillation ε_0 at a fixed frequency (FIG. 4.29c). The yield strain ε_y is the point of intersection between the plateau in G' (for $\varepsilon_0 < \varepsilon_y$) and the power law $G' \propto \varepsilon_0^\alpha$ (for $\varepsilon_0 > \varepsilon_y$). σ_y can then be found from $\sigma_y = \varepsilon_y |G^*(\varepsilon_y)|$. Similarly, we could deduce σ_y from the variation in G' in response to an oscillating stress, σ_0.

4.2.3 Non-linear viscoelasticity The response to an oscillatory perturbation of small amplitude σ_0 or ε_0 (less than the yield strain) is linear and dominated by elastic behaviour: $G' \gg G''$, and G' is independent of the amplitude (FIG. 4.29c). On the other hand, above the yield strain the response becomes non-linear. The complex modulus is thus defined in terms of the fundamental frequency according to eq. (4.11), and the response is viscoplastic: $G' \ll G''$ and G' decreases as $\varepsilon_0^{-3/2}$, as shown in FIG. 4.29c.

This visco-elasto-plastic behaviour can be described (FIG. 4.2) by a spring, a dashpot, and a sliding friction element (schematized in FIG. 4.29c). The spring represents the elastic response: its stiffness corresponds to the elastic modulus G. The sliding friction element describes the plasticity: it is rigid if the force applied to it is lower than the yield stress σ_y, but it slides freely beyond it. The dashpot represents the viscous dissipation, with Newtonian viscosity η given by the loss modulus in the linear regime, such that $G'' = \eta\omega$ [44]. This phenomenological model gives a good prediction of the variations of G^* with strain amplitude shown in FIG. 4.29c.

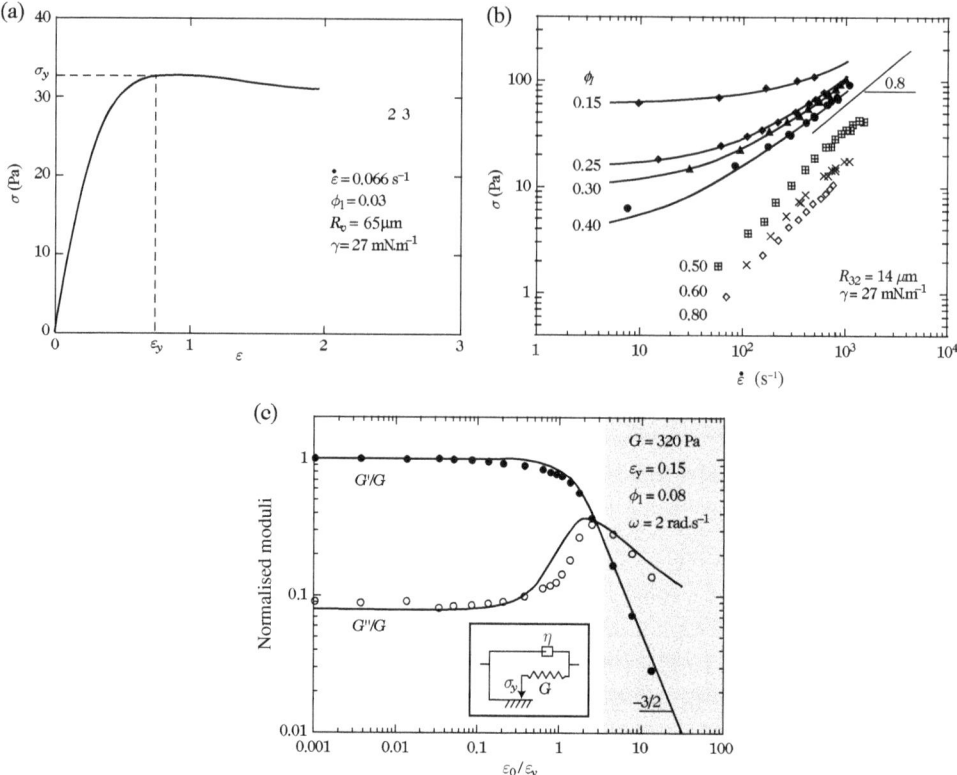

FIG. 4.29: Measuring the yield stress and yield strain of a foam using the methods described in §1.2.7, chap. 5. (a) Stress as a function of strain during start-up flow at constant shear rate $\dot{\varepsilon}$ (parallel-plate rheometer). Data from [32]. (b) Flow curves $\sigma(\dot{\varepsilon})$ of foams used in oil-well drilling, measured at steady state using a capillary rheometer under an applied pressure drop of 50 bars. The continuous lines represent fits to a Herschel–Bulkley relation (eq. (4.6)). Note the absence of a yield stress for $\phi_l > 0.40$. Data from [29]. (c) G' and G'' as a function of the strain amplitude ε_0 measured in oscillatory mode (cylindrical Couette rheometer). The continuous curves represent fits to the elasto-visco-plastic model illustrated, with $\eta\omega/G = 0.08$. The grey area corresponds to the region where the strain is localized (see p. 214). Data from [44].

4.2.4 Effect of liquid fraction

As for the shear modulus, the values of σ_y and ε_y are greatest in dry foams and decrease when ϕ_l increases, reaching zero at ϕ_l^*. In effect, the strain (and therefore the stress) necessary to reduce to zero the length of a Plateau border or the area of a film is smaller when the liquid content is greater (FIG. 4.30).

For disordered 3D foams, measurements in oscillatory mode (FIG. 4.31) are summarized in two empirical expressions [54]:

$$\varepsilon_y = a_1 \left(\phi_l^* - \phi_l\right) \tag{4.62}$$

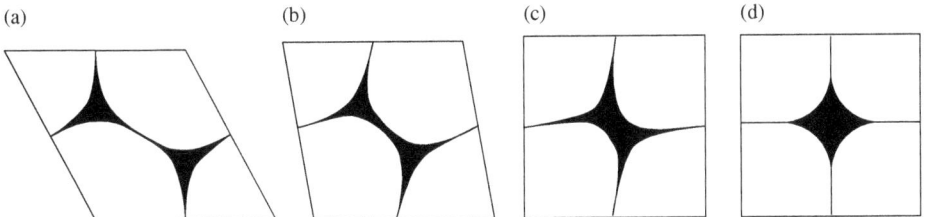

FIG. 4.30: Start of a shear-induced T1 (from **a** to **d**) in an ordered wet 2D foam ($\phi_l = 0.06$). The process is similar to a T1 in a dry foam until the point at which the Plateau borders touch (**b**). The wet vertex is stable at (**c**), but at (**d**) it becomes unstable and drives the completion of the topological change. From [49].

and

$$\sigma_y = a_2 \frac{\gamma}{R_V} (\phi_l^* - \phi_l)^2, \tag{4.63}$$

where $\phi_l^* = 0.36$ is the volume fraction of close-packed spheres (FIG. 4.17), a_1 is a constant between 0.5 and 1, and a_2 is a constant beween 0.2 and 0.5. In 2D, the yielding behaviour is similar [50]. Moreover, the flow curves (FIG. 4.29b) indicate a transition from a viscoelastic liquid (the absence of a yield stress in a dispersion of bubbles) to a yield stress fluid ($\sigma_y \neq 0$) in which the yield stress increases as the foam becomes drier. The transition takes place close to $\phi_l = 0.40$, which is compatible with $\phi_l^* = 0.36$.

4.3 Foam flow

We now turn to the steady flow of a 3D foam resulting from an applied stress greater than σ_y (or an applied strain greater than ε_y), but nevertheless sufficiently weak so as not to induce coalescence or bubble break-up. Current understanding of 3D foam flow is based on experimental results, complemented by numerical simulations and models, which constitute the pieces of an as yet unfinished puzzle. We will review some of them, though not exhaustively.

4.3.1 Effective viscosity
At a macroscopic scale, both experiments and bubble model simulations (§2.2.2, chap. 5) suggest that steady shear of a foam is characterized by the phenomenological Herschel–Bulkley law (eq. (4.6)). The effective viscosity in steady flow depends on the shear rate $\dot\varepsilon$ according to

$$\eta_{\text{eff}} \equiv \frac{\sigma}{\dot\varepsilon} = \frac{\sigma_y}{\dot\varepsilon} + \eta_p \dot\varepsilon^{\beta-1}, \tag{4.64}$$

with an exponent $\beta < 1$ (FIG. 4.32), corresponding to shear-thinning behaviour [17, 26, 29, 32]. Note that the effective viscosity of a foam η_{eff} is several orders of magnitude greater than the solution from which it is made (§3.3 and FIG. 4.32). The exponent

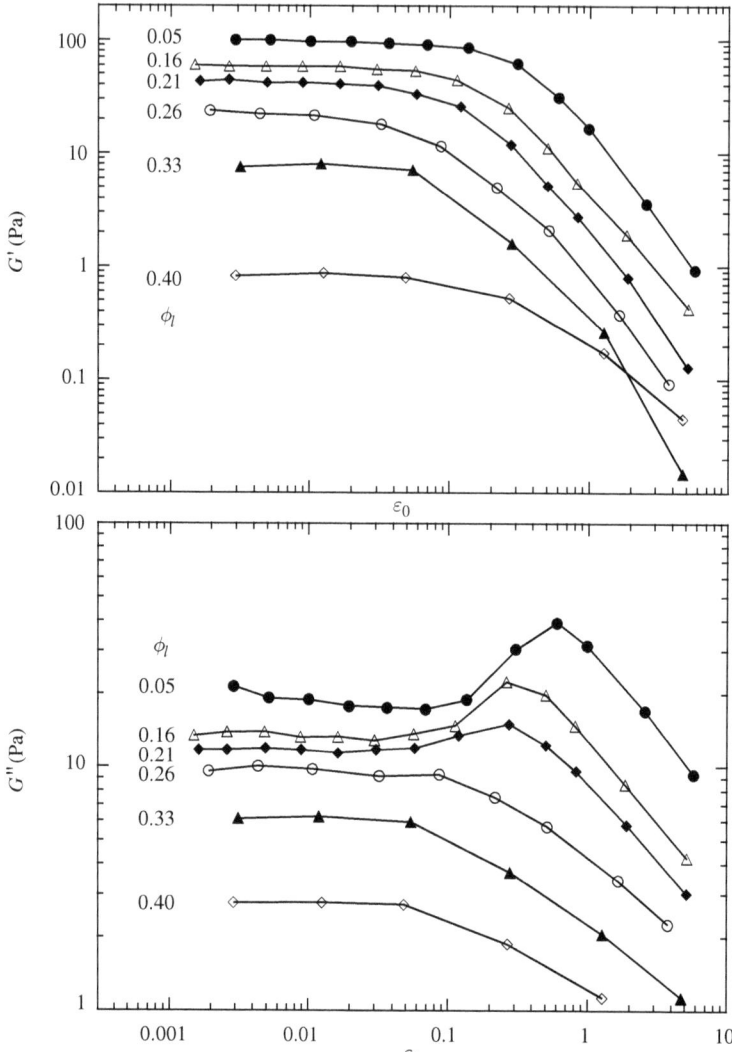

FIG. 4.31: The storage modulus G' and loss modulus G'' as a function of strain amplitude ε_0 in oscillatory mode ($\omega = 1$ rad.s^{-1}, cone-plate or cylindrical Couette geometry, see §1.2.7, chap. 5). The sample is a polydisperse foam ($R_{32} \approx 66$ μm and $\gamma \approx 20$ mN.m^{-1}). At the two highest liquid fractions, the measurements are affected by drainage. Data from [54].

β depends on the rheological properties of the gas/liquid interfaces [17] (§3.3.1): if the interfaces have a low surface dilatational elasticity (like a solution of SDS above the cmc), then the dissipation is localized inside the films, and $\beta \approx 0.5$. In the case of incompressible interfaces (as for a shaving foam or a fatty acid solution), the dissipation

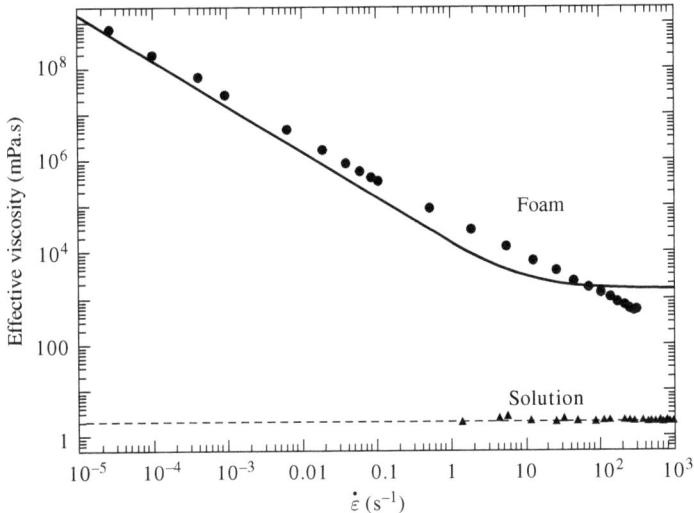

FIG. 4.32: Effective viscosity (eq. (4.64)) as a function of the shear rate for a shaving foam ($R_V = 30$ µm and $\phi_l = 0.08$) and for the (Newtonian) solution from which it is made ($\eta = 1.8$ mPa.s), measured in steady state flow using a cylindrical Couette geometry (internal radius = 20.0 mm, gap-width = 4.1 mm). The continuous curve shows a Bingham law (eq. (4.5)) with $\sigma_y = 15$ Pa and $\eta_p = 1.5$ Pa.s. Data from [26].

at the interfaces is dominant, and $\beta \approx 0.2$. In practice, the change of β from 0.5 to 0.2 occurs at a surface dilatational modulus (eq. (3.13)) of the order of several tens of mN.m^{-1} [17]. Note that the value of β is different from the exponents which arise due to dissipation at a solid wall (§3.3.3).

The plastic viscosity or consistency η_p also depends on the physico-chemical characteristics of a foam [17]. For a given surface tension, bubble size, and liquid fraction, η_p is much greater for incompressible interfaces than for perfectly compressible ones. In other words, the rate of dissipation of mechanical energy increases with the rigidity of the interfaces.

If the shear rate is not too great, the contribution of the yield stress term in eq. (4.64) dominates the effective viscosity, which thus varies as $\gamma/(R_V \dot\varepsilon)$; we call this the plastic regime. In the limit of large $\dot\varepsilon$, the flow is dominated by viscous effects. Recent experimental and theoretical work suggests that the quantity $\sigma - \sigma_y$ depends on the capillary number $Ca = \eta R_V \dot\varepsilon / \gamma$ through the relationship [17]

$$\frac{\sigma - \sigma_y}{\gamma/R_V} = f(\phi_l)\, Ca^{0.5}, \qquad (4.65)$$

where the function $f(\phi_l)$ depends not only on the liquid fraction but also on the disorder of the foam. In steady state flow the bubbles slide past each other and the contacts between bubbles are thus constantly renewed. The contact dynamics introduce local shear flows which controls the film thickness. This process is governed by the competition between the viscous stress $\eta\dot{\varepsilon}$ and the Laplace pressure γ/R_V, which determines the capillary number.

Macroscopic phenomenological laws like eq. (4.64) or the scalar model represented in FIG. 4.29c do not include all the physical properties that govern foam flow. It is also necessary to try and capture spatial variations [3, 55].

4.3.2 Is the flow homogeneous?

Experiments and simulations show that foam flow is not always homogeneous. For an experiment in a rheometer, it is therefore useful to introduce a local shear rate $\dot{\varepsilon}_{\text{local}}$, defined over a length-scale smaller than the gap-width, but larger than the bubble size. At low shear rates, if the stress σ is only slightly greater than the yield stress σ_y, "solid" ($\dot{\varepsilon}_{\text{local}} = 0$) and "liquid" ($\dot{\varepsilon}_{\text{local}} > 0$) regions may coexist within the sample. When the shear rate varies discontinuously at the interface between the two, the flowing region is called a shear band.

Other complex fluids consisting of jammed packings, like concentrated emulsions, soft pastes, gel microbeads, or colloidal dispersions, also exhibit shear banding in steady flow. Shear bands are not a universal characteristic of complex fluids, insofar as they depend on physico-chemical parameters—for example the volume fraction of the dispersed phase or the nature of the interactions between its elements. Moreover, they are occasionally superimposed upon thixotropic effects[16] which further complicates our understanding of them [28]. We next describe the conditions necessary for shear bands to appear in a 3D foam (the 2D case is shown in FIG. 2.29).

Flow under homogeneous stress Flows in a parallel-plate or cone-plate rheometer, or even a cylindrical Couette rheometer if the gap is narrow (cf. §1.2.7, chap. 5), are characterized by a constant stress across the gap. We therefore expect that the shear rate is constant across the gap too, whatever the constitutive law. However, this is not always found experimentally. Imaging by optical tomography (§1.2.4, chap. 5) shows that in a dry foam, above the yield strain, the velocity profile does not vary linearly with position, but that shear bands appear, at a variable depth [53]. Heterogeneous flows have also been found in *Surface Evolver* simulations of ordered dry 3D foams [52]. On the other hand, bubble model simulations of disordered wet 3D foams don't show this [24]. So the physical parameters governing foam flow are still not well-understood, as the following paragraph confirms.

Flow under heterogeneous stress Flows in a cylindrical Couette rheometer are characterized by a shear stress which decreases across the gap as $1/r^2$, where r is the distance from the axis of rotation (§1.2.7, chap. 5). Consequently, when a yield stress fluid is sheared at a rate $\dot{\varepsilon}$ that is so low that viscous stresses are negligible compared to σ_y, only a narrow region close to the inner cylinder experiences a stress greater than σ_y and flows. The rest of the foam sample then behaves as an elastic material (FIG. 4.33). MRI measurements of the local velocity profiles (§1.2.4, chap. 5) in the steady shear of foams of various bubble sizes, liquid fractions (ϕ_l between 0.05 and 0.12), and interfacial rigidities [48] in a large-gap Couette cell have not detected any shear banding; instead, the shear rate was found to decrease continuously to

---

[16] A fluid is thixotropic if its effective viscosity changes over time.

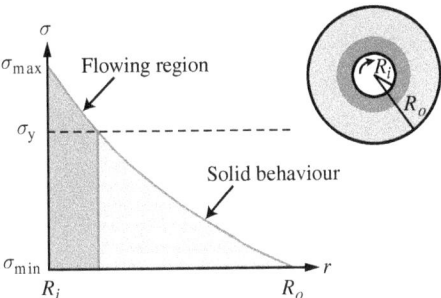

FIG. 4.33: Shear of a yield stress fluid in a cylindrical Couette rheometer with internal radius R_i and external radius R_o (cf. FIG. 5.11). A view from above and the profile of stress aross the gap (σ decreases as $1/r^2$) showing that only the dark grey region close to the inner cylinder flows.

zero at the interface between the flowing and solid-like regions as σ tended toward σ_y (FIG. 4.33). These results, combined with torque measurements, suggest that in steady flow these foams behave like Herschel–Bulkley fluids (eq. (4.6)). It is therefore necessary to distinguish the heterogeneous flow in a large-gap Couette geometry from the shear banding described above, a distinction which is also apparent in 2D (see [10] and §5.4, chap. 2).

5 Appendix: From the discrete to the continuous

The elastic response of a few bubbles, discussed in §3.1, and of a foam with many bubbles, given by eq. (4.49), are directly comparable. On the other hand, the plastic response is very different and going from small to large scale requires further consideration. In a dry hexagonal 2D foam, i.e. the periodic repetition of a single bubble (FIG. 4.10), the stress is completely relaxed after the first T1 (FIG. 4.11a). If the shear continues, we observe periods of elastic loading followed by plastic unloading, and the stress oscillates between 0 and the yield stress. In contrast, FIG. 4.29a shows that in a large *disordered* foam the stress converges to a constant value in the plastic regime. An intermediate regime is observed in a sample of about 1,000 bubbles, shown in FIG. 4.22a, with partial relaxations of the stress due to T1s, and small, though not negligible, stress fluctuations around the average stress. What is significant is that the relative amplitude of the fluctuations decreases with the number of bubbles and with disorder.

To be more quantitative, we consider the Stokes experiment in a foam (FIG. 4.34a), in which a small spherical bead of radius R_{bead} is pulled slowly through a foam at constant speed. The bead is suspended from a balance by a thin rigid rod, enabling the force on the bead to be measured. FIG. 4.34b shows an initial period of elastic loading before the force fluctuates around an average value. In this second regime, linear elastic loading with constant slope is interrupted by quasi-instantaneous unloading due to T1s. The T1s are localized around the sphere (FIG. 4.34c), and the amplitude

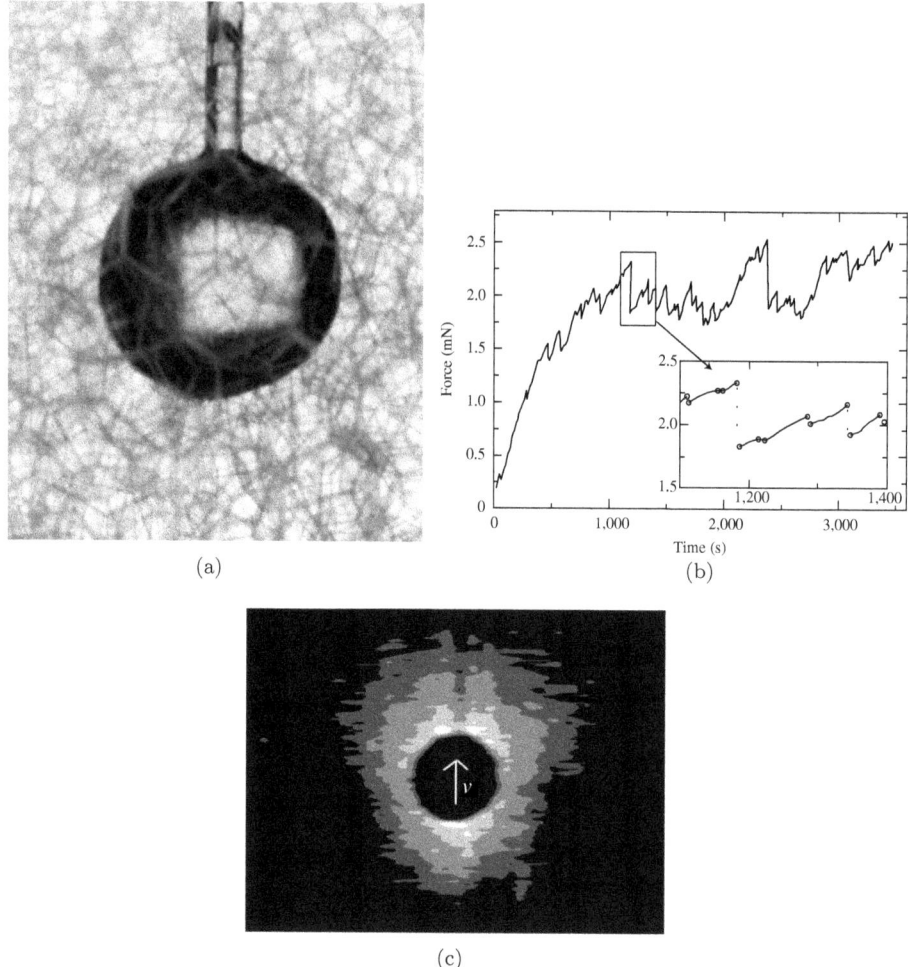

FIG. 4.34: Stokes experiment in a 3D foam. (a) A spherical bead ($R_{bead} = 0.5$ cm) pulled at constant speed (5 µm.s^{-1}) through a disordered monodisperse dry foam ($R_V = 1.5$ mm). Photograph courtesy of I. Cantat and O. Pitois. (b) Force exerted on the bead as a function of position. (c) Projected map of the T1 density (cf. the result in the 2D case in FIG. 2.28d). Data from [7]. Image courtesy of I. Cantat and O. Pitois.

of the jump in the force is of the same order of magnitude as if a bubble in direct contact with the sphere were detached from it: $\delta F \sim \gamma R_V$. On dimensional grounds, the average force is approximated by the product of the yield stress γ/R_V and the surface area of the obstacle R_{bead}^2. The relative fluctuations are therefore of the order of R_V^2/R_{bead}^2, and so are smoothed out when the bead becomes very large compared to the bubbles [7].

6 Experiments

6.1 Observation of T1s

difficulty level: ☻☻	cost of materials: $$
preparation time: 4 hrs	experimental time: 4 hrs

<u>Phenomena demonstrated</u>: *T1s and flow.*

<u>Reference in the text</u>: §3.2.

<u>Materials</u>:

- equipment for making a 2D foam (see §6.2, chap. 2)
- an obstacle which is the same thickness or thicker than the foam, for example a slice cut from a plastic cork
- an aquarium air-pump and a flexible tube.

1. Position the obstacle at the centre of the device used to create the foam (centre of the bowl or CD case). We recommend the assembly between two solid plates (i.e. the CD case), or assembly between the liquid and the solid plate.
2. In order to create a flow in one direction, insert the end of the tube at one end of the device and connect it to the pump using the flexible tubing to create air bubbles continuously at constant gas flow rate. Make the 2D foam according to the instructions given in §6.2, chap. 2.
3. Observe the T1s, primarily upstream and downstream of the obstacle (FIGS. 2.28d and 4.34c).

The presence of the obstacle perturbs what would otherwise be a plug flow, induces shearing of the bubbles, and causes some edges to disappear, triggering T1s.

6.2 Visualization of the yield stress

difficulty level: ☻☻	cost of materials: $
preparation time: 1 hr	experimental time: 1 hr

<u>Phenomenon demonstrated</u>: *yield stress of a foam.*

<u>Reference in the text</u>: §4.2.1.

<u>Materials</u>:

- fairly wet shaving foam ($\phi_l \sim 0.20$)
- a flat, rough, rigid plate (around 10 cm × 10 cm). To make it rough, sandpaper can be stuck onto the plate
- something to incline the plate: one end of the plate should rest on a horizontal plane and the other end should be progressively raised using a wedge or laboratory jack
- a square or rectangular frame of around 5 cm in length and width and a height of around 2 cm.

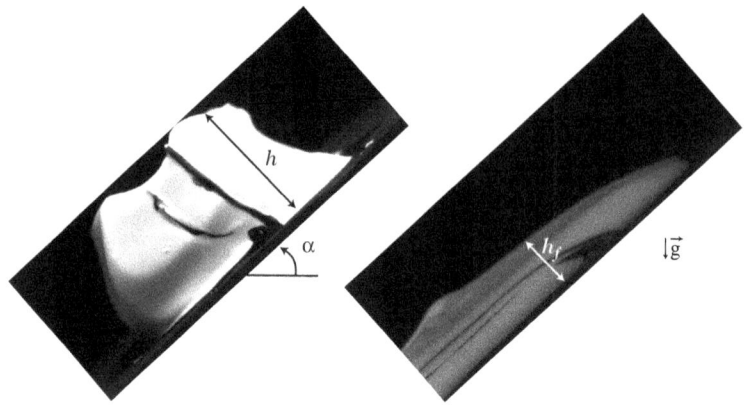

FIG. 4.35: Flow of a shaving foam down a plane inclined at an angle α to the horizontal. The image on the left is taken at the start of the flow; the image on the right is taken at the end of the flow. The black lines are drawn on the foam with a liquid containing a dispersion of carbon particles (printer toner or Indian ink for example), allowing the flow to be visualized. Photographs courtesy of F. Rouyer.

1. Place the flat plate horizontally.
2. Position the frame in the middle of the plate.
3. Fill the frame with foam and make the upper surface level.
4. Carefully pull the frame away without changing the shape of the foam.
5. Slowly (but not so slowly that the foam coarsens) tilt the plane at an angle $\alpha \sim 10°$.
6. If the foam doesn't move, increase the angle slightly.
7. Now leave the foam to spread towards the bottom of the plate and measure its final thickness h_f.

Under the effect of its own weight, the foam is subjected to a shear stress $\sigma = \phi_l \rho_l g h \sin \alpha$. The foam flows when this shear stress exceeds its yield stress σ_y, and stops flowing at a height h_f for which $\sigma_y = \phi_l \rho_l g h_f \sin \alpha$.

7 Exercises

Solutions to the exercises are available on request through the following webpage: http://www.oup.co.uk/academic/physics/admin/solutions.

7.1 The Young–Laplace law and the stress in a spherical bubble

(a) Use Batchelor's expression for the stress tensor (eq. (4.18)) to recover the Young–Laplace law (eq. (2.3)), giving the pressure inside a spherical gas bubble of radius R in a liquid at pressure P_0.
(b) Now calculate the contribution of this bubble to the stress tensor and show that its contribution to the average deviatoric stress is zero. What is the contribution of the bubble to the average pressure in the foam?

7.2 Elasticity of a dry 2D foam

(a) Find the yield strain ε_y and yield stress σ_y (eq. (4.37)) of a dry hexagonal 2D foam using the diagram in FIG. 4.36. Now deduce the shear modulus $\hat{G}$ of the foam, assuming linear elasticity.

(b) Eq. (4.26) shows that as the yield strain is approached, the relationship between stress and strain becomes non-linear. Demonstrate the linear elastic behaviour at small strains by an expansion of eq. (4.26) in ε. Deduce from it the shear modulus $\hat{G}$.

(c) Compare the approximate (part (a)) and exact (part (b)) expressions for the modulus.

(d) In a dry hexagonal 2D foam at equilibrium, each vertex is a symmetrical junction of three films. By placing on each vertex the same small quantity of liquid, we obtain, according to the decoration theorem, the structure of a wet foam at equilibrium. The shape of the vertices is illustrated in FIG. 4.18. Develop a symmetry argument to show that the contribution of the decorated vertices to the average shear stress is zero. Use the fact that the off-diagonal elements of the stress tensor must remain invariant when a vertex is turned through an angle of 120°.

(e) Show that, for the same reason, the linear elastic response of a 2D hexagonal foam does not change if we decorate its vertices with small, three-sided bubbles.

7.3 Poynting's law

In a non-linear elastic solid, the stress as a function of strain can be expressed in terms of the Finger tensor (eq. (4.13)):

$$\sigma_{ij} = f(B_{ij}). \tag{4.66}$$

In the following, consider an isotropic undeformed solid.

(a) Show that for the shear strain ε illustrated in FIG. 4.1, the Finger tensor is

$$B = \begin{pmatrix} 1+\varepsilon^2 & \varepsilon & 0 \\ \varepsilon & 1 & 0 \\ 0 & 0 & 1 \end{pmatrix}. \tag{4.67}$$

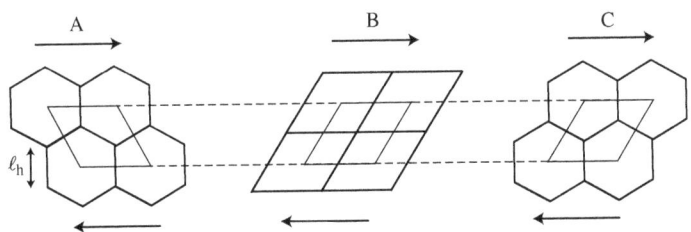

FIG. 4.36: A sheared hexagonal 2D foam, as in FIG. 4.10. The basic unit of the network is the region inside the thin-lined rhombus. (A) Foam at rest before the shear; (B) the strain reaches the yield strain; (C) T1s occur.

(b) Show that the tensor B is diagonal in a frame of reference rotated through an angle θ about the z axis, with

$$2\cot(2\theta) = \varepsilon. \tag{4.68}$$

In a strained isotropic elastic material, the principal axes of B and of the stress tensor are the same, so in the new frame of reference the stress tensor must be diagonal:

$$\sigma = \begin{pmatrix} \sigma_1 & 0 & 0 \\ 0 & \sigma_2 & 0 \\ 0 & 0 & \sigma_3 \end{pmatrix}. \tag{4.69}$$

Find an expression for σ in the (x, y, z) frame of reference, and use it to show that

$$2\cot(2\theta) = N_1/\sigma_{xy}. \tag{4.70}$$

(c) Deduce Poynting's law (eq. (4.50)) from eqs. (4.68) and (4.70).

7.4 Stress and strain in a square lattice

Consider a lattice of squares, with side-length a at rest, oriented in the x and y directions, and assume that each edge of the lattice transmits a force 2γ. This lattice could represent a 2D foam in which Plateau's laws are not respected. Assume that all pressures are zero.

(a) Subject the lattice to an elongational strain $\varepsilon_{xx} = -\varepsilon_{yy} = \varepsilon$, $\varepsilon_{xy} = 0$. Calculate the stress tensor, either using Batchelor's expression (eq. (4.18)) or by determining directly the force per unit length exerted on a line parallel to the x or y direction. Note that Hooke's law in 2D (eq. (4.15)) may also be expressed as $\sigma_{xx} = 2G\varepsilon_{xx} + \text{Tr}(\sigma)/2$, where $\text{Tr}(\sigma) = \sigma_{xx} + \sigma_{yy}$ is the trace of the stress tensor. Hence deduce that $G = \gamma/a$.

(b) Subject the lattice to an affine shear by shifting the point $M(x, y)$ to the point $M'(x + \varepsilon y, y)$. Using the same methods as in part (a), show that the stresses σ_{xx} and σ_{yy} are quadratic functions (to leading order in ε) of the strain, and give their difference as a function of σ_{xy}.

7.5 Elasticity and plasticity

This exercise is a continuation of exercise 7.3, chap. 2.

(a) Elasticity

Consider again the equilibrium shape of the system defined in exercise 7.3, chap. 2, changing a and b whilst preserving $ab = S$.

(1) Write the energy of the system in terms of the ratio a/b for the stable (and possibly metastable) configuration, and give the value of a/b for which the energy is minimized.

(2) Show that for this ratio $F_x/b = F_y/a$, with F_x the resultant force in the x direction, exerted on the two right-hand films, and F_y the resultant force in the y direction exerted on the two upper films.

(b) Plasticity

When the ratio b/a exceeds the limit of existence of the metastable state, a T1 occurs; determine the energy lost during the relaxation towards a new stable state.

7.6 Compressibility of a foam

(a) The isotherm of compressibility for a dry foam, in equilibrium with a reservoir of pressure P_{ext}, is $\chi_T = -(1/V) \left|\partial V/\partial P_{\text{ext}}\right|_T$. Assuming that a change in volume may be made without changing the structure, i.e. by a homothetic transformation, use one of the methods below to show that [33]

$$\chi_T^{-1} = P_{\text{ext}} + \frac{4\gamma S_{\text{int}}}{9V}. \tag{4.71}$$

Method 1 (pressure argument): differentiate the equation of state (eq. (2.18) or exercise 7.6, chap. 2), or use Batchelor's expression (eq. (4.18)).

Method 2 (energy argument): use the thermodynamic relationship $\left|\partial^2 \mathcal{F}/\partial V^2\right|_T = 1/(V\chi_T)$, where $\mathcal{F}$ is the free energy of the foam. (See exercise 7.6, chap. 2 for the terms that must be taken into account in $\mathcal{F}$.)

(b) Make a qualitative comparison of the foam and gas compressibilities, either at atmospheric pressure P_{ext}, or at the foam pressure $\langle P_{\text{int}}\rangle$. Hence find the bulk modulus of a dry foam and compare it to the shear modulus.

(c) The isothermal bulk compression modulus of an elastic material is the inverse of the isotherm of compressibility, defined by $K_T = -V \left|\partial P/\partial V\right|_T$, where V is the volume of the material, P its pressure, and T its temperature. By considering an infinitesimal variation of pressure δP, show that the bulk compression modulus of a wet foam with liquid fraction ϕ_l is

$$K_T(1-\phi_l) \approx \frac{K_{T,\text{gas}}}{1-\phi_l}, \tag{4.72}$$

where $K_{T,\text{gas}}$ is the bulk compression modulus of the gas.

References

[1] D.J. ACHESON, *Elementary Fluid Dynamics*, Clarendon Press, Oxford, 1990.
[2] G. BATCHELOR, *J. Fluid Mech.*, **41**, 545, 1970.
[3] S. BÉNITO, C.-H. BRUNEAU, T. COLIN, C. GAY, and F. MOLINO, *Eur. Phys. J. E*, **25**, 225, 2008; S. BÉNITO, F. MOLINO, C.-H. BRUNEAU, T. COLIN, and C. GAY, *Eur. Phys. J. E*, **35**, 51, 2012.
[4] S. BESSON and G. DEBRÉGEAS, *Eur. Phys. J. E*, **24**, 109, 2007.
[5] A.-L. BIANCE, S. COHEN-ADDAD, and R. HÖHLER, *Soft Matter*, **5**, 4672, 2009.
[6] D. BUZZA, C.-Y. LU, and M.E. CATES, *J. Phys. II*, **5**, 37, 1995.
[7] I. CANTAT and O. PITOIS, *Phys. Fluids*, **18**, 083302, 2006.
[8] I. CANTAT, *Phys. Fluids*, **25**, 031303, 2013.
[9] I. CHEDDADI, P. SARAMITO, B. DOLLET, C. RAUFASTE, and F. GRANER, *Eur. Phys. J. E*, **34**, 1, 2011; I. CHEDDADI, P. SARAMITO, F. GRANER, *J. Rheol.*, **56**, 213, 2012.

[10] I. CHEDDADI, P. SARAMITO, C. RAUFASTE, P. MARMOTTANT, and F. GRANER, *Eur. Phys. J. E* **27**, 123, 2008.
[11] S. COHEN-ADDAD, H. HOBALLAH, and R. HÖHLER, *Phys. Rev. E*, **57**, 6897, 1998.
[12] S. COHEN-ADDAD, R. HÖHLER, and Y. KHIDAS, *Phys. Rev. Lett.*, **93**, 028302, 2004.
[13] S. COHEN-ADDAD, R. HÖHLER, and O. PITOIS, *Ann. Rev. Fluid Mech.*, **45**, 241, 2013.
[14] S.J. COX, F. GRANER, and M.F. VAZ, *Soft Matter*, **4**, 1871, 2008.
[15] P.-G. DE GENNES, F. BROCHARD-WYART, and D. QUÉRÉ, *Capillarity and Wetting Phenomena: Drops, Bubbles, Pearls, Waves*, Springer, New York, 2004.
[16] N.D. DENKOV, V. SUBRAMANIAN, D. GUROVICH, and A. LIPSA, *Colloids Surf. A.*, **263**, 129, 2005.
[17] N.D. DENKOV, S. TCHOLAKOVA, K. GOLEMANOV, K.P. ANANTHPADMANABHAN, and A. LIPS, *Soft Matter*, **5**, 3389, 2009.
[18] B. DERJAGUIN, *Kolloid Z.*, **64**, 1, 1933.
[19] F. DIAS, D. DUTYKH, and J.-M. GHIDAGLIA, *Comput. Fluids*, **39**, 283, 2010.
[20] M. DURAND and H.A. STONE, *Phys. Rev. Lett.*, **97**, 226101, 2006.
[21] D.J. DURIAN, *Phys. Rev. E*, **55**, 1739, 1997.
[22] D.A. EDWARDS, H. BRENNER, and D.T. WASAN, *Interfacial Transport Processes and Rheology*, Butterworth Heinemann, Stoneham, 1991.
[23] J. FERRY, *Viscoelastic Properties of Polymers*, Wiley, New York, 1980.
[24] B.S. GARDINER, B.Z. DLUGOGORSKI, and G.J. JAMESON, *J. Non-Newton. Fluid Mech.*, **92**, 151, 2000.
[25] B.S. GARDINER, B.Z. DLUGOGORSKI, and G.J. JAMESON, *J. Rheol.*, **42**, 1437, 1998.
[26] A.D. GOPAL and D.J. DURIAN, *J. Colloid Interface Sci.*, **213**, 169, 1999.
[27] A.D. GOPAL and D.J. DURIAN, *Phys. Rev. Lett.*, **91**, 188303, 2003.
[28] B. HERZHAFT, *J. Colloid Interface Sci.*, **247**, 412, 2002.
[29] B. HERZHAFT, S. KAKADJIAN, and M. MOAN, *Colloids Surf. A*, **263**, 153, 2005.
[30] R. HÖHLER and S. COHEN-ADDAD, *J. Phys.: Condens. Matter*, **17**, R1041, 2005.
[31] R. HÖHLER, S. COHEN-ADDAD, and A. ASNACIOS, *Europhys. Lett.*, **48**, 93, 1999.
[32] S.A. KHAN, C.A. SCHNEPPER, and R.C. ARMSTRONG, *J. Rheol.*, **32**, 69, 1988.
[33] A.M. KRAYNIK and D.A. REINELT, *J. Colloid Interface Sci.*, **181**, 511, 1996.
[34] A.M. KRAYNIK and D.A. REINELT, in *Proc. XIVth Int. Congr. Rheology*, Seoul, The Korean Society of Rheology, 2004.
[35] A.M. KRAYNIK, D.A. REINELT, and F. VAN SWOL, *Phys. Rev. E*, **67**, 031403, 2003.
[36] A.M. KRAYNIK, D.A. REINELT, and F. VAN SWOL, *Phys. Rev. Lett.*, **93**, 208301, 2004.
[37] N.P. KRUYT, *J. Appl. Mech.*, **74**, 560, 2007.
[38] V. LABIAUSSE, R. HÖHLER, and S. COHEN-ADDAD, *J. Rheol.*, **51**, 479, 2007.
[39] M.-D. LACASSE, G.S. GREST, and D. LEVINE, *Phys. Rev. E*, **54**, 5436, 1996.
[40] R.G. LARSON, *J. Rheol.*, **41**, 365, 1997.

[41] R.G. LARSON, *The Structure and Rheology of Complex Fluids*, Oxford University Press, New York, 1999.
[42] A.J. LIU, S. RAMASWAMY, T.G. MASON, H. GANG, and D.A. WEITZ, *Phys. Rev. Lett.*, **76**, 3017, 1996.
[43] C.W. MACOSKO, *Rheology: Principles, Measurements, and Applications*, Wiley-VCH, Weinheim, 1994.
[44] P. MARMOTTANT and F. GRANER, *Eur. Phys. J. E*, **23**, 337, 2007.
[45] S. MARZE, A. SAINT-JALMES, and D. LANGEVIN, *Colloids Surf. A*, **263**, 121, 2005.
[46] T.G. MASON, J. BIBETTE, and D.A. WEITZ, *Phys. Rev. Lett.*, **75**, 2051, 1995.
[47] N. MUJICA and S. FAUVE, *Phys. Rev. E*, **66**, 021404, 2002.
[48] G. OVARLEZ, K. KRISHAN, and S. COHEN-ADDAD, *EPL*, **91**, 68005, 2010.
[49] H.M. PRINCEN, *J. Colloid Interface Sci.*, **91**, 160, 1983.
[50] C. RAUFASTE, B. DOLLET, S. COX, Y. JIANG, and F. GRANER, *Eur. Phys. J. E*, **23**, 217, 2007.
[51] D.A. REINELT and A.M. KRAYNIK, *J. Fluid Mech.*, **311**, 327, 1996.
[52] D.A. REINELT and A.M. KRAYNIK, *J. Rheol.*, **44**, 453, 2000.
[53] F. ROUYER, S. COHEN-ADDAD, M. VIGNES-ADLER, and R. HÖHLER, *Phys. Rev. E*, **67**, 021405, 2003.
[54] A. SAINT-JALMES and D.J. DURIAN *J. Rheol.*, **43**, 1411, 1999.
[55] P. SARAMITO, *J. Non-Newtonian Fluid Mech.*, **145**, 1, 2007; P. SARAMITO, *J. Non-Newtonian Fluid Mech.*, **158**, 154, 2009.
[56] L.W. SCHWARTZ and H.M. PRINCEN, *J. Colloid Interface Sci.*, **118**, 201, 1987.
[57] D. STAMENOVIĆ, *J. Colloid Interface Sci.*, **145**, 255, 1991.
[58] E. TERRIAC, J. ETRILLARD, and I. CANTAT, *Europhys. Lett.*, **74**, 909, 2006.
[59] S. VINCENT-BONNIEU, R. HÖHLER, and S. COHEN-ADDAD, *Europhys. Lett.*, **74**, 533, 2006.
[60] D. WEAIRE, *Phil. Mag. Lett.*, **60**, 27, 1989.
[61] D. WEAIRE and S. HUTZLER, *The Physics of Foams*, Clarendon Press, Oxford, 1999.

5
Experimental and numerical methods

The study of foams is interdisciplinary, both in terms of the questions that foams give rise to and the methods (experimental, simulations, and theoretical) used to study them.

1 Experimental methods

We present some of the experimental techniques used in research laboratories in order to make and study foams. The appropriate method depends on the length-scale that should be probed, i.e. liquid/gas interface, single liquid film, bubble, or macroscopic sample. We will also show how the optical and electrical properties of a foam depend on its physical parameters.

1.1 Methods used to study interfaces and isolated films

There are numerous techniques used to study gas/liquid interfaces in the presence of surfactants. We choose to concentrate on those for measuring the surface tension, the interfacial viscoelasticity, and the properties of thin liquid films. Other techniques such as ellipsometry, Brewster angle microscopy, and infrared spectroscopy are described elsewhere [1].

1.1.1 Surface tension
There are several different ways to measure surface tension; we briefly introduce the three that are used most often, and the reader is referred to [15] for more detailed descriptions.

One possibility is to measure the capillary force exerted on an object immersed in a solution. In *Wilhelmy's* method a thin strip of platinum or filter paper (to enforce total wetting) is used, while in *de Nouy's* method a ring is used. The precision is of the order of ± 0.1 mN.m^{-1}, and the method is most suited to measuring equilibrium values.

In the *pendant drop* method [1, 15], a liquid drop (typically 1 mm in diameter) is suspended from the end of a needle and its shape determined. Since the pressure difference at every point of the interface is given by the Young–Laplace law (§2.1.3, chap. 2), the shape of the drop and its variation in time depend on the surface tension. The method gives $\gamma(t)$ for times t from one second to several hours, and the accuracy of the measurement is again ± 0.1 mN.m^{-1}. The *bubble* configuration [1, 15], in which a bubble is held at the end of a needle in a tank filled with solution, and the *sessile drop* configuration, a liquid drop placed on a substrate, are similar.

Finally, the maximum-bubble-pressure method gives the surface tension by measuring the pressure at which a bubble detaches from a submerged capillary [43]. A capillary of radius R_c (typically of the order of several hundred micrometres) is immersed in a foaming solution and gas injected to inflate a bubble with known pressure at each instant. This pressure is $2\gamma/R_c$, by the Young–Laplace law, and is maximal when the bubble radius is equal to R_c. On further inflation, the bubble detaches from the capillary, and by rapidly repeating the inflation and detachment (usually at frequencies of a few kHz), it is possible to measure $\gamma(t)$ on a time-scale of the order of 1 ms to 1 s. This precision is necessary for surfactants of low molecular mass (see §6.1, chap. 3) for which the surface tension is significantly lowered immediately after the formation of an interface; for example, for SDS, above the cmc, the tension drops from 72 mN.m^{-1} to around 38 mN.m^{-1} in less than a second.

1.1.2 Surface viscoelasticity

Rheological methods for measuring dilatational viscoelasticity In order to study the response to an expansion or a compression, it is necessary to vary the area of a gas/liquid interface. The in-phase and out-of-phase response of the surface tension to oscillations in interface area over a range of frequencies (experimentally ranging from a fraction of one Hz to several thousand Hz) gives both the elastic E'_s and viscous E''_s 2D moduli (§2.2.3, chap. 3).

The pendant drop or bubble method for measuring surface tension (§1.1.1) can also be used in oscillatory mode. A sinusoidal oscillation of the volume of the drop induces a change in the area of the interface and then by tracking the changing shape of the drop the complex modulus E^*_s can be found, typically for frequencies between 0.01 and 100 Hz.

The detection of capillary waves (whether produced by thermal fluctuations or induced by an oscillator) allows the dilational viscoelastic properties of an interface to be measured at high frequency, typically several hundred Hz [52]. In practice, this is carried out by electro-capillary methods, in which a wave of controlled frequency and amplitude, for example an alternating voltage of several hundred volts, is applied to a strip positioned 1 mm above the interface. This raises and lowers the surface of the liquid without ever touching it. Then the wavelength and the attenuation of the wave that propagates at the surface are measured optically [52]. The appeal of this method for measuring interfacial viscoelasticity lies in the fact that the dispersion relation for a surface wave is different from that in pure liquid, because of the coupling between the wave in the liquid and the viscoelastic properties of the film in compression. The propagation of the wave creates regions of expansion and compression of the interface (FIG. 5.1a). The observed wavelength always depends strongly on the surface tension, whilst the attenuation is linked to the surface dilatational viscoelasticity.

Finally, it is also possible to make measurements in a Langmuir trough (see shaded section on p. 58). A solution is poured into a shallow container containing a mobile barrier that allows the area of the interface to be varied (FIG. 5.1b). Using a Wilhelmy plate (§1.1.1) allows variations in surface tension to be measured while the barrier is translated or slowly oscillated (over a period of seconds or minutes).

Rheological methods for measuring shear viscoelasticity Measuring the 2D shear moduli is a delicate procedure because it requires simultaneous control of the

(a)

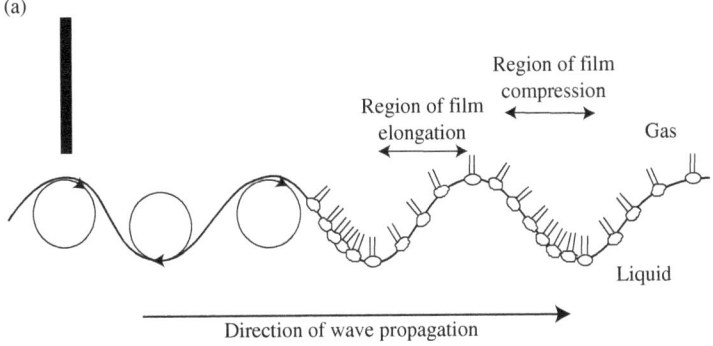

(b)

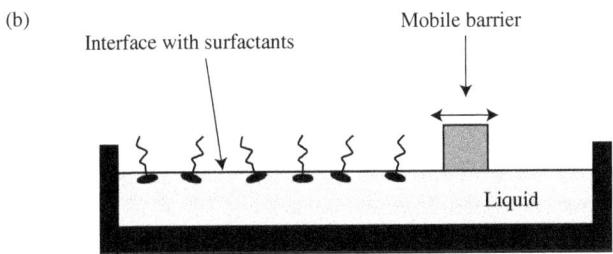

FIG. 5.1: Methods for measuring interfacial viscoelasticity in compression: (a) Coupling between a capillary wave and expansion/compression of the film. Counter-rotating circles represent trajectories of fluid particles as the wave passes. The thick vertical bar represents a metal strip to which an alternating current is applied to induce a surface wave. (b) Langmuir trough with an oscillating barrier.

shear applied to an interface, measurement of the interfacial stresses and strains, and accounting for the coupling with flow in the subphase. However, it is possible, both in oscillatory and continuous modes.

A number of methods are based on generating shear by moving a floating object such as a magnetized needle (FIG. 5.2a) [6].

Other methods include a 2D analogue of a 3D Couette rheometer (§1.2.7), in which a disc or horizontal bicone is suspended from a vertical thread and placed at the interface and rotated relative to the outside wall (FIG. 5.2b). Measuring (usually by optical means) its angle of rotation gives the strain, whilst measuring the torque gives the 2D shear stress [45, 61].

A third group of methods, more suitable for insoluble surfactants, uses a Langmuir trough (FIG. 5.1b). The compression barrier is used to push the free surface of the liquid into a narrow channel and the interfacial velocity field measured by tracking markers (e.g. talc) placed on the surface.

Finally, it is possible to observe the flow in a 3D channel such as a Plateau border: as described in §4, chap. 3, the coupling between the flow at the surface and in the

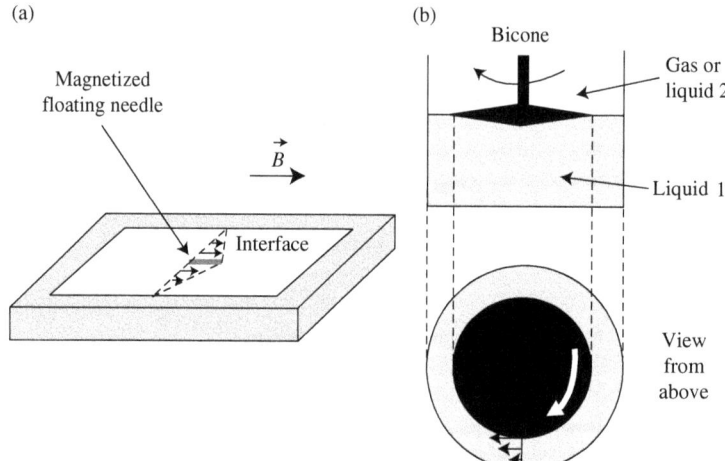

FIG. 5.2: Methods for measuring interfacial shear viscoelasticity. (a) A floating magnetized needle is oriented by a magnetic field and moved by a magnetic field gradient: it shears the surface. (b) Shear of the interface using a bicone device: once positioned precisely at the interface, its circumference shears the interface. The velocity field is represented by arrows, here in the case where the inner cylinder is turned and the outer wall is fixed.

bulk is controlled by the surface shear viscosity. By measuring the speed of macroscopic drainage, we obtain the surface mobility and therefore the shear viscosity [17, 47].

1.1.3 Reflectometry of liquid films The colourful iridescence so easily observed on the surface of a bubble is the result of interference between the waves of light that are reflected by each of the two interfaces of a film. This is possible because the films are so thin: for a film of thickness h illuminated at normal incidence, the intensity of the reflected light, ignoring multiple reflections, is

$$I = RI_0 \left[1 - \cos\left(\frac{4\pi n_0 h}{\lambda}\right) \right], \qquad (5.1)$$

where n_0 is the refractive index of the soap solution, λ the wavelength of light used, I_0 the intensity of the light source, and R the reflection coefficient of the interface ($R \approx 4\%$ for an air/water interface).

For a given wavelength λ, the intensity is minimal when h is a multiple of $\lambda/2n_0$ and maximal when h is an odd multiple of $\lambda/4n_0$. In effect, if one ray is reflected at a gas/liquid interface and another at a liquid/gas interface, between the two rays there is a phase shift of π; this is taken into account in eq. (5.1).

If white light is used, each wavelength gives rise to its own particular interference pattern, and the intensity minimum for each colour occurs for a different film thickness. All the intensities sum incoherently, which gives a progression of colours (cf. FIGS. 2.33 and 3.55). A particular colour therefore corresponds to a certain thickness (although the converse is not true: different thicknesses can show the same colour), and gradients of thickness give rise to gradients of colour. The colour and the intensity also depend strongly on the direction from which they are observed.

As the thickness increases, the colour gradations become blurred: green and pink are obtained first (typically corresponding to thicknesses of several hundred nanometres) then a mixture of numerous colours called *high-order white*.

At the other extreme, if the film is very thin, the difference in the optical path length between the two rays becomes very small compared to λ. Then the only possible phase shift between the two rays is π and so there can only be nearly destructive interference and the film becomes almost transparent, known as a *black film*.[1] The thickness of a black film can be up to 100 nm, so they can be found in a foam (see §2.3.2, chap. 3).

In order to determine the precise thickness of a film as it drains and its thickness changes, a monochromatic source is used to find the minima and maxima of reflected intensity (eq. (5.1)). For an absolute measurement, it is necessary to identify the order of the interference colours (starting with the black film, for example) [7, 54].

Interferometry is also possible with X-rays or neutron beams [3, 20, 50], allowing structures of the order of nanometres to be probed. X-rays are able to show not only the thickness but also the structure of the film. Neutron sources have lower flux than X-rays, but because it is possible to *label* parts of the material (by replacing hydrogen atoms with deuterium), they enable precise study of the structure [20].

1.1.4 Film balance technique The *film balance* technique (also called the *porous plate technique* or the *thin-film balance*) is based on an idea of Mysels [44]. It enables the study of the properties of a single liquid film held in a frame and the interactions between interfaces at the scale of thin films [7, 54].

A film is suspended horizontally in a hole (of typically 1 mm in diameter) pierced in a disc-shaped glass frit (of typical thickness 3 mm; see FIG. 5.3a). The frit and the capillary tube to which it is soldered act as a reservoir of solution and are kept at constant pressure. The frit is enclosed in a chamber in which the pressure P applied to the film can be varied (FIG. 5.3b). If the repulsive forces between the two interfaces of the film are large enough, they create a disjoining pressure Π_d sufficient to equilibrate the pressure P. In this case a stable film is obtained, and by measuring the equilibrium thickness h for each value of P by monochromatic interferometry (§1.1.3), we can construct the curves $\Pi_d(h)$ (§2.3.2, chap. 3). In the example given for a micellar solution in FIG. 5.4, several branches are found, illustrating the stratification of the micelles in the film (§2.3, chap. 3). The film doesn't thin continuously, but its thickness jumps from one value to another as the pressure is increased. On the other hand, by reducing the pressure, the curve returns to zero along the same branch each time, indicating that the organization of the micelles into layers only occurs under confinement.

The film can be visualized directly using a microscope, making it possible to observe its homogeneity, any instabilities, and even the dynamics associated with a jump in thickness. Several examples are presented in chap. 3 (FIGS. 3.17, 3.18, 3.40, and 3.44). Such direct visualization of the film structure, and thus of possible non-trivial effects

---

[1] Newton called it "black film" because he observed it against a black background.

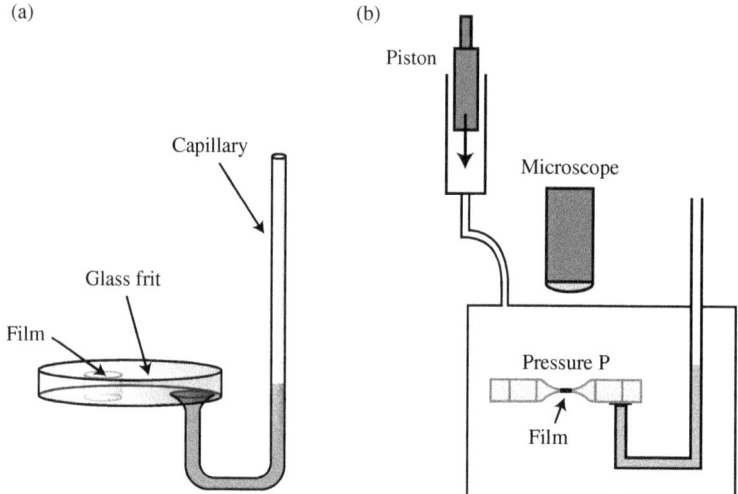

FIG. 5.3: The film balance technique. (a) A glass frit is soldered to a U-shaped capillary tube; the film is formed in a hole in the frit. (b) The frit is placed in a sealed chamber within which the pressure is controlled, and the film is observed through a microscope.

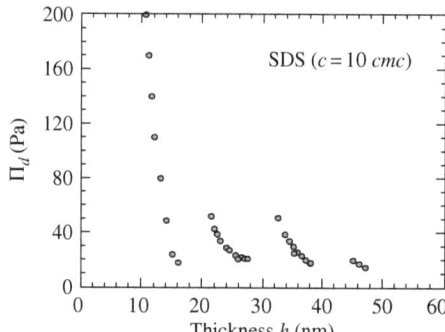

FIG. 5.4: Disjoining pressure $\Pi_d(h)$ as a function of the thickness h, measured with the film balance technique [7]. The different branches of the curve illustrate the effect of the confinement described in FIG. 3.16.

(thickness heterogenities, aggregation), enables a better understanding of the origin of the foamability of a solution.

1.2 Methods for studying foams

1.2.1 Making foams The basic principle of producing a foam is to find a way of incorporating a large volume of gas in a tiny volume of liquid (cf. §5.1, chap. 1).

It is possible to incorporate the gas by beating or whisking a solution (either manually, or with an electric whisk). The gas/liquid interface is stretched and folded in such a way as to force air bubbles to enter the liquid. The duration and speed of the beating controls the homogeneity of the foam obtained. With electric mixers, large quantities of homogeneous foam may be obtained, with a range of possible liquid fractions (depending on the amount of solution) between 0.03 and 0.5. In general, the bubble size distribution is large.

By bubbling gas into a solution it is possible to generate bubbles between typically 0.2 and 10 mm in size. These foams are usually dry (liquid fraction of the order of 10^{-2} to 10^{-3}), because drainage occurs while they are being produced (see §4, chap. 3). The rate of foam production is limited by the frequency of bubble detachment (typically 1 L.min^{-1}). To generate foam more quickly (more than 10 L.min^{-1}), it is possible to simultaneously inject gas and liquid at high pressure through a porous frit, a mesh, or a constriction, inducing a turbulent flow of the resulting mixture. This process of foam production is therefore known as turbulent mixing, and, since it requires no moving parts, it is remarkably robust. Micro-fluidic devices are similar, but do not require turbulent flow [38, 53]; instead, they produce equal-sized bubbles (or droplets) one at a time at a rate of the order of several mL.min^{-1}.

Inside a can of shaving foam, a gas (usually a hydrocarbon or a chlorofluorocarbon) is liquified under high pressure and emulsified with the surfactant solution. When the button is pressed, the pressure is released slightly and each hydrocarbon drop becomes a bubble that is much larger. The excess volume then leaves the can in the form of a foam. This method produces very homogeneous foams, with bubble diameters of about 40 µm. The properties of the gas are crucial: in particular it must be liquifiable under modest pressures (a few atmospheres).

Other techniques exist, for example *in situ* gas production through a chemical reaction (as in the case of solid foams made from polyurethane or aluminium [5], or bread).

1.2.2 Foamability tests In order to test the foamability of a solution, the simplest method is to shake a flask containing a small amount of the solution and to see if a foam forms and, if so, for how long it lasts. This is often formalized in a tumbling tube: a sealed vertical cylinder rotated perpendicular to its axis for a fixed time. These methods are an easy way to distinguish those solutions which don't foam from those that foam very well.

In order to obtain reproducible and quantitative results, a common laboratory test consists of injecting a constant and controlled flow of gas into the solution (FIG. 5.5a). The quantities measured are the volume of foam produced, and the characteristic time with which the foam disappears when the bubbling stops. If V_g denotes the volume of gas injected, poor foamability is characterized by a low value of the ratio V_foam/V_g (whatever the temperature T), indicating that not much gas is trapped in the foam. When this ratio is close to 1, almost all of the gas has been incorporated into the foam, indicating quasi-optimal foamability. The size of the bubbles can be changed by varying the geometry of the gas injector, making it possible to study foamability as a

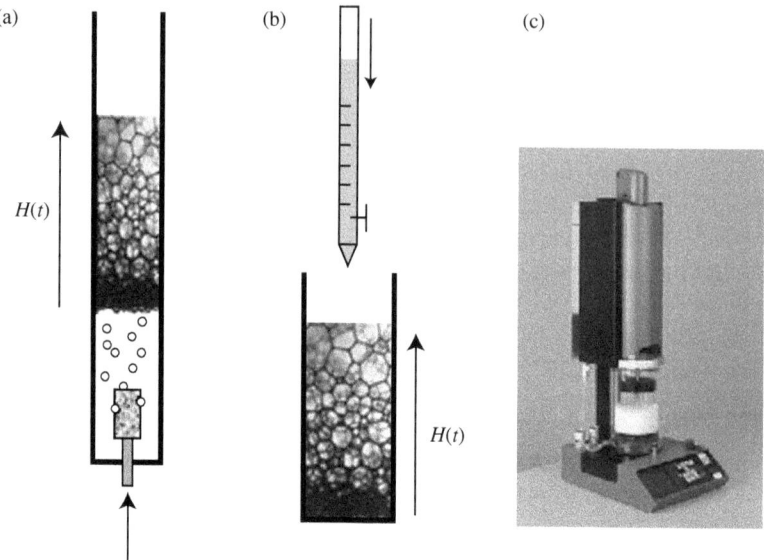

FIG. 5.5: Methods for measuring foamability: (a) bubbling; (b) Ross–Miles method; (c) commercial apparatus with a mechanical mixer (reproduced by kind permission of SITA Messtechnik GmbH, Germany).

function of bubble size. It is often observed that, for a given surfactant concentration, smaller bubbles lead to better foamability.

In the popular Ross–Miles method (FIG. 5.5b), foaming solution is allowed to flow into a beaker from a graduated pipette, fixed at a given height. The foam volume is measured as a function of time while the solution flows, after it stops, and for different drop heights.

Difference commercial devices are also now available (FIG. 5.5c), which make a foam with a rotating blade or rotor (often very rapidly, at several hundred revolutions per minute) in a reservoir containing the solution. The quantity and homogeneity of the foam produced can be characterized by optical (§1.2.3, §1.2.5) or electrical (§1.2.6) techniques.

1.2.3 Surface imaging The bubbles at the surface of a foam are easy to observe, with a magnifying glass or a microscope if necessary. By measuring the diameter of these surface bubbles, we can characterize the distribution of their sizes and their average diameter, and by carrying out such measurements at different places, we can evaluate the homogeneity of the sample.

A photograph of a foam in contact with a transparent wall shows the network of Plateau borders around the surface bubbles (FIG. 5.6a). Changing the lighting conditions (for example using a diffuse or a point source, either in reflection or in transmission) changes the apparent width of the Plateau borders and the apparent diameter of the bubbles. Geometric optics and ray tracing can be used to explain this by linking

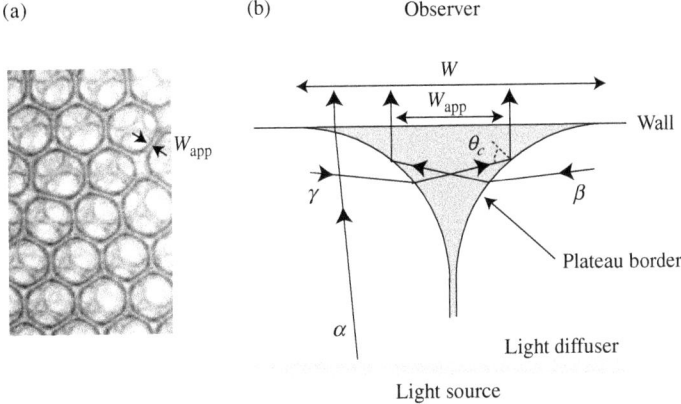

FIG. 5.6: Surface imaging. (a) Photograph of a monodisperse 3D foam under a glass plate, illuminated by a uniform source of diffuse transmitted light. The diffuser is the foam itself. Each Plateau border in contact with the plate appears demarcated by two dark lines. Photograph courtesy of S. Cohen-Addad. (b) Schematic diagram of a Plateau border in contact with a transparent wall, illuminated in transmission by a uniform diffuse light source; α, β, and γ represent several possible optical paths (see text). The apparent width W_{app} of a Plateau border on an image is less than its actual width W (cf. eq. (5.2)).

the characteristics of the images to the actual dimensions of the Plateau borders and the bubbles [56]. Then illumination with a uniform source of diffuse transmitted light shows each Plateau border in contact with the wall as a clear band, demarcated by two dark lines (FIG. 5.6a).

In order to understand a photograph such as this, we construct optical paths to the observer that obey the laws of geometric optics; see, for example, α, β, and γ in FIG. 5.6b. By traversing these paths in reverse, i.e. from the observer, it is possible to identify those which originated at the light source. These rays thus establish the regions that the observer perceives as illuminated. For example, ray α passes far from the middle of the Plateau border and crosses the interfaces at an angle of incidence close to 90°. It is therefore only weakly deflected and effectively links the observer to the light source. The paths which cross the wall close to the centre of the Plateau border are more complex, since they may be reflected or refracted. If, like rays β and γ, they meet the gas/liquid interface at an angle of incidence equal to or slightly greater than the critical angle θ_c for total reflection, they cannot have originated in the light source. These paths define, on the image viewed by the observer, two dark lines on either side of the centre of the Plateau border. A geometric calculation gives the expression for the apparent width of the Plateau border W_{app} as a function of its actual width W [56]:

$$W_{\text{app}} = W(1 - \sin \theta_c). \tag{5.2}$$

At an air/water interface, θ_c is $48.8°$, giving $W_{\text{app}} = 0.75\,W$. In the presence of surfactants, θ_c can vary slightly depending on their nature and their concentration.

In the same way, the image of a spherical soap bubble (for example in a very wet foam), when illuminated by a uniform source of diffuse transmitted light, shows a dark ring of apparent outer diameter $d_{\text{app}} = d \cos(\theta_c/2)$, where d is the actual diameter of the bubble [56].

1.2.4 Bulk imaging

Optical methods When a foam is sufficiently dry that it appears transparent, it is possible to observe bubbles deep within a sample using a microscope or magnifying glass. Remarkably, Matzke [41] counted the number of faces and edges of a large number of bubbles in this way, and deduced statistical information about the structure of a foam (cf. §3, chap. 2). Microscopy is sometimes associated with optical tomography [42] or even stereoscopy [28] (FIG. 5.7).

Multiple light scattering, due to reflections and refractions at the liquid/gas interfaces (cf. §1.2.5), prohibits observation of the internal structure of a foam sample. The different types of 3D microscopy (confocal, two-photon, or structured illumination microscopy, for example), although common in biology, are still not widely used for foams.

X-ray tomography Foams scatter X-rays much less than visible light, and this is exploited to image the bulk of a foam by tomography. Synchrotrons provide a source of X-rays that is even bright enough to image wet foams [36]. A small volume of foam

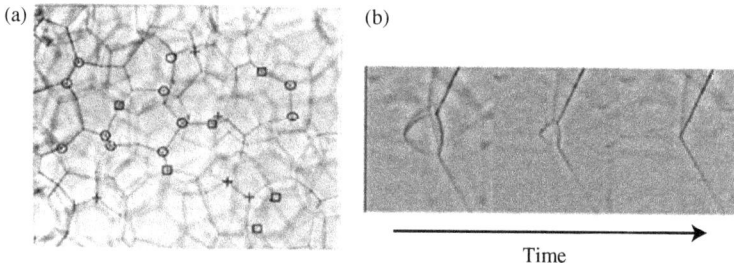

FIG. 5.7: 3D optical imaging. **(a)** Structure deep within the foam, obtained by optical tomography. The foam is illuminated by a laser sheet scanner and imaged in successive slices, and then its structure can be reconstructed in 3D. The image shows one of these slices, with circles indicating the vertices in this slice and squares/crosses indicating the vertices in the slice above/below, respectively. Photograph courtesy of C. Monnereau, reproduced by kind permission of [42], © 2001 The American Physical Society. **(b)** Coarsening tracked by optical stereoscopy, showing the disappearance of a bubble with four faces over time. Photographs taken from two different angles of observation (only one of which is shown here) allow the 3D coordinates of the bubble's four vertices to be located precisely. Photograph courtesy of S. Hilgenfeldt, reproduced from [28], © 2001 The American Physical Society.

(of the order of a cubic centimetre) is placed in a parallel horizontal beam of X-rays in front of a detector with typically 1,000 × 1,000 pixels or more. The resulting 2D image is similar to a medical X-ray: the X-rays are slightly absorbed by the liquid phase but the gas phase has no effect on them. The grey level of a pixel is therefore proportional to the total amount of liquid crossed by the beam along its straight path through the sample. The sample is turned about a vertical axis and roughly one thousand 2D images are recorded at different angles. From these images a mathematical algorithm is applied to reconstruct the 3D structure of the foam. The time required to create a high-resolution 3D image (FIG. 5.8) is currently a fraction of a second.

The 3D images consist of elementary cubes, called voxels (in analogy with pixels in a digital 2D picture), with sides of the order of 10 μm in length, which is sufficient to clearly resolve the Plateau borders. The thin films must be reconstructed a posteriori after segmentation (see §3.1.1). It is then possible to record for each bubble its faces, its volume, and its evolution in time, for example to follow foam coarsening (§3.2, chap. 3).

Imaging by nuclear magnetic resonance Nuclear magnetic resonance (NMR) probes a material by analysing the electromagnetic signal emitted by the spins of atomic nuclei when they are placed in an intense stationary magnetic field and excited by an appropriate radio-frequency wave. The signal emitted depends both on the nature of the material probed and the excitation characteristics. With magnetic field gradients, it is possible to excite each point in space differently, and deduce the nature and the position of each element of material from the signal: this is the principle of medical magnetic resonance imaging (MRI), which has been used to study foam coarsening [24].

If the material moves, it "experiences" different values of the magnetic field, and the signal received depends on the field experienced by each material element along

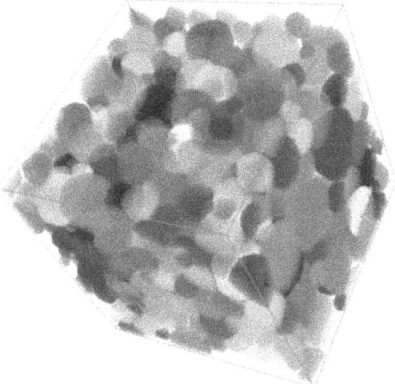

FIG. 5.8: X-ray tomography: 3D image of a foam with liquid fraction $\phi_l = 0.15$. Each bubble is identified (see FIG. 5.18) and coloured in a different shade of grey. Image courtesy of J. Lambert; data from [36]. See PLATE 12.

the whole of its trajectory, and therefore, *inter alia*, its velocity. The exploitation of this information allows us to generate a 3D map of velocities within the material, and thus to probe the flow of foams at the local scale [49] (cf. §4, chap. 4). To reduce the acquisition time, only a single component of the velocity is measured, or the region studied is limited to a single plane.

1.2.5 Multiple light-scattering We describe here how to take advantage of the fact that, due to reflection and refraction at the many liquid/gas interfaces, foams scatter light strongly. By characterizing the propagation of light through a foam (eq. (5.3) below), we can determine the average bubble size if the liquid fraction is known, and vice versa (eq. (5.6)). Furthermore, a coherent light source creates interference patterns which provide details about the dynamics of topological changes (eq. (5.7)).

Transport mean free path Geometric optics describe the propagation of light within a foam consisting of bubbles that are larger than the wavelength of light. When a light ray meets a gas/liquid interface, it is partly refracted and partly reflected. By following each of the two possible trajectories until they meet another interface, and repeating, we can construct a complex tree of rays. The "memory" of the initial direction of propagation is lost after a few refractions and/or reflections, over a characteristic distance known as the *transport mean free path*, denoted ℓ^*. We model the propagation of a light ray through a foam as an effective random walk, consisting of paths ℓ^* in statistically independent directions.

In a very wet foam, the distance ℓ^* is comparable to the size of a bubble: in fact, the liquid between two bubbles acts as a lens which strongly deflects a ray. On the other hand, the thin films between neighbouring bubbles in a dry foam do not deflect the transmitted rays, and ℓ^* greatly exceeds the bubble size. In this description, the only characteristic length-scale is the bubble size and so we expect that, at given liquid fraction ϕ_l, ℓ^* varies like the average bubble radius $\langle R_V \rangle$. In fact, experimentally [57],

$$\ell^* = 2 \frac{\langle R_V \rangle}{\sqrt{\phi_l}}. \tag{5.3}$$

The idea of a random walk occurs in other physical phenomena, such as the Brownian motion of a molecule in solution. At the macroscopic scale, this motion is described in terms of a concentration which varies in space and time according to Fick's law [1]. By analogy, we will show that light transport across an opaque medium is described by a light energy density obeying the same laws, which enables the propagation of light across a sample of multiply scattering foam to be modelled.

Diffuse light transmission Diffuse transmission spectroscopy (DTS) measures the diffuse light intensity transmitted through a foam by illuminating a strip of foam with a collimated light beam of intensity I_0, and measuring the total light intensity I which crosses the sample. The transmission factor $Tr = I/I_0$ is linked to the thickness of the strip L and to the transport mean free path ℓ^*. In the absence of optical absorption in the sample, and in the multiple scattering regime ($L \gg \ell^*$), we have [57]

$$Tr = \frac{1 + z_e}{L/\ell^* + 2z_e}. \tag{5.4}$$

The extrapolation parameter z_e, of order one, characterizes the conditions at the optical limits between the sample and the transparent plates which delimit it. For a foam confined between two glass plates, we have [57]

$$z_e = 0.74 + 3.44\,\phi_l. \tag{5.5}$$

In the limit $L \gg \ell^*$, eqs. (5.3) to (5.5) imply that, omitting prefactors,

$$Tr \sim \frac{\ell^*}{L} \sim \frac{\langle R_V \rangle}{L\sqrt{\phi_l}}. \tag{5.6}$$

A foam becomes more opaque as it becomes wetter, as the bubbles become smaller, or as the sample becomes thicker. A measurement of optical transmission enables the average bubble size to be determined for a known liquid fraction, and vice versa. As this measurement is instantaneous and non-destructive, it is appropriate for measuring the temporal evolution of the average bubble size during coarsening (FIG. 3.27) or of the liquid fraction during drainage (FIG. 3.29). Relative measurements, compared to a given bubble size or a given initial liquid fraction, are easy to perform; for absolute measurements, the technique must be calibrated using a sample with known diffusivity.

Dynamic light scattering Very often, we seek statistical information about the dynamics of topological changes within a foam, whether induced by coarsening (§3.2.1, chap. 3) or by an imposed mechanical perturbation (see §4.2.1, chap. 4 or §4.2.3, chap. 4). This information can be obtained by a dynamic light-scattering technique known as diffusing-wave spectroscopy (DWS), which probes the local dynamics in turbid media such as foams, concentrated emulsions, and pastes (FIG. 5.9) [60].

In DWS, the sample is illuminated with a coherent light beam and either the transmitted or back-scattered light is recorded on a screen placed close to the sample. A pattern of "speckles" (regions of different light intensity) is observed (FIG. 5.9a). The speckles are the result of interference between all the light rays arriving at the screen.

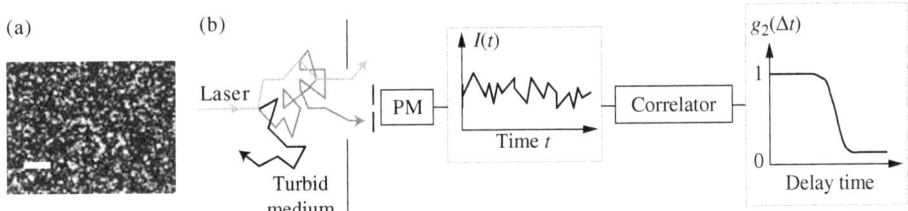

FIG. 5.9: Diffusing-wave spectroscopy (DWS). (a) Speckle pattern. The scale bar corresponds to an angle of 0.02 mrad. Photograph courtesy of S. Cohen-Addad. (b) DWS (in transmission) is used to probe the dynamics in a turbid medium: the speckle intensity is measured using a photomultiplier, and the autocorrelation function $g_2(\Delta t)$ is calculated using a signal correlator.

A change of structure in the sample, for example a T1, modifies the phase of those rays which cross the rearranged zone, and changes the intensity of the speckles, which appear to "twinkle". To quantify this, we measure the intensity of the speckle pattern $I(t)$ as a function of time, and calculate its (normalized) autocorrelation,

$$g_2(\Delta t) = \frac{\langle I(t)I(t+\Delta t)\rangle - \langle I(t)\rangle^2}{\langle I(t)^2\rangle - \langle I(t)\rangle^2}. \tag{5.7}$$

By taking an average, shown with angle brackets, we avoid having to choose a particular speckle or a particular bubble structure with fixed disorder. There are two equivalent ways of taking the average: if the internal dynamics of the sample (due to coarsening or flow, for example) modifies the structure over time, we use a photomultiplier to measure the intensity of a single speckle, and we average its autocorrelation over time t (FIG. 5.9b); if, on the other hand, the dynamics are slow or non-ergodic, we use a "multispeckle" technique instead, in which a large number of speckles is recorded simultaneously with a CCD camera [4, 12, 13], and the correlation between their respective intensities is averaged at instants t and $t + \Delta t$.

In the limit $\Delta t \to 0$, $g_2(\Delta t)$ tends to 1, indicating that over short time intervals the structure of the foam is unchanged. As Δt increases, the internal dynamics increasingly modify the structure, due for example to T1s, and the correlation between the structures at times t and $t + \Delta t$ is lost.

Statistical models of light propagation and topological changes permit a more quantitative analysis. For example, the characteristic time over which $g_2(\Delta t)$ decreases in a coarsening foam is proportional to the average time interval τ between T1s in a given volume [19], which increases as coarsening proceeds (FIG. 3.25). It is also possible to infer the fraction of the sample that undergoes T1s per unit time, for example during shear [27], to measure the duration of a T1 [4], or even to characterize the intermittency of the dynamics [12].

1.2.6 Conductivity The electrical properties of a foam are strongly dependent on its liquid content. Given a relationship between the electrical conductivity and ϕ_l, measuring the former would allow us to infer the latter.

In the very wet limit, the bubbles separate into isolated spheres. The relative conductivity Λ of this bubbly liquid, defined as the conductivity of the sample divided by that of the foaming solution, is given by the Maxwell relationship [21]:

$$\Lambda = \frac{2\phi_l}{3 - \phi_l}. \tag{5.8}$$

In the dry limit, the foam consists of a network of randomly oriented, conducting Plateau borders, in which the resistance varies as the inverse of ϕ_l according to eq. (2.38). The relative conductivity of the foam is now given by Lemlich's relationship [21]:

$$\Lambda = \frac{\phi_l}{3}. \tag{5.9}$$

Experimental methods 239

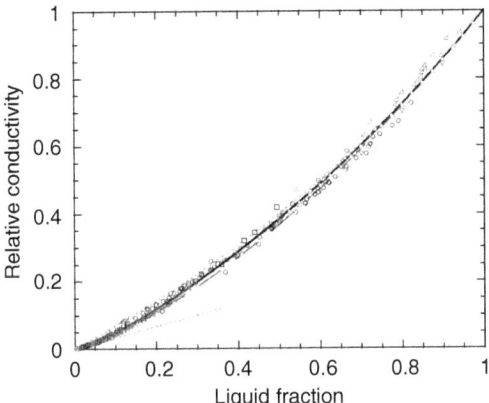

FIG. 5.10: Relative electrical conductivity as a function of liquid fraction [21]. The experimental measurements (symbols) obtained for foams and suspensions are well-described by eqs. (5.10) and (5.11) (thick line) over the whole range of liquid fraction. Eq. (5.9) (thin dotted line) is a linear approximation valid for liquid fractions of up to a few percent. See [21] for further details.

Measurements of the conductivity of foams and suspensions for liquid fractions between 0.02 and 0.9 suggest that it is continuous over the whole range of ϕ_l, as shown in FIG. 5.10. The empirical expression (a one-parameter best-fit to the data in FIG. 5.10)

$$\Lambda = \frac{2\phi_l(1 + 12\phi_l)}{6 + 29\phi_l - 9\phi_l^2} \qquad (5.10)$$

describes all the data well and agrees with the predictions of Lemlich and Maxwell in the two limits [21].

Conversely, the liquid fraction can be expressed in terms of the relative conductivity as

$$\phi_l = \frac{3\Lambda(1 + 11\Lambda)}{1 + 25\Lambda + 10\Lambda^2}. \qquad (5.11)$$

Neither average bubble size, polydispersity, or interfacial properties affect these relationships. Thus, no matter what the system (foam, emulsion, suspension of a non-conducting phase) or its physico-chemistry, the relative conductivity depends only on the degree of dispersion. Consequently, given the conductivity of the solution, which depends on the electrolyte concentration and the electrode geometry, a conductivity measurement of a sample of foam gives direct access to its liquid fraction (see FIG. 3.29 for example).

This conductivity measurement is simple to use: from the electrical point of view, the foam behaves as a circuit consisting of a resistor and a capacitor in parallel. The measurements are carried out with an alternating electrical voltage, typical 1 V oscillating at 1 kHz (see experiment §7.4, chap. 3).

1.2.7 Rheometry To study the rheological behaviour of a 3D foam it should be probed with the simplest and most well-defined perturbations possible. Once a constitutive law relating stress, strain, strain-rate, and structural parameters (cf. chap. 4) is established, it should be possible to predict the flow in any given geometry.

Methods To impose a controlled deformation on a foam, we inject the foam into the cell of a rheometer, which measures the strain under imposed stress or the stress under imposed strain. The choice of cell geometry depends on the information required; the characteristics of the strains and stresses which the sample undergoes in each possible geometry are described in FIG. 5.11. A detailed presentation of the methods of rheometry can be found in [39], for example.

The four geometries "sliding plate", "parallel plate", "cone-plate", or "cylindrical Couette" are all a priori suitable for measuring the linear response, under the condition that the gap size is much greater than the bubble size (typically at least 20 bubbles). To study the non-linear regime, a geometry for which the stress is the same everywhere within the gap is preferable, as in the sliding plate, cone-plate (if the cone angle is small), or even the cylindrical geometry (if the gap is small compared to the radius of the cylinder). By measuring the force exerted by the sample perpendicular to the plate, we obtain the first normal stress difference which accompanies the shear (cf. §4.1.2, chap. 4). Finally, to study the flow of a foam under the effect of a pressure gradient, we use a capillary rheometer.

We have seen in §4.3.2, chap. 4 and FIG. 2.29 that the flow of a foam is sometimes heterogeneous, or localized, whether or not the stress is homogeneous. Under these conditions, it is not possible to deduce the local strain from the motion of the walls, and it is necessary to observe the sample itself, for example optically or with MRI (cf. §1.2.4).

Wall slip A frequent problem in foam rheometry is the slip of the sample at the walls of the rheometer [40], which are covered by a film of the foaming solution (§3.3.3, chap. 4). The degree of slip depends on the material from which the wall is made: plastic is often more slippery than steel. Wall slip is distinct from shear banding, which can occur in the bulk, far from the walls. It is often straightforward to detect wall slip experimentally, since the parameters in the constitutive law appear to depend on the thickness of the sample or the size of the rheometer. It distorts the analysis of the rheological measurements since the motion of the walls or the pressure drop is greater than what is actually experienced by the foam. To eliminate it, it is necessary to roughen the walls with (water-resistant) sandpaper or by etching grooves in the surface, perpendicular to the direction of flow; the grain of the paper or the depth of the grooves must be greater than the bubble diameter.

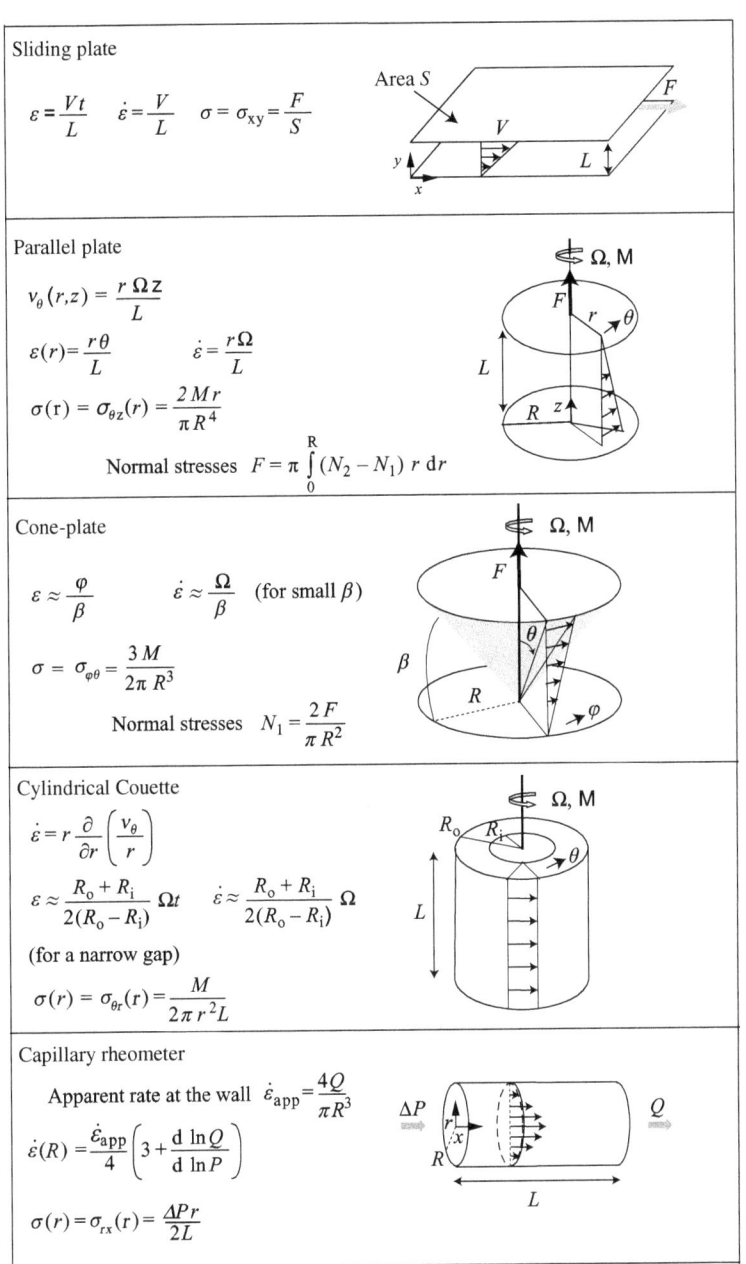

FIG. 5.11: Schematic diagrams of the different rheometer geometries used in foam rheology. The variables are strain ε, strain rate $\dot{\varepsilon}$, shear stress σ, force **F** and torque **M**, velocity V and angular velocity Ω, tangential fluid velocity v_θ, pressure drop ΔP, and volumetric flow rate Q.

2 Numerical simulations

In this section, we describe the main numerical tools used to simulate the static structure and rheology of foams. Some have high levels of precision while others generate statistics for large numbers of bubbles.

Numerical simulations are an important tool for both the engineer and the researcher. Whilst an analytical approach is often useful for ordered foams, simulations are usually necessary to understand disordered foams, given the larger number of parameters involved: vertex positions, edge lengths, and curvatures, and, in 3D, the curvature of the faces. They enable foam behaviour to be predicted without actual experimentation, and allow the different physical parameters to be varied independently, something which is often impossible experimentally, for example to ignore gravity or gas diffusion, and even control polydispersity. Moreover, some quantities, such as the fields of pressure or local velocity, and film curvatures, which are difficult or impossible to measure experimentally, are accessible and it becomes possible to explore parameter ranges beyond the reach of experiments.

The simulation techniques presented here are adapted either to predicting the foam's static (equilibrium) state, or its flow. Even then, the choice of simulation tool can depend on the significance of dissipation in the problem studied, the range of liquid fraction, the dimension (2D or 3D), and the compromise sought between obtaining the fine detail of the structure, the number of bubbles to be considered, and the calculation time.

2.1 Predicting static structure

To obtain the structure of a foam at equilibrium, i.e. the shape and pressure of the bubbles, it is necessary to satisfy Plateau's laws and the Young–Laplace law everywhere [8], or, as we describe here, find a local minimum of energy.

Starting from an arbitrary initial configuration in which the bubbles are connected and tessellate space, for example a Voronoi tesselation [58] of randomly scattered points, the interfaces are deformed so as to minimize the energy, respecting the volume of each bubble and any further constraints at the boundaries. During the search for a minimum, the topology of the initial tessellation may be modified by one or more T1s.

We present below the two most popular methods for predicting the equilibrium shape of bubbles and their pressures. Although designed with dry foams in mind, they can be adapted to the wet case. They are in principle equivalent and both are freely available on the Internet, with each having different advantages.

2.1.1 Potts model The Potts model [30] searches *randomly* for an energy minimum. This enables a minimum, which may be very different from the initial configuration, to be found rapidly, even for a large number of bubbles. Developed by R.B. Potts in 1952 for statistical physics, it has been used to simulate the grains (microcrystals) in crystalline materials [51] as well as foams [29]. It forms the basis of the free software CompuCell [30], for which extensive training is available.

It is a lattice model: as in a digital photograph, a foam is represented by small surface elements (pixels in 2D) or volume elements (voxels in 3D) with fixed positions.

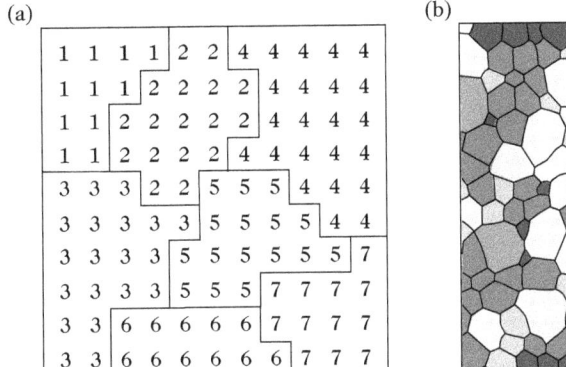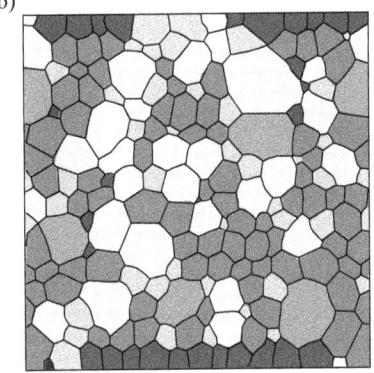

FIG. 5.12: Potts model simulation [26], here in 2D. (**a**) Bubble i consists of all pixels which are labelled with the number i, with i running here from 1 to 7. (**b**) Polydisperse foam at equilibrium. The shading indicates the number of sides of a bubble. 589 bubbles on a lattice of 256×256 pixels, with periodic boundary conditions to left and right and hard walls at the top and bottom.

Bubble i is defined as all those pixels labelled with the same number (or index) i. In a dry foam, each pixel belongs to only one bubble, so the bubbles tessellate space (FIG. 5.12). The interface between bubbles i and j is exactly those sites where a pixel labelled i neighbours a pixel labelled j. The energy of the foam is equal to the contact area (i.e. the area of the interface between sites with different indices) multiplied by 2γ. Additional terms in the energy functional (see exercise 4.2) ensure conservation of the bubble volumes.

The energy minimization is *sequential*: a pixel located at the edge of bubble i is chosen at random and associated with the neighbouring bubble j, i.e. it takes the index of one of its neighbours. If this operation lowers the energy, it is retained (otherwise the pixel is returned to bubble i). In this case, it reduces the size of bubble i and increases the size of bubble j and moves the interface between them (a small variation in bubble volume is therefore permitted). It can happen that this puts bubble j into contact with a new bubble k: in other words, it may cause a T1. This operation is repeated thousands of times, and the total energy gradually decreases to a local minimum.

The model also allows the effect of an external force to be incorporated. It also permits, with a certain probability (which decreases when the energy cost increases), that a pixel can change bubble even if this increases the energy. This allows the exploration of different energy minima and means that the model is appropriate for non-deterministic systems with fluctuations, for example physical systems at a temperature which is significantly greater than their energy variations (divided by Boltzmann's constant) or biological cells [30]. Finally, it allows the simulation of a million 2D or 3D bubbles [55] (see FIG. 3.24), with a degree of precision of the order of one percent of the value of the interface area (and thus of the energy) [29].

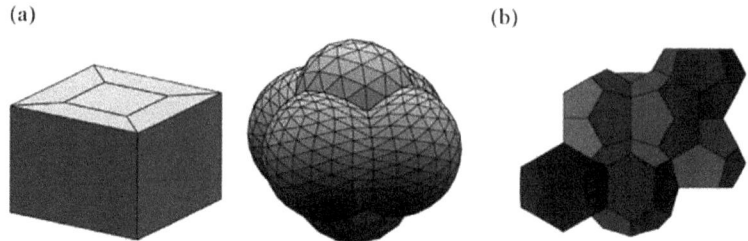

FIG. 5.13: *Surface Evolver* simulation [31]. **(a)** Starting from an initial configuration (on the left), each face of each bubble is divided into small triangles that move in order to decrease the total energy, until a local minimum is found (on the right). **(b)** Example of an ordered monodisperse foam in 3D, the Weaire–Phelan structure. This is the structure with the least surface area per unit volume known (see shaded section on p. 37) [59]. Six barrel-shaped bubbles touch each other on their hexagonal faces. They are therefore aligned in columns along three perpendicular axes. The gaps are filled with pentagonal dodecahedra (the two bubbles at the far right of the figure).

2.1.2 *Surface Evolver* software The *Surface Evolver* (FIG. 5.13) seeks an energy minimum in a *deterministic* way [9]. Developed with grain growth and minimal surfaces in mind, it has also been used successfully in other fields in which surface energy is important. Its creator, K. Brakke, has been providing updates and documentation for free on his Internet site [31] since 1992.

The surfaces of the bubbles are discretized into a mesh of small triangles which determines the precision; it can be refined as the calculation proceeds, allowing the shape of each bubble to be described with a level of precision as high as desired (see for example FIG. 2.17). In the case of 2D foams, the surface elements are the edges themselves, which can be represented as arcs of circles and do not need to be subdivided.

The variables of the problem are the attributes of these surface elements (vertex position, edge curvature, etc.), generating a parameter space of very large dimension. The software calculates the energy gradient in this vector space and determines the direction of greatest energy decrease. All the elements are then moved in this direction together: the entire structure is affected simultaneously. If during this process the interface between two bubbles becomes smaller than some given critical size, a T1 can be triggered, the new surface tesselated, and the minimization continued.

The calculation converges when the energy varies less than some specified amount. It is necessary to strike a compromise between the required precision (mesh size, convergence criterion, critical interface size before triggering a T1) and the calculation time. The software is used for simulations of up to a thousand bubbles in 3D [35].

2.2 Predicting dynamics

When the dynamics are slow, a foam passes through a succession of equilibrium states (§2.2.1); we can study this using the static foam simulations described above. If the flows are more rapid, then the foam is not at equilibrium. Its structure is determined by the competition between three forces: surface tension, pressure forces, and viscous forces, which are functions of the local velocity gradient. For very rapid flows or wet foams, inertia must also be taken into account. We present below a model that works in both 2D and 3D (§2.2.2), and two models limited to 2D (§2.2.3, 2.2.4).

2.2.1 Quasi-static models Potts model and *Surface Evolver* can both be used in 2D or in 3D to simulate slow dynamics, either in coarsening (§3.2, chap. 3), where the bubble volumes evolve slowly, or quasi-static flows (FIG. 5.14).

The foam structure is first driven to an equilibrium state determined by the initial conditions, which may include, for example, the position of the walls which contain the foam. To simulate the flow induced by the movement of these walls, for example, the walls are moved step by step and a new minimum of energy is sought at each step (FIG. 4.22b). The evolution of the structure towards equilibrium during each relaxation is of no physical significance: only the successive equilibrium states are relevant.

In *Surface Evolver*, a T1 is triggered when a face separating two bubbles, or an edge, becomes smaller than a critical value, chosen in advance. With a higher critical value, the T1s happen more easily, providing an indirect way of taking into account the liquid fraction [48] (cf. FIG. 4.30 and §3.3.2). In the Potts model, it is the pixel size which determines the critical value [48].

2.2.2 Bubble model The bubble model, first described by Durian in 2D [18], and later extended to 3D [23], enables viscous dynamics to be probed. This approach is more phenomenological than the preceding ones, giving a fast method of calculation which does not respect the exact shape of the bubbles. The structure is determined by the position of the centres of spherical bubbles of given, constant, diameter. The bubbles can overlap, and interact via a repulsive contact force, proportional to their overlap, and through frictional forces between neighbouring bubbles, proportional to

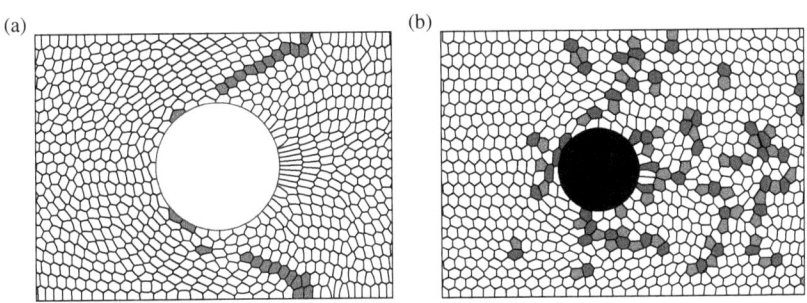

FIG. 5.14: A dry monodisperse 2D foam flowing from left to right around an obstacle. Good agreement is obtained between these two quasi-static simulations based on different algorithms [48]. (a) *Surface Evolver*, with an initially disordered foam; the grey bubbles are those which initially form a straight line at the entrance to the channel. (b) Potts model, with an initially ordered foam; the grey bubbles are those which have 5 or 7 sides.

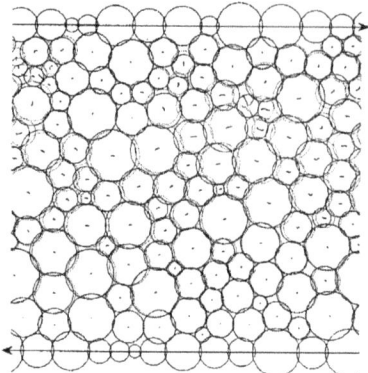

FIG. 5.15: Bubble model simulation of a 2D foam under shear. The bubbles are represented by slightly (in this wet foam) overlapping circles: greater overlap represents greater bubble deformation for given liquid fraction, and lower liquid fraction in general. The bubbles cut by horizontal lines are fixed and move in the direction of the arrows to impose a shear on the rest of the foam. The dotted and continuous circles correspond to two successive instants in time, and the line segment at the centre of each bubble represents its motion. In this example, a T1 takes place towards the top right corner, where significant motion can be seen, even though the movement of the walls is imperceptible (reproduced from [18], © 1997 The American Physical Society).

their relative velocity (FIG. 5.15). These frictional forces do not take into account the flow in the liquid phase (see §3.3, chap. 4).

From a given initial state, the forces exerted on each bubble, which depend uniquely on the (known) positions of the centres and the (unknown) relative velocities, are calculated. Since inertia is neglected, the sum of these forces must be zero, which enables the velocity of each bubble to be determined. The bubbles are then moved, and the process repeated. It is a similar process to molecular dynamics, which is used to predict the motion of a group of atoms [2].

2.2.3 Viscous froth Model The flow of a 2D foam between two plates is special because of the additional contribution to the dissipation from the plates (see §5.4, chap. 2, FIG. 2.29, and also [11]): the friction forces are functions of the velocity (and not the velocity gradient as in 3D) and are relatively well understood (see §3.3, chap. 4).

There are many different simulations of the dynamic behaviour of a 2D foam. The most precise ones exploit a full discretization of the structure (viscous froth model [34]): each edge is represented by a certain number of elementary line segments, acted on by the pressure in adjacent bubbles, by the surface tension, and by a friction force at the plates, proportional to velocity. As for the bubble model, the sum of forces on each segment is zero, which gives the velocity of each segment. If this motion leads to the disappearance of an edge, a T1 is triggered (see §2.2.1), allowing the prediction of the effect of frictional forces on T1s (FIG. 5.16).

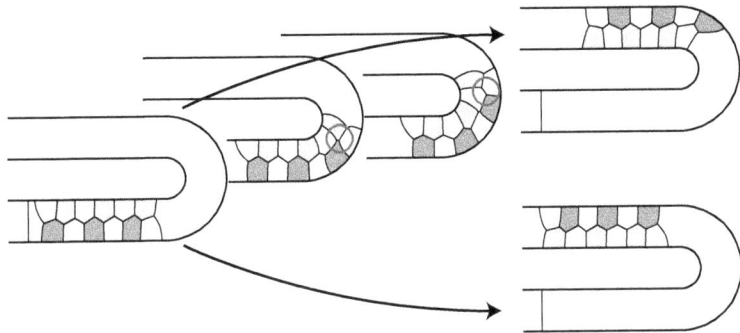

FIG. 5.16: Movement of a 2D foam along a U-shaped channel simulated with the viscous froth model, which includes the friction with the plates bounding the foam. Lower arrow: at low velocity the foam moves as a plug and the bubbles keep the same neighbours. Upper arrow: above a critical driving velocity, the bubbles on the inside of the bend move more quickly and a T1 occurs [34].

2.2.4 Vertex model For 2D flows with a large number of bubbles, further simplification is necessary. One possibility is to describe the foam by the vertex positions and their connectivity (vertex model [46]). The edges are approximated by straight lines (FIG. 5.17) and the dynamics are reproduced with an algorithm similar to the one above.

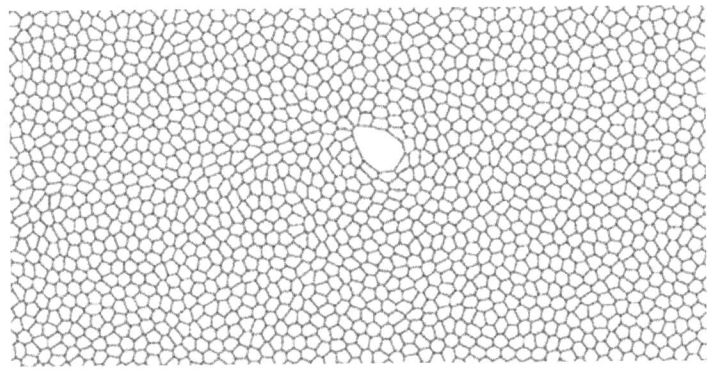

FIG. 5.17: Simulation of the flow of a foam between two plates using the vertex model. The foam is monodisperse and disordered with one defect: one bubble larger than the others. There are periodic boundary conditions and the foam moves to the right at a given rate. This simulation validated the theoretical prediction of the velocity of the large bubble, which is higher than the average (see §5.4, chap. 2) [10].

3 Methods of image analysis

An image of a foam, particularly in 2D, is remarkably informative as it contains almost all the physical information required (except for the viscous dissipation). In general, it is necessary to start by treating the image in some way (§3.1), identifying the bubbles and their boundaries for example, although images from numerical simulations can be used directly, without treatment. The procedures for gathering data from the treated images (§3.2) are often the same for experiments and simulations. The 2D case is simpler (§3.3).

3.1 Image treatment

The image obtained from a camera is a measure of the grey level in each pixel. If a foam is illuminated and observed from the same side (reflection), the liquid appears brighter than the gas (FIG. 5.18a); if illuminated from behind (transmission), the reverse is true.

To make the most of the images, it is necessary to first segment them, that is, to identify individual bubbles. Various methods, adapted to either 2D or 3D, allow the recognition of bubbles in spite of low contrast, noise, poor resolution, or variations in intensity. Here we consider only the case of high-quality 2D images, the easiest to treat and analyse (§3.3.1).

3.1.1 Segmentation For an image with good contrast and many pixels per bubble, segmentation can be automated with the following steps:

- Pre-treatment: compensate for slight variations in the lighting and in the background, for example by filtering the image to reduce this noise.
- Thresholding: grey-levels greater than a given value (dependent on the image being treated), called the "threshold", are turned white (the liquid phase in FIG. 5.18b); the others become black (the gas phase in FIG. 5.18b).
- Dilation (predominantly for dry foams): sometimes a bubble is not completely surrounded by "white" and therefore appears to be connected to a neighbour.

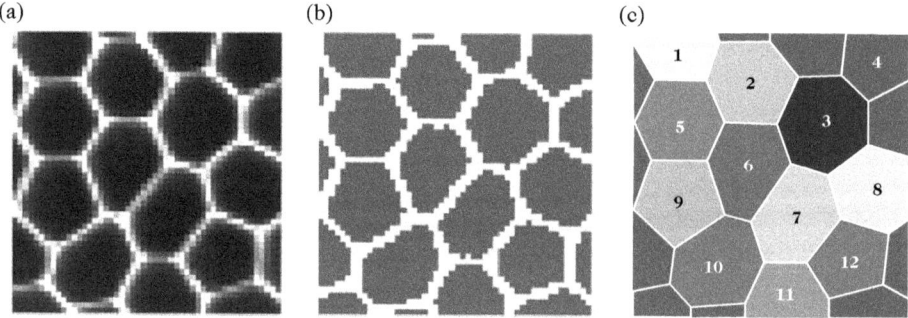

FIG. 5.18: Main steps in image treatment, here in 2D: **(a)** raw image (courtesy of I. Cantat); **(b)** image after thresholding and dilation; **(c)** image after skeletonization and segmentation. The liquid appears brighter than the gas.

In this case, we assign the "white" value to the n nearest neighbours of each of the white pixels; this disconnects the bubbles, and those bubbles smaller than n pixels are lost.
- Skeletonization of the continuous (white) phase: the pixels at the edge of the white region are changed to black as long as this doesn't alter the topology. The final image then consists of black bubbles surrounded by a white border exactly one pixel thick (FIG. 5.18c).
- Segmentation: identification of each individual bubble. The same number—the number of the bubble—is attributed to all touching black pixels (FIG. 5.18c); this allows each bubble to be coloured and easily identified (see FIG. 5.8).

3.1.2 Bubble tracking To characterize the kinematics of the foam (cf. §3.2.5), a succession of images is required (a video), at time intervals Δt sufficiently short to capture the evolution of interest. For example, an image every 30 min is sufficient to follow slow coarsening, an image every second is enough for a quasi-static flow, and a high-speed camera (up to 1,000 images per second) allows the study of rapid flows.

Even if the image is not segmented (i.e. the bubbles have not been individually identified), subtracting one image from the next in the sequence gives a qualitative visualization which is often sufficient. In this case, it is possible to shift the image so as to find the optimal correlation with the next one, which allows the average velocity field to be determined.

If we segment the image, we can also determine the individual speed of each bubble. This requires that we follow each bubble from one image to the next, thus matching a bubble k in image m with a bubble k' in image $m+1$. If each bubble advances less than one diameter between two successive images, one method is to determine if the centre of mass $(x_{k,m}, y_{k,m})$ of the bubble k in the image m lies in bubble k' in image $m+1$. If this is the case, then k and k' are the two successive indices of this particular bubble.

If the average velocity field $v(x, y)$ is known, we can even follow the bubbles which advance more than one diameter between two successive images by seeking the bubble in image $m+1$ that contains the point $(x_{k,m} + v_x \Delta t, y_{k,m} + v_y \Delta t)$. Conversely, if we follow each bubble, we can measure the velocity of its centre of mass, and thus average these speeds, weighted by the bubble volumes (or by the mass of the liquid films) in order to obtain $v(x, y)$.

3.1.3 Basic data Ideally, for each bubble, a 3D image gives access to raw data on:

(i) all its interior pixels, its centre of mass, its volume, and the local liquid fraction;
(ii) its number of faces, the face areas, and curvatures;
(iii) the number of edges per face, edge length, curvature, and cross-section;
(iv) the number of vertices and their position.

In all of this we have assumed that we know all the voxels of each bubble. In practice, this is the case for a segmented image from X-ray or optical tomography or NMR (see §3.1.1), and for *Surface Evolver* or Potts simulations. However, there are occasions when this does not hold, for example tomography or NMR images of foams that are too dry for the liquid films separating the bubbles to be seen, or vertex model simulations.

3.2 Image analysis

How can we exploit all of the information available in an image? From the basic data (§3.1.3) we can measure many quantities. We give here a non-exhaustive list.

Each measurement is carried out in response to a certain question. Depending on the problem posed, we might measure the complete distribution of a quantity, or content ourselves with its average or standard deviation.

3.2.1 Topology, disorder
A bubble is characterized by its number of neighbours, i.e. faces, f. This is used to determine the topological disorder of the foam, which is a measure of how closely the different bubbles resemble each other. It is characterized by the variance (dispersion of the values around the average) of the distribution of the number of faces in the foam, $\mu_2^f = \langle (f - \langle f \rangle)^2 \rangle$, or equivalently $\mu_2^f = \langle f^2 \rangle - 2\langle f \rangle \langle f \rangle + \langle f \rangle^2 = \langle f^2 \rangle - \langle f \rangle^2$. The resulting number is positive, and we often take its square root, the standard deviation $\delta f = \sqrt{\mu_2^f}$, or its dimensionless form

$$\frac{\delta f}{\langle f \rangle} = \sqrt{\frac{\langle f^2 \rangle}{\langle f \rangle^2} - 1}. \tag{5.12}$$

Around its perimeter each face has a number of edges n, whose distribution also has a standard deviation. It is another measure of topological disorder, this time for a single bubble: the more irregular the bubble is, the greater $\delta n / \langle n \rangle$ becomes.

This sort of topological analysis is evident in the empirical Aboav–Weaire relationship, which links the number of faces of a bubble f to the average number of faces of its neighbours, $\langle f \rangle_{\text{neigh}}$ (§3.1.4, chap. 2). It is observed that if $f \langle f \rangle_{\text{neigh}}$ is plotted as a function of f, the result is approximately a straight line (FIG. 2.11):

$$f \langle f \rangle_{\text{neigh}} = (\langle f \rangle - a_f) f + a_f \langle f \rangle + \mu_2^f. \tag{5.13}$$

Note that if the number of faces f of a bubble is large, then the average number of faces of its neighbours is small. The parameter a_f is a characteristic of the foam, which can be measured by multiplying eq. (5.13) by f and averaging over all bubbles in the foam [33]:

$$a_f = \langle f \rangle - \frac{1}{\mu_2^f} \left\langle f^2 \left(\langle f \rangle_{\text{neigh}} - \langle f \rangle \right) \right\rangle. \tag{5.14}$$

3.2.2 Bubble size, polydispersity
A further characteristic of a bubble is its size. We have access to the volume V of each bubble from MRI, optical or X-ray tomography, 3D microscopy, simulations, and in 2D foams (in which case we measure the area A, see §3.3). The bubble radius is available from other experiments, including light scattering and 2D images of 3D foams (projection, radiography, slices, microscopy).

The radius has several possible definitions (as opposed to the volume V, which is uniquely defined) and different experiments measure different radii. We can in principle define several characteristic sizes of a bubble, for example the value $R_S = (S/4\pi)^{1/2}$ associated with the surface (the radius of a sphere with the same surface area S) or $R_V = (3V/4\pi)^{1/3}$ associated with the volume (the radius of a sphere with the same volume V). For a bubble obeying Plateau's laws, eq. (2.19) gives $R_S \approx 1.05\, R_V$.

Given a definition of the radius R, there are several ways to define a characteristic length based on an average over the whole foam, using the moments of the distribution of R. Thus, $R_{\text{avg}} = \langle R_V \rangle$ is the average radius and $R_{\text{rms}} = \langle R_V^2 \rangle^{1/2}$ is the mean square radius. More generally, we construct a family of characteristic lengths:

$$R_{n,n-1} = \frac{\langle R_V^n \rangle}{\langle R_V^{n-1} \rangle} \qquad (5.15)$$

where n is a whole number (see FIG. 5.19). For example, $R_{10} = R_{\text{avg}}$ and $R_{21} = R_{\text{rms}}^2/R_{\text{avg}}$; R_{32}, the Sauter mean radius, is used in rheological constitutive models (§4.1.1, chap. 4) as well as models of osmotic pressure (§4.2, chap. 2). Their values are close, and for a monodisperse foam (i.e. in which all the bubbles have the same volume) they are equal.

The correspondence between measurements of radius and volume is not direct. In particular, in a 2D slice of a 3D foam the apparent size of the bubble depends on the position of the cut, so that bubbles of the same volume can appear to have different radii. A monodisperse foam may thus show a polydisperse cross-section. Likewise, if we see only the free surface of the foam or observe those bubbles that touch a (transparent)

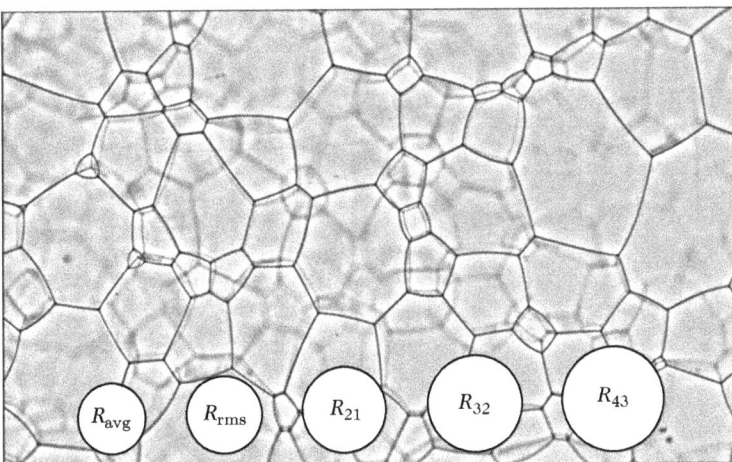

FIG. 5.19: Different values of the average bubble size. The radius of each circle represents the value of the quantity indicated in the circle, defined according to eq. (5.15) and measured for the polydisperse foam shown in the background. Photograph courtesy of K. Feitosa, D. Durian [22], reproduced by kind permission of Springer Science + Business Media.

wall (FIG. 5.6), the apparent distribution of bubble sizes doesn't directly reflect the actual volumes.

The geometric disorder is the (dimensionless) standard deviation of the distribution of the bubble volumes, analogous to eq. (5.12):

$$\mu_2^V = \langle (V - \langle V \rangle)^2 \rangle \quad \text{and} \quad \frac{\delta V}{\langle V \rangle} = \sqrt{\frac{\langle V^2 \rangle}{\langle V \rangle^2} - 1}. \qquad (5.16)$$

It is almost equivalent to considering the polydispersity in the bubble radii [35]:

$$p = \frac{R_{32}}{\langle R_V^3 \rangle^{1/3}} - 1 = \frac{\langle R_V^3 \rangle^{2/3}}{\langle R_V^2 \rangle} - 1 = \frac{\langle V \rangle^{2/3}}{\langle V^{2/3} \rangle} - 1. \qquad (5.17)$$

The size disorder is given by $\delta R_V / \langle R_V \rangle = (\langle R_V^2 \rangle / \langle R_V \rangle^2 - 1)^{1/2}$. If $\delta V / \langle V \rangle$ and $\delta R_V / \langle R_V \rangle$ are small compared to 1, they are approximately related by $\delta V / \langle V \rangle = 3 \delta R_V / \langle R_V \rangle$. These three measures, $\delta V / \langle V \rangle$, $\delta R_V / \langle R_V \rangle$, and p, are zero for a monodisperse foam and positive otherwise.

With regard to correlations, a large bubble often has smaller neighbours. This negative correlation is characterized in a similar way to the topology (eq. (5.14)) by the parameter

$$a_V = \langle V \rangle - \frac{1}{\mu_2^V} \langle V^2 (\langle V_{\text{neigh}} \rangle - \langle V \rangle) \rangle. \qquad (5.18)$$

3.2.3 Surface area, deformation

It is important to know the total surface area of a foam, especially because this gives its elastic energy (eq. (2.15)). It is difficult to exploit however, as the differences in energy between the different possible arrangements of a given group of bubbles are often too small to be detectable experimentally. In order to resolve this, *Surface Evolver* simulations are often used to calculate the ratio $S/V^{2/3}$ for each bubble. Then the regularity of the bubble can be assessed by comparing it with a sphere or a regular bubble (eqs. (2.16) and (2.19)).

The deformation of a bubble can be characterized by the "inertia matrix":

$$\begin{pmatrix} \langle x^2 \rangle & \langle xy \rangle & \langle xz \rangle \\ \langle yx \rangle & \langle y^2 \rangle & \langle yz \rangle \\ \langle zx \rangle & \langle zy \rangle & \langle z^2 \rangle \end{pmatrix} \qquad (5.19)$$

where the average is over the coordinates (x, y, z) of all the pixels of the bubble (with the origin taken at its barycentre). This matrix, with units of m^2, describes the principal directions of deformation of the bubble and its anisotropy; the eigenvalues give the directions in which a bubble is elongated.

The deformation of a foam (or a subregion) is characterized by the "texture" [25]:

$$\begin{pmatrix} \langle X^2 \rangle & \langle XY \rangle & \langle XZ \rangle \\ \langle YX \rangle & \langle Y^2 \rangle & \langle YZ \rangle \\ \langle ZX \rangle & \langle ZY \rangle & \langle Z^2 \rangle \end{pmatrix} \tag{5.20}$$

where the average is over all vectors, with components (X, Y, Z), which connect the barycentres of two neighbouring bubbles. This matrix, also expressed in m^2, describes the principal directions of deformation and the anisotropy of a foam; there is an equivalent definition in 2D (FIG. 5.20).

When a dry foam is deformed, each bubble undergoes a similar deformation, so the (intra-cellular) inertia and the (inter-cellular) texture are correlated. In a wetter foam, each bubble becomes more rounded, so that its inertia is more isotropic while the texture continues to reflect the deformation of the whole foam.

3.2.4 Forces, pressure, curvature Bubbles exert forces on each other which originate from three sources: gas pressure, film tension, and viscous friction. Dynamic quantities, like viscous stresses, cannot be measured directly on the image.

On the other hand, the image does allow access to geometric quantities, like the contribution of pressures and films to the stress (eq. (4.18)), for which it is necessary to know the surface tension γ. In 2D, it is sometimes possible to make direct measurements (see §3.3.3), although this is usually difficult, because of sensitivity to the pixellization of the images.

To determine the average pressure in a region, we measure the average of the bubble pressures, weighted by their volume. The contributions of the films is determined by integrating the last term of eq. (4.18) over each face, which requires knowledge of the exact shape of the face (and therefore its curvature).

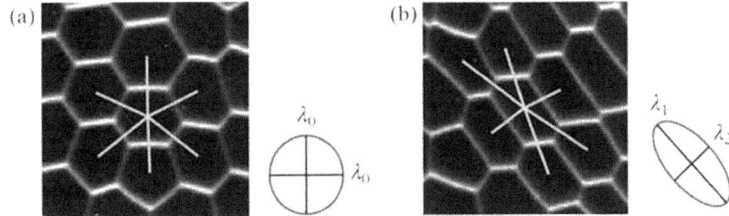

FIG. 5.20: Deformation of a foam, here in 2D. In (a), the foam is almost isotropic: the texture of a bubble, calculated from its links with neighbouring bubbles, can be represented by a circle (the two eigenvalues of the matrix in eq. (5.20) are equal, denoted λ_0). In (b), the foam is deformed: the texture is represented by an elongated ellipse in which the axes are the two distinct eigenvalues, $\lambda_1 \neq \lambda_2$. Images courtesy of C. Raufaste, P. Marmottant, reproduced by kind permission of The European Physical Journal [25].

To measure the curvature (which is also useful for determining the geometric charge, eq. (2.23)), it is necessary to fit to the image a surface of constant mean curvature. The fit may be improved by noting that the curvature arises from the pressure difference between bubbles (eq. (2.4)), which implies relationships between the curvature of adjacent faces (eqs. (2.5) and (2.24)). In other words, by determining simultaneously the pressures and the curvatures best approximating the faces, we can improve the accuracy. For N bubbles, and therefore N unknown pressures, there are about $N\langle f\rangle/2$ faces, and the same number of measured curvatures. Then there are $\langle f\rangle/2$ times more givens than unknowns, which improves the approximation to the pressures and curvatures.

The structure of a foam observed in an experiment can be reconstructed in the *Surface Evolver*. Using the fact that each face has constant mean curvature, the pressures and curvatures can then be calculated.

3.2.5 Evolution in time Following individual bubbles over time allows, for example, measurement of their velocity (cf. §3.1.2) and their growth rate (see §3.1, chap. 3) defined at a fixed number of faces, because we can only track the bubble in the absence of T1s.

To identify the topological changes, we generate the list of bubbles that share a face (neighbours). If this list changes between two successive images, it is because there has been a topological change, allowing us to count the number of T1s, T2s, or film ruptures per unit time and relate them to the number of faces.

We can also measure how the texture changes (eq. (5.20)) between two images, allowing us to treat all the topological changes simultaneously, and to study their direction as well as their characteristic size [16, 25].

3.3 Image analysis, liquid fraction, and stress in 2D

The quantities defined in §3.1 and §3.2 each have an equivalent in a 2D foam. For the topology and the geometry, it is sufficient to replace f with n, and V with A. There are some features, however, which are specific to 2D foams.

3.3.1 2D image analysis In 2D, images are easy to obtain experimentally (certainly more so than in 3D), and especially to segment. On an image which shows good contrast and homogeneous lighting (for example with a diffuse light source), the bubbles can be individually identified (FIG. 5.18). Even if the foam is very dry (between two glass plates) the edges of the bubbles, and thus the faces, are visible because of the surface Plateau borders (FIG. 5.6a). In simulations, the data is again often more precise. At equilibrium the edges are circular arcs, which facilitates numerical or analytical treatment.

3.3.2 2D liquid fraction Defining the liquid fraction in 2D poses considerable problems (cf. §5.3, chap. 2). In a Hele–Shaw cell (FIG. 2.25a) it is possible to define the liquid fraction as in 3D, i.e. as the fraction of the volume between the plates that the liquid occupies. But this depends on whether the walls are wetted or not by the soap solution, and this definition is only weakly related to the physical properties of the foam. In the bubble raft (FIG. 2.25c), or the liquid–glass arrangement (FIG. 2.25b and 2.25d), the total volume of the foam is not precisely defined, and therefore neither is the liquid fraction.

In fact, in all realizations of a 2D foam, it is necessary to define and measure an *effective* liquid fraction, appropriate to the problem under investigation. Thus for measurements of flow, it is possible to establish a link between the gas and the liquid flows. Or for the study of mechanical properties, we measure the distance between two vertices at the moment when a T1 is initiated (§2.2.1) [48]. Finally, if we are interested in what happens in the vicinity of a glass wall (friction, for example), there are optical methods [56] for estimating the shape and size of surface Plateau borders (FIG. 5.6).

3.3.3 2D stress The 3D surface tension depends only on the composition of the soap solution. In 2D it is replaced by the line tension, which also depends on the bubble geometry (cf. §5.2, chap. 2): it must be measured directly on each occasion, which is not always easy [14, 48].

Foams in a liquid–glass experiment probably constitute the unique (experimental) situation where the pressure in each bubble can be measured directly. If the pressure increases in a bubble, it descends slightly into the liquid phase without noticably changing its volume. The change in surface area, viewed from above, is small but it is nonetheless sufficient to deduce by image analysis the change in pressure from its relaxed state (FIG. 2.28c) [16].

In a very dry 2D foam in a Hele–Shaw cell it is possible to measure the stresses due to the bubble edges. Their curvature can be measured by fitting them to circular arcs, but if the foam is at, or close to, equilibrium (quasi-static case), the stress depends only weakly on the curvatures, and it is sufficient to determine the position of the vertices and treat the edges as if they were straight [32].

4 Exercises

Solutions to the exercises are available on request through the following webpage: http://www.oup.co.uk/academic/physics/admin/solutions.

4.1 Measurement of the average liquid fraction of a foam

A U-tube can give an idea of the average liquid fraction of a foam, using Archimedes' principle. If the bottom of the tube is filled with the foaming solution and foam is placed on one side of the tube, the liquid on the other side rises.

For a foam with volume $V_{\text{foam}} = 1$ mL in a tube of diameter 3 mm, what is the liquid fraction ϕ_l if the liquid on the other side rises to a height of 8.5 mm?

4.2 Pressure in the Potts model

The free enthalpy expression $\mathcal{G}$ on which the Potts model is based (called for simplicity *the energy* in §2.1.1) for the whole foam is

$$\mathcal{G} = \gamma d_p^2 \sum_{\text{pixel } i} \sum_{\text{neighbours } j} [1 - \delta(n_i, n_j)] + \sum_{\text{bubbles } k} K \frac{(V_k - V_k^o)^2}{V_k^o},$$

where d_p is the edge length of a pixel, n_i is the index of pixel i, $\delta(n_i, n_j) = 1$ if $n_i = n_j$ and 0 otherwise, $V_k^o = n_k RT / P_{\text{ext}}$ is the reference volume of bubble k, and K is a physical constant to be determined.

(a) Show that this expression does correspond, in the limit of small variations in volume, to the free enthalpy of the foam, and give the value of K. You should write the free enthalpy of the foam as the sum of the gas and liquid enthalpy and the interfacial energy. Then consider a transformation in which the volume of the bubbles changes from V_k^o to V_k at constant external temperature and pressure, assume that the gas in the foam is a perfect gas, and calculate the change in free enthalpy during the transformation.

In practice, the foam structure that is obtained doesn't depend on the value of K chosen, although its value must be sufficiently large that the differences $(V_k - V_k^o)/V_k^o$ are much less than 1 (barely compressible foam). For numerical reasons, we are limited to a value of K much smaller than the value found previously and we are able to confirm that $(V_k - V_k^o)/V_k^o \ll 1$.

(b) Give the pressure P_k in bubble k as a function of V_k, V_k^o, and K, and show that $P_k = -\partial \mathcal{G}/\partial V_k$ in the limit of small variations in volume.

References

[1] A.W. ADAMSON and A.P. GAST, *Physical Chemistry of Surfaces*, Wiley, New York, 6th ed., 1997.
[2] M.P. ALLEN and D.J. TILDESLEY, *Computer Simulation of Liquids*, Oxford University Press, Oxford, 2002.
[3] M. AXELOS and F. BOUÉ, *Langmuir*, **19**, 6598, 2003.
[4] R. BANDYOPADHYAY, A.S. GITTINGS, S.S. SUH, P.K. DIXON, and D.J. DURIAN, *Rev. Sci. Instr.*, **76**, 093110, 2005.
[5] J. BANHART, *J. Metals*, **52**, 22, 2000.
[6] G.B. BANTCHEV and D.K. SCHWARTZ, *Langmuir*, **19**, 2673, 2003.
[7] V. BERGERON, *J. Phys: Condens. Matter*, **11**, R215, 1999.
[8] F. BOLTON and D. WEAIRE, *Phil. Mag. B*, **65**, 473, 1992.
[9] K. BRAKKE, *Exp. Math.*, **1**, 141, 1992.
[10] I. CANTAT and R. DELANNAY, *Eur. Phys. J. E*, **18**, 55, 2005.
[11] I. CHEDDADI, P. SARAMITO, C. RAUFASTE, P. MARMOTTANT, and F. GRANER, *Eur. Phys. J. E*, **27**, 123, 2008.
[12] L. CIPELLETTI, H. BISSIG, V. TRAPPE, P. BALLESTA, and S. MAZOYER, *J. Phys: Condens. Matter*, **152**, S257, 2003.
[13] S. COHEN-ADDAD and R. HÖHLER, *Phys. Rev. Lett.*, **86**, 4700, 2001.
[14] S. COURTY, B. DOLLET, F. ELIAS, P. HEINIG, and F. GRANER, *Europhys. Lett.*, **64**, 709, 2003; B. DOLLET, F. ELIAS, F. GRANER, *Europhys. Lett.*, **88**, 69901, 2009.

[15] P.-G. DE GENNES, F. BROCHARD-WYART, and D. QUÉRÉ, *Capillarity and Wetting Phenomena: Drops, Bubbles, Pearls, Waves*, Springer, New York, 2004.
[16] B. DOLLET and F. GRANER, *J. Fluid Mech.*, **585**, 181, 2007.
[17] W. DRENCKHAN, H. RITACCO, A. SAINT-JALMES, D. LANGEVIN, P. MAC-GUINESS, A. SAUGEY, and D. WEAIRE, *Phys. Fluids*, **19**, 102101, 2007.
[18] D.J. DURIAN, *Phys. Rev. E*, **55**, 1739, 1997.
[19] D.J. DURIAN, D.A. WEITZ, and D.J. PINE, *Science*, **252**, 686, 1991.
[20] J. ETRILLARD, M. AXELOS, I. CANTAT, F. ARTZNER, A. RENAULT, and F. BOUÉ, *Langmuir*, **21**, 2229, 2005.
[21] K. FEITOSA, S. MARZE, A. SAINT-JALMES, and D.J. DURIAN, *J. Phys.: Condens. Matter*, **17**, 6301, 2005.
[22] K. FEITOSA and D.J. DURIAN, *Eur. Phys. J. E*, **26**, 309, 2008.
[23] B.S. GARDINER, B.Z. DLUGOGORSKI, and G.J. JAMESON, *J. Non-Newton. Fluid Mech.*, **92**, 151, 2000.
[24] C.P. GONATAS, J.S. LEIGH, A.G. YODH, J.A. GLAZIER, and B. PRAUSE, *Phys. Rev. Lett.*, **75**, 573, 1995.
[25] F. GRANER, B. DOLLET, C. RAUFASTE, and P. MARMOTTANT, *Eur. Phys. J. E*, **25**, 349, 2008.
[26] F. GRANER, Y. JIANG, E. JANIAUD, and C. FLAMENT, *Phys. Rev. E*, **63**, 011402, 2001.
[27] P. HÉBRAUD, F. LEQUEUX, J.P. MUNCH, and D.J. PINE, *Phys. Rev. Lett.*, **78**, 4657, 1997.
[28] S. HILGENFELDT and A. VAN DOORNUM, *APS Meeting Abstracts*, 13014, 2003.
[29] E.A. HOLM, J.A. GLAZIER, D.J. SROLOVITZ, and G.S. GREST, *Phys. Rev. A*, **43**, 2662, 1991.
[30] http://www.compucell3d.org/
[31] http://www.susqu.edu/brakke/evolver/
[32] E. JANIAUD and F. GRANER, *J. Fluid Mech.*, **532**, 243, 2005; S. ISHIHARA and K. SUGIMURA, *J. Theor. Biol.*, **313**, 201, 2012; S. ISHIHARA, K. SUGIMURA, S.J. COX, I. BONNET, Y. BELLAÏCHE, and F. GRANER, *Euro. Phys. J. E*, in press, 2013.
[33] S. JURINE, S. COX, and F. GRANER, *Colloids Surf. A*, **263**, 18, 2005.
[34] N. KERN, D. WEAIRE, A. MARTIN, S. HUTZLER, and S.J. COX, *Phys. Rev. E*, **70**, 041411, 2004.
[35] A.M. KRAYNIK, D.A. REINELT, and F. VAN SWOL, *Phys. Rev. Lett.*, **93**, 208301, 2004.
[36] J. LAMBERT, I. CANTAT, A. RENAULT, F. GRANER, J.A. GLAZIER, I. VERETENNIKOV, and P. CLOETENS, *Colloids Surf. A*, **263**, 295, 2005.
[37] J. LAMBERT, R. MOSKO, I. CANTAT, P. CLOETENS, J.A. GLAZIER, F. GRANER, and R. DELANNAY, *Phys. Rev. Lett.*, **104**, 248304, 2010.
[38] E. LORENCEAU, Y. YIP CHEUNG SANG, R. HÖHLER, and S. COHEN-ADDAD, *Phys. Fluids*, **18**, 097103, 2006.
[39] C.W. MACOSKO, *Rheology: Principles, Measurements, and Applications*, Wiley-VCH, Weinheim, 1994.
[40] S. MARZE, D. LANGEVIN, and A. SAINT-JALMES, *J. Rheol.*, **52**, 1091, 2008.
[41] E.B. MATZKE, *Am. J. Botany*, **33**, 58, 1946.

[42] C. MONNEREAU and M. VIGNES-ADLER, *Phys. Rev. Lett.*, **80**, 5228, 1998.
[43] K.J. MYSELS, *Langmuir*, **2**, 428, 1986.
[44] K.J. MYSELS and M.N. JONES, *Discuss. Faraday Soc.*, **42**, 42, 1966.
[45] R. NAGARAJAN, S.I. CHUNG, and D.T. WASAN, *J. Colloid Interface Sci.*, **204**, 53, 1998.
[46] T. OKUZONO and K. KAWASAKI, *Phys. Rev. E*, **51**, 1246, 1995.
[47] O. PITOIS, C. FRITZ, and M. VIGNES-ADLER, *J. Colloid Interface Sci.*, **282**, 458, 2005.
[48] C. RAUFASTE, B. DOLLET, S. COX, Y. JIANG, and F. GRANER, *Eur. Phys. J. E*, **23**, 217, 2007.
[49] S. RODTS, J.C. BAUDEZ, and P. COUSSOT, *Europhys. Lett.*, **69**, 636, 2005.
[50] D. SENTENAC and J.-J. BENATTAR, *Phys. Rev. Lett.*, **81**, 160, 1998.
[51] D.J. SROLOVITZ, M.P. ANDERSON, G.S. GREST, and P.S. SAHNI, *Scripta metall.*, **17**, 241, 1983.
[52] C. STENVOT and D. LANGEVIN, *Langmuir*, **4**, 1179, 1988.
[53] H.A. STONE, A.D. STROOCK, and A. AJDARI, *Ann. Rev. Fluid Mech.*, **36**, 381, 2004.
[54] C. STUBENRAUCH and R. VON KLITZING, *J. Phys.: Condens. Matter*, **15**, R1197, 2003.
[55] G.L. THOMAS, R.M.C. DE ALMEIDA, and F. GRANER, *Phys. Rev. E*, **74**, 021407, 2006.
[56] A. VAN DER NET, L. BLONDEL, A. SAUGEY, and W. DRENCKHAN, *Colloids Surf. A*, **309**, 159, 2007.
[57] M.U. VERA, A. SAINT-JALMES, and D.J. DURIAN, *Appl. Opt.*, **40**, 4210, 2001.
[58] G.F. VORONOÏ, *J. Reine Angew. Math.*, **133**, 97, 1908; **134**, 198, 1908; **136**, 67, 1909.
[59] D. WEAIRE and R. PHELAN, *Phil. Mag. Lett.*, **69**, 107, 1994.
[60] D.A. WEITZ and D.J. PINE, "Diffusing-wave spectroscopy", in *Dynamic Light Scattering*, Chap. 16, Clarendon, Oxford, 1993.
[61] C. ZAKRI, A. RENAULT, and B. BERGE, *Physica B*, **248**, 208, 1998.

Notation

Note that page numbers are not provided for symbols that are used throughout the text

c	concentration	
cmc	critical micelle concentration	146
c_s	bulk concentration below the interface	85
d	average bubble diameter	
d_m	typical size of a surfactant molecule	146
e	thickness of a 2D foam (distance between the plates)	57
e_x, e_y, e_z	reference unit vectors in the laboratory frame	
f	number of faces of a bubble	32
$g = 9.8$ m.s^{-2}	gravity, in the direction $-e_z$	
h	thickness of a film	20
$k_B = 1.38 \times 10^{-23}$ J.K^{-1}	Boltzmann's constant	
ℓ	length of a Plateau border	46
ℓ_{ij}	length of the edge separating bubbles i and j in 2D	35
ℓ^*	transport mean free path of light	236
n	number of edges of a face (in 3D); number of sides of a bubble (in 2D)	31
$\mathbf{n}$	unit vector normal to a surface	
$\hat{n}$	unit vector normal to a line, in the plane of the interface	
p	pressure in the liquid phase	48
q	geometric charge	41
q_t	topological charge	32
r	radius of curvature of the cross-section of a Plateau border	47
r_1, r_2	radii of curvature of a Plateau border	47
s	surface area of the cross-section of a Plateau border	48
u	local velocity in the liquid phase	
v	velocity of a bubble	
x, y	horizontal directions	
z	vertical direction, oriented upwards	
A	surface area of a 2D bubble	38
B_{ij}	Finger tensor	174
$Bo = \eta_s/(\eta r)$	Boussinesq number	120

$Ca = \eta u/\gamma$	capillary number	190
$C^r = (dV/dt)/V^{1/3}$	relative growth rate of a bubble	102
D_v	bulk diffusion coefficient of a surfactant	85
D_f	bulk diffusion coefficient of a gas through a liquid	109
D_{eff}	effective diffusion coefficient of gas across a film	109
E	energy of a foam	35
E_{GM}	Gibbs–Marangoni elastic modulus	87
$E_s^* = E_s' + iE_s''$	complex surface dilational elastic modulus	88
F_{ij}	deformation gradient tensor	174
G	static shear modulus	169
$\hat{G}$	static shear modulus in 2D (in N.m^{-1})	180
$G^* = G' + iG''$	complex shear modulus	173
$G_s^* = G_s' + iG_s''$	complex surface shear modulus	88
$H = 1/R_1 + 1/R_2$	mean curvature of a film	26
He	Henry's coefficient	109
J	compliance	172
K_p	reduced permeability of a Plateau border	117
L_{int}	total length of interface in a 2D foam $(E = \lambda L_{\text{int}})$	59
N	number of bubbles in a foam	
N_1, N_2	normal stress differences	175
P	gas pressure in a bubble	
$P_{\text{atm}} = 10^5$ Pa	atmospheric pressure	
P_c	capillary pressure	48
R	radius of curvature of a film	25
R_{ij}	radius of curvature of the edge separating bubbles i and j in 2D	40
R_1 and R_2	principal radii of curvature of a film	25
R_{32}	Sauter mean radius	251
R_S	equivalent spherical radius of a bubble with surface area S	38
R_V	equivalent spherical radius of a bubble with volume V	38
S_i	surface area of bubble i	35
S_{ij}	area of the face separating bubbles i and j $(E = \sum_{i<j} 2\gamma S_{ij})$	27
S_{int}	total area of air/water interface in a foam $(E = \gamma S_{\text{int}})$	35
T	temperature	
T1	type 1 topological transformation	78
T2	type 2 topological transformation	81
V	bubble volume	
V_m	molar volume of gas	109
V_r	typical volume of a topological change	205
$\mathbf{X}$	displacement field	174

Symbol	Description	Page
α	foam permeability	118
$\delta_a = \sqrt{3} - \pi/2$	geometric coefficient	124
$\delta_b = 1.74$	geometric coefficient	124
γ	surface tension of a single interface	23
$\gamma_0 = 72 \times 10^{-3}$ N.m^{-1}	surface tension of pure water in air	23
ε	shear strain	168, 174
$\dot{\varepsilon} = d\varepsilon/dt$	strain rate	169
ε_{ij}	strain tensor	174
ε_y	yield strain	171
η	bulk viscosity, $\eta = 10^{-3}$ Pa.s of water at 20 °C	169
η_d	surface dilatational viscosity	87
η_p	plastic viscosity	171
η_s	surface shear viscosity	88
$\kappa_{ij} = 1/R_{ij}$	curvature of the edge separating bubbles i and j in 2D	40
$\lambda \sim \gamma e$	line tension of an interface in 2D	59
$\lambda_c = \sqrt{\gamma/\rho_l g}$	capillary length	54
λ_D	Debye length	93
μ	chemical potential	83
$\mu_2^f = \langle f^2 \rangle - \langle f \rangle^2$	variance in the number of faces	250
$\mu_2^V = \langle V^2 \rangle - \langle V \rangle^2$	variance in the bubble volumes	252
ρ	density of a foam	21
ρ_l	density of a liquid	21
σ	shear stress	168
$\hat{\sigma}$	shear stress in 2D (in N.m^{-1})	180
σ_{ij}	stress tensor	174
σ_{ij}^s	surface stress tensor	87
σ_y	yield stress	171
ϕ_l	liquid fraction (volume fraction of the liquid phase of a foam)	21
ϕ_l^*	critical liquid fraction	195
Γ	surface concentration of surfactant	82
Γ_∞	saturated surface concentration of surfactant	85
Π_o	osmotic pressure of a foam	51
Π_d	disjoining pressure	48
$\Pi_l = \gamma_0 - \gamma$	Langmuir surface pressure	83

Index

2D (foam) 55, 65, 254

A

adsorption 18, 82, 84, 90, 98, 148, 161
antifoam/antifoaming agent 140
avalanche 140

B

Batchelor 175, 218
biological tissue 12
black film 137, 155, 229
  common 98
  Newton 98
Boussinesq number 120–123
bubbly liquid 21, 76

C

capillary
  length 54
  number 190–193, 213
  pressure 48, 97, 98, 135
  suction 20, 48, 97
charge
  electrostatic 94
  geometric 41–44, 100, 103
  topological 32, 43, 161
close packing 21, 46
cmc (*see* micelle)
coalescence (*see* rupture)
coarsening 99
  Mullins 44, 102, 104
  Lifshitz–Slyozov–Wagner 76
  Ostwald ripening 76
  von Neumann 42, 100, 151, 161
  characteristic time 104, 108
  coupling with drainage 133
colour in films (*see* iridescence)
conductivity 115, 158, 238
convective instability 128
creep 172, 202, 205
curvature
  mean 27, 30, 102
  radius of 27, 48

D

Darcy's law 117
Debye length 93
diffusion
  coefficient 85, 102, 109, 152
  of gas 100, 102, 109
  of surfactants 85
dilation (*see* expansion)
dimple 92, 136
dissipation 126, 151, 186–190, 209, 211, 246
drainage
  coupled with coarsening 133
  film 135, 152
  equation 127
  forced 115, 129, 157
  free 114, 132, 156, 158, 162
  characteristic time 132
DTS (*see* multiple light scattering)
DWS (*see* multiple light scattering)

E

edge (*see* Plateau border)
elasticity
  foam 178, 194, 198, 202
  surface/interfacial 86, 88, 91, 136
  Gibbs (*see* Gibbs)
  Langmuir (*see* Langmuir)
elastic limit 183
elongation 62, 178, 220
emulsion 11, 30, 51, 76, 77, 147, 237
energy
  dissipated 151, 187
  elastic 169, 187, 208
  local minimum 22, 36, 180, 184
  surface 17, 23, 35, 72, 197
Euler's formula 31, 32, 43, 71, 81
expansion/compression 87, 211, 226

F

ferrofluid 149
film balance technique 136, 229
fluctuations 26, 91, 137, 215
foamability 82, 98, 231

G

Gibbs
  elasticity 87
  equation 83

H

Hamaker (constant) 92
hexagonal structure 12, 61, 179, 184, 196, 219

hole (opening in film) 138
honeycomb (*see* hexagonal structure)
hydrophilic/hydrophobic 18, 147
hysteresis 184

I

idealized bubble (with zero mean curvature) 37, 44, 103
ideal foam 26, 30, 31, 59, 100
interface
  mobile/immobile 90, 120, 122, 130, 136
  compressible/incompressible 188–193, 213
iridescence 20, 67, 129, 228

K

Kelvin cell 33, 50, 68, 181, 183

L

Langmuir
  monolayer 58
  trough 227
  equation 83
  foam 56, 58
Laplace Law (see Young–Laplace Law)
Lewis relation 38
liquid fraction
  2D 60
  3D 21, 110–115, 238
localization of flow 214, 215
length
  Debye (*see* Debye)
  capillary (*see* capillary)
local minimum (*see* energy)

M

magnetic resonance 215, 235
Marangoni effect 91
marginal regeneration 136, 155
micelle 23, 83, 146, 229
MRI (*see* magnetic resonance)
Mullins (*see* coarsening)
multiple light scattering 236

N

NMR (*see* magnetic resonance)
node (*see* vertex)

O

Ostwald ripening (*see* coarsening)

P

permeability 118, 119
plasticity 62, 170, 183, 209, 215, 220
Plateau
  border 21, 47, 188, 233
  laws 20, 26, 30, 46, 59, 70, 180, 183

Poiseuille flow 117
polymer 10, 147
pressure
  capillary (*see* capillary)
  disjoining 19, 20, 49, 92, 94, 97, 112, 230
  osmotic 21, 51
  surface 24, 83, 162

Q

quasi-static 62, 170, 178, 183, 186, 198, 245

R

rheometer 210, 214
Reynolds
  number 116
  velocity 135
rupture 78, 134, 140

S

Sauter mean radius 194, 251
SDS 19, 69, 110, 146, 211, 225
self-similar (growth) 105, 161
shear 169, 172, 240
  experiment 172, 210, 228, 241
  modulus 89, 88, 169, 173, 180–183
  surface 86, 88, 120, 228
  rate 88, 169, 172, 211
solid foam 8
stability (*see* foamability)
strain 168, 170, 173, 181, 240
  surface 87, 88, 121
  rate 171
  yield 171, 183, 210, 217
stratification 95
stress 168, 174, 177, 215, 240
  deviatoric 174
  normal 174, 198, 198
  surface 87, 121
  yield 171, 208–211, 218
suction (*see* capillary)
surface
  elasticity (*see* elasticity)
  energy (*see* energy)
  pressure (*see* pressure)
  strain (*see* strain)
  stress (*see* stress)
  tension (*see* tension)
  viscosity (*see* viscosity)
surface area
  effective 112
  minimal 19, 244
  specific 6, 119, 127
surfactant/surface-active molecule 18, 24, 63, 82, 131, 145, 153, 190, 225

T

T1 36, 78, 183, 186, 205, 208, 211, 215, 217, 245, 246, 254
T2 81, 104, 151, 254
tension
  film 26, 69
  line 42, 59, 178
  surface 17, 23, 27, 64, 82, 91, 225
(characteristic) time
  of drainage (*see* drainage)
  of coarsening (*see* coarsening)
tissue (*see* biological)
tomography 2, 21, 234, 235
topological
  change (*see* T1)
  charge (*see* charge)
  rearrangement (*see* T1)
topology 31, 38, 100, 250

V

van der Waals 19, 92, 94
vertex 21, 30, 46, 125, 126, 129, 132, 194, 219, 234, 247
viscosity
  foam 169, 171, 211
  surface 86, 87, 120, 152, 190, 204
von Neumann's law (*see* coarsening)

W

Weaire–Phelan 37, 244

Y

yield
  stress (*see* stress)
  strain (*see* strain)
Young–Laplace law 17, 25, 26, 69, 218

The manufacturer's authorised representative in the EU for product safety is Oxford University Press España S.A. of el Parque Empresarial San Fernando de Henares, Avenida de Castilla, 2 – 28830 Madrid (www.oup.es/en or product. safety@oup.com). OUP España S.A. also acts as importer into Spain of products made by the manufacturer.

www.ingramcontent.com/pod-product-compliance
Ingram Content Group UK Ltd.
Pitfield, Milton Keynes, MK11 3LW, UK
UKHW051532240326
469240UK00017B/131